Science and Fiction

More information about this series at http://www.springer.com/series/11657

Science and Fiction – A Springer Series

This collection of entertaining and thought-provoking books will appeal equally to science buffs, scientists and science-fiction fans. It was born out of the recognition that scientific discovery and the creation of plausible fictional scenarios are often two sides of the same coin. Each relies on an understanding of the way the world works, coupled with the imaginative ability to invent new or alternative explanations - and even other worlds. Authored by practicing scientists as well as writers of hard science fiction, these books explore and exploit the borderlands between accepted science and its fictional counterpart. Uncovering mutual influences, promoting fruitful interaction, narrating and analyzing fictional scenarios, together they serve as a reaction vessel for inspired new ideas in science, technology, and beyond.

Whether fiction, fact, or forever undecidable: the Springer Series "Science and Fiction" intends to go where no one has gone before!

Its largely non-technical books take several different approaches. Journey with their authors as they

- Indulge in science speculation – describing intriguing, plausible yet unproven ideas;
- Exploit science fiction for educational purposes and as a means of promoting critical thinking;
- Explore the interplay of science and science fiction – throughout the history of the genre and looking ahead;
- Delve into related topics including, but not limited to: science as a creative process, the limits of science, interplay of literature and knowledge;
- Tell fictional short stories built around well-defined scientific ideas, with a supplement summarizing the science underlying the plot.

Readers can look forward to a broad range of topics, as intriguing as they are important. Here just a few by way of illustration:

- Time travel, superluminal travel, wormholes, teleportation
- Extraterrestrial intelligence and alien civilizations
- Artificial intelligence, planetary brains, the universe as a computer, simulated worlds
- Non-anthropocentric viewpoints
- Synthetic biology, genetic engineering, developing nanotechnologies
- Eco/infrastructure/meteorite-impact disaster scenarios
- Future scenarios, transhumanism, posthumanism, intelligence explosion
- Virtual worlds, cyberspace dramas
- Consciousness and mind manipulation

Time Machine Tales

The Science Fiction Adventures and Philosophical Puzzles of Time Travel

 Springer

Paul J. Nahin
Department of Electrical Engineering
University of New Hampshire
Durham, NH, USA

ISSN 2197-1188 ISSN 2197-1196 (electronic)
Science and Fiction
ISBN 978-3-319-48862-2 ISBN 978-3-319-48864-6 (eBook)
DOI 10.1007/978-3-319-48864-6

Library of Congress Control Number: 2016957723

Printed on acid-free paper

This Springer imprint is published by Springer Nature
The registered company is Springer International Publishing AG
The registered company address is: Gewerbestrasse 11, 6330 Cham, Switzerland

Also By Paul J. Nahin

Oliver Heaviside (1988, 2002), Johns Hopkins
Time Machines (1993, 1999), Springer
The Science of Radio (1996, 2001), Springer
An Imaginary Tale (1998, 2007, 2010), Princeton
Duelling Idiots (2000, 2002), Princeton
When Least Is Best (2004, 2007), Princeton
Dr. Euler's Fabulous Formula (2006, 2011), Princeton
Chases and Escapes (2007, 2012), Princeton
Digital Dice (2008, 2013), Princeton
Mrs. Perkins's Electric Quilt (2009), Princeton
Time Travel (1997, 2011), Johns Hopkins
Number-Crunching (2011), Princeton
The Logician and the Engineer (2013), Princeton
Will You Be Alive Ten Years From Now? (2014), Princeton
Holy Sci-Fi! (2014), Springer
Inside Interesting Integrals (2015), Springer
In Praise of Simple Physics (2016), Princeton

Frontispiece: The Pioneers of Time Travel

The scientific pioneers were Albert Einstein (1879–1955) and Kurt Gödel (1906–1978), good personal friends who are shown here in 1954 at the Institute for Advanced Study in Princeton, New Jersey, in a photo taken by Richard Arens. It was Einstein's 1916 general theory of relativity (theory of gravity) that Gödel used as the basis for his 1949 paper that was the first to show that the general theory does **not** forbid time travel into the past.

The literary pioneer of time travel was of course Herbert George Wells (1866–1946), who is shown here as a college freshman cut-up around 1885. The photograph was taken as a prank by an unknown friend while Wells was a student in a biology course given by Thomas Huxley, at the Normal School of Science in

South Kensington (a branch of the University of London). A far too thin and impoverished Wells was then still a teenager, and *The Time Machine* lay a distant 10 years in the future.

Einstein/Gödel photograph courtesy of the American Institute of Physics Emilio Segré Visual Archives of the AIP Niels Bohr Library. Wells photograph courtesy of the rare Books and Special Collections Department of the Library of the University of Illinois at Urbana-Champaign.

A Note on the Story Citations and Science Fiction History

"You will find it a very good practice always to verify your references, sir."
—advice given in 1847 to a young scholar by Martin Joseph
Routh, President of Magdalen College, Oxford

Most of the pulp science fiction stories I've cited in this book, in their original form as ink on paper, have long since vanished from our region of spacetime and exist today only (alas) on microfilm reels in scholarly vaults. I am especially indebted to Texas A & M, the Claremont Colleges, the California State Universities at Northridge and Fullerton, Mount Holyoke College, the New York City Public

Library, and the University of Delaware, for giving me access via Inter-Library Loan (through my home institution, the University of New Hampshire) to their extensive archives of ancient science fiction magazines.

A number of the really good stories *have* been anthologized, however, and so are still readily available today in book form. In essentially all cases, though, for historical reasons, I've given the original publication information (magazine and date). You can find which of the stories cited are available in one or more anthology reprints by going to an immensely useful, searchable database on the Web, at: http://www.isfdb.org, and I gratefully thank all those in the science fiction community responsible for creating and maintaining that database.

The following two books by science fiction historian Sam Moskowitz (1920–1997), who lived through what Isaac Asimov called the 'Golden Age of [magazine] Science Fiction,' may be difficult to find today but, if you are interested in the early history of *magazine* science fiction (beyond simply the subgenre of time travel), the hunt for them will be well worth your time:

Science Fiction by Gaslight: a history and anthology of science fiction in the popular magazines, 1891–1911 (World Publishing Company 1968);
Under the Moons of Mars: a history and anthology of "The Scientific Romance" in the Munsey Magazines, 1912–1920 (Holt, Rinehart and Winston 1970).

Some First Words

Is time travel in principle (never mind the difficulties) a possibility? It has received some thought in the past and deserves some more.
—David Park, in his 1980 book *The Image of Eternity*

He used to have quite a reputation, but the last couple of years he's been working on time ... You know, time travel, that sort of rot. An A-1 crackpot.
—a character (discussing a colleague) disagrees with Park, in Mack Reynolds' "Advice from Tomorrow," *Science Fiction Quarterly*, August 1953

In 1993 the first edition of my book *Time Machines* was published by the Press of the American Institute of Physics. In 1999, after Springer acquired AIP Press, the second edition of that book appeared. So, is this the third edition? Well, yes *and* no. It *is* because large chunks of the 1999 edition are still here, along with new discussions of the advances by physicists and philosophers that have appeared in the intervening 18 years. The prime example of that centers on the time travel paradoxes. Those discussions contain mostly what is in the second edition, but they have also been brought up to date with the latest thinking on the paradoxes, by physicists and philosophers.

And yet this book is *not quite* the third edition because the emphasis is now on the philosophical and on science fiction, rather than on physics as it was when written for AIP Press. In that spirit there are, for example, no Tech Notes filled with algebra, integrals, and differential equations, as there are in the first and second editions of *Time Machines*. That's because I wish to avoid having this book seem to be simply a long physics treatise. I have, in fact, some sympathy with the following views, expressed by two philosophers:

"There is one metaphor in the physicist's account of space-time which one would expect *anyone* to recognize as such, for metaphor is here strained far beyond the breaking point, i.e., when it is said that time is 'at right angles to each of the other three dimensions.' Can anyone really attach any meaning to this—except as a recipe for drawing diagrams?"[1]

and

"This is from the outset a study in descriptive metaphysics. In consequence, I shall have nothing to say about twice-differentiable Lorentzian manifolds, Minkowski diagrams, world-lines, time-like separations, space-time worms [a 'thick' world-line], or temporal parts."[2]

I don't *completely* endorse these sentiments, however, and so please understand that I am not denying the ultimate importance of *physics* when it comes to achieving a deep understanding of time travel. To quote yet another philosopher,

"Arm chair reflections on the concept of causation [are] not going to yield new insights. The grandfather paradox is simply a way of pointing to the fact that if the usual laws of physics are supposed to hold true in a chronology violating space-time, then consistency constraints emerge. *[To understand these constraints] involves solving problems in physics, not armchair philosophical reflections* [my emphasis]."[3]

I could not agree more. So, in *Time Machine Tales* you *will* find some physics. In support of time travel to the future (and in how to make a wormhole time machine for travel into the past), for example, I'll show you a high school level derivation of the famous time dilation formula from special relativity. There are some spacetime diagrams, some simple algebraic manipulations, and here and there just a touch of freshman calculus; even the metric tensor gets a few words, too. But it is, admittedly, pretty light-weight stuff.

So, while certainly saluting the premier position of physics, *Time Machine Tales* is not a scholarly, in-depth treatment of time travel physics. Rather, it is an examination of how science fiction writers (and many philosophers, too) have viewed time travel. (Even in the physics discussions, science fiction will regularly appear.) Those views, by their very nature, are far more romantic than are those of hardcore theoretical physicists. History has shown, of course, that the results of the work of theoretical physicists may, in the end, prove to actually be far more astonishing than anything fiction writers cook-up—and if there is any scientific subject for which that may again prove to be true it's time travel—but for us, here, it will be the fiction writer who has center stage.

The philosophers will be only slightly less important in this book. While much of the early philosophical literature on time travel and backwards causation reads like imaginative fairy tales spun out of vacuous vapors (more on this soon), many modern philosophers have shown themselves to be quite sophisticated. What they

---

[1]C. W. K. Mundle, "The Space-Time World," *Mind*, April 1967, pp. 264–269.

[2]J. F. Rosenberg, "One Way of Understanding Time," *Philosophia*, October 1972, pp. 283–301.

[3]John Earman, "Recent Work on Time Travel," in *Time's Arrows Today: recent physical and philosophical work on the direction of time* (Steven F. Savitt, editor), Cambridge University Press 1995, pp. 268–310. We'll discuss the idea of *consistency constraints* in some detail later in the book. Earman is Professor Emeritus of History and Philosophy of Science at the University of Pittsburgh.

have written deserves serious consideration by anyone interested in time travel, and that includes physicists. However, while the time travel interests of philosophers and physicists have a lot of overlap, those interests are *not* in total agreement. For example, while both groups talk of the grandfather paradox, the philosophers worry in particular about motivation (*why* the murderous mission?), while physicists have *never* to my knowledge asked themselves that question[4] (other than to figure out how to avoid it!). After all, philosophers talk of flesh-and-blood humans as time travelers, while the physicists send only billiard balls (with no personal identities or memories) on time trips into the past for the *expressed purpose* of avoiding the messy human issues of 'motivation' and free will. This approach by physicists isn't because they are cold and emotionless. It is a useful strategy because, if it can be shown that a mere billiard ball can travel into the past then, as one *philosopher* pointed out long ago, "It is implausible that it should be possible for some physical systems to travel back in time, and not others. Thus, if we suppose that simple objects can time-travel ... then we must suppose that more complicated systems, e.g., human beings, can also time-travel."[5]

For the most part, philosophers and physicists have worked at the extreme, opposite points of the time travel spectrum. Much better, I think, would be to adopt the following, more balanced position advocated recently: "The study of time machines is a good opportunity for forging a partnership between philosophy and physics. Of course, philosophers have to recognize that in this particular instance the partnership is necessarily an unequal one since the mathematical physicists have to do the heavy lifting. But it seems clear that a little more cooperation with philosophers of science in attending to the analysis of what it takes to be a time machine could have led to some helpful clarifications in the physics literature."[6]

In the past, philosophers gained a reputation for being just a bit too 'unconstrained by the facts' for scientific tastes—as the English mathematician Augustus De Morgan (1806–1871) wrote in an 1842 letter, "There are no writers who give us so much *must* with so little *why*, as the metaphysicians"[7]—but I do think today's physicists would do well to reexamine that harsh opinion.

Philosophers of the 'old school' may look askance at a non-philosopher (me!) leveling criticism at them, and so let me step aside and quote from a member of the

---

[4]Nicholas J. J. Smith, "Why Would Time Travelers Try to Kill Their Younger Selves?" *The Monist*, July 2005, pp. 388–395. As Smith writes, "[Motivation] does not impact upon the possibility, or even the likelihood of backwards time travel. Yet it is deeply puzzling, and we will have no idea what time travel would actually be *like* until we explore it." See also Peter B. M. Vranas, "Can I Kill My Younger Self? Time Travel and the Retrosuicide Paradox," *Pacific Philosophical Quarterly*, December 2009, pp. 520–534.

[5]P. Horwich, "On Some Alleged Paradoxes of Time Travel," *Journal of Philosophy*, August 1975, pp. 432–444.

[6]John Earman, Christopher Smeenk, and Christian Wüthrich, "Do the Laws of Physics Forbid the Operation of Time Machines?," *Synthese*, July 2009, pp. 91–124.

[7]D. J. Cohen, *Equations from God: pure mathematics and Victorian faith*, The Johns Hopkins University Press 2007, p. 119.

'modern school' of philosophical thought: "Space-time is the basic spatiotemporal entity. Many philosophers have mouthed this truth, but few have swallowed it, and very few have digested it … An appreciation of this truth is crucial to what is commonly referred to as the philosophy of space and time … In large measure the lack of progress in this area can be traced to the fact that philosophers have not taken seriously the corollary that talk about space and time is really talk about the spatial and temporal aspects of spacetime."[8] This is a polite way of telling philosophers that they had better learn some physics!

What provoked those harsh words was that 'modern' philosopher's perception that 'old school' philosophers were not talking science when they wrote of space and time, but rather were in the business of telling each other irrelevant stories and myths, a curious philosophical approach involving the 'telling of tales' that reached its peak in the early and mid-1960s. Spacetime story telling seems to have started with a paper by the Oxford philosopher Anthony Quinton (1925–2010), who argued[9] that although there can be multiple, disjointed spaces, there can only be a single time that is the same for everyone, everywhere. The issue is *not* the truth or not of that assertion (Newton believed it, modern physicists don't), but rather Quinton's technique for arriving at it: myth construction.

Myth construction strikes those trained in the technical sciences as, while perhaps interesting—even physicists, after all, can enjoy a good fairy tale now and then—something quaint and totally beside the point. In his paper Quinton tells a fairy tale about how he thinks someone can live continuously in time and yet, via dreaming, be in two different spatial worlds; when awake he is in one world, while when the person is asleep he is in the other. Quinton argues that this multispatial myth is plausible, but that a search for an analogous multitemporal myth is doomed. This prompted a reply[10] from another 'old school' philosopher who rebutted Quinton with an even more outlandish counter-myth involving "the warring tribes of Okku and Bokku"!

It was this back-and-forth spinning of hypothetical tales that caused the 'modern' philosopher to write in his paper (note 30) that "the procedure for arriving at answers to these questions [about space and time] adopted by Quinton and most other ['old school' philosophers] is, to say the least, a curious one: a story is told about a mythical land—usually called something like the land of Okkus-Bokkus [which is now seen to an outrageous pun]—and then we are asked what we would say if confronted by experiences like those of the Okkus-Bokkusians. As often happens with such a question, people have said all sorts of things, not all of which are interesting or enlightening."

Another modern philosopher was even less gentle in his rejection of the fairy tale approach to spacetime physics: "Quinton [and others of a similar approach invite

---

[8] J. Earman, "Space-Time or How to Solve Philosophical Problems and Dissolve Philosophical Muddles Without Really Trying," *Journal of Philosophy*, May 1970, pp. 259–276.

[9] A. Quinton, "Spaces and Times," *Philosophy*, April 1962, pp. 130–147.

[10] R. G. Swinburne, "Times," *Analysis*, June 1965, pp. 185–191.

us] to say what we should think in certain strange circumstances which they describe within common-sense language [as opposed to scientific terminology]. I must say that if I found myself in the circumstances which they describe I just would not know what to think. Probably I should simply conclude that I had gone mad ... It looks as though these writers are inviting us to consider *what we should say if we knew no science* [my emphasis]."[11]

Even before the modern philosopher (note 30) wrote in 1970 to complain about myth-making, another had already done so: "Whenever a human being produces an argument which opens 'Suppose I had 23 senses ...,' 'Suppose I were God ...,' 'Suppose I experienced objects extended in four spatial dimensions ...,' we can protest that the argument is worthless. For in supposing that he has transcended our human point of view, he has also transcended the limits of our understanding."[12] As this author concluded his very funny paper, such opening sentences are the signatures of myths from "The Philosopher's Fairy Tale Book."

The strained relationship between myth-making philosophers and physicists, especially concerning time travel, has a historically interesting antecedent in the 1920s negative reaction among many over Einstein's theories of relativity (the very theories that give apparent life to time travel). To illustrate my point, consider the October 1913 letter Oskar Kraus (1872–1942), a philosophy professor at the German University in Prague, sent to Ernst Gehrcke (1878–1960), a physics professor at the Reich Institute of Physics and Technology in Berlin. Both men were opponents of Einstein but, as Kraus wrote in his letter, it was only Gehrcke among the physicists he considered to be sympathetic to him: "[I] would not know ... anyone else but you who as a specialist would not reject the intervention of a philosopher from the start."[13]

So, I think Earman's proposal a sound one, an echo in fact of similar words that the physicist Kip Thorne wrote (in the Foreword to the second edition of *Time Machines*) concerning science fiction writers: "Smart physicists seek insight everywhere, including from clever science fiction writers who long ago began probing seriously the logical consequences that would ensue if the laws of physics permitted time travel."[14]

To emphasize this new, combined, diversified focus (but also to retain some connection with my earlier books) is the reason I have altered the title, just a bit. In addition, each chapter now concludes with several open-ended questions, suitable for motivating either classroom discussions or more extensive essay responses.

---

[11]J. J. C. Smart, "The Unity of Space-Time: Mathematics Versus Myth Making," *Australasian Journal of Philosophy*, (no. 2) 1967, 214–217.

[12]M. Hollis, "Times and Spaces," *Mind*, October 1967, pp. 524–536. Hollis ends by saying he is prepared to accept the failure of his paper to convince many of his colleges to change their ways, and he is waiting for one of them to write a paper opening with "Twice upon a time in another space no distance in any direction from here ..."!

[13]Quoted from the Introduction to Milena Wazeck, *Einstein's Opponents: the public controversy about the theory of relativity in the 1920s*, Cambridge 2014 (published in German in 2009).

[14]Thorne is Professor Emeritus of Physics at the California Institute of Technology.

Teachers, in particular, may find this a useful feature if using the book in an academic setting. The book ends with reprints of two of my own published time travel stories (one from *Analog* and the other from *Omni*), with each serving as an illustration of technical issues raised in the book. From my own teaching of an undergraduate honors class in time travel at the University of New Hampshire, I think the assigning of story writing to be an excellent tool for teachers to use. I found reading student stories to be a lot of fun, and students may well surprise teachers with innovative ideas.

Now that I've mentioned story writing, let me say something about the heavy presence of time travel science fiction stories in this book, the majority of which originated in the often maligned pulp magazines of the 1920s through the 1950s. 'Pulp' has long been burdened with a bad literary reputation. As the editor of one anthology of pulp fiction bluntly put it, "Pulp equated with rubbish. Crap of the basest nature."[15] Part of the reason for that was cosmetic; as I wrote in an earlier book, "The term *pulp* came from the use of inexpensive wood-pulp—you could *feel* the lumpy wood chips in each ragged, untrimmed page—to make paper that was far too crummy for the use by any publisher of 'words meant to last.' Such paper quickly yellowed, turned brittle, and finally, amid billowing clouds of bits and pieces, entered into eternal oblivion. Think of the paper used in your newspaper before its final contribution to civilization in the bottom of your cat's litter box; pulp was worse."[16]

And then a little later, in the same book, "The stories in *Amazing* [*Stories* magazine] were 'read it in the morning, forget it by dinnertime' adventure fiction, the stuff you'd put inside a newspaper if on a crowded train or bus so fellow passengers wouldn't know what a low-grade mind you had. The transient nature of pulp fiction was independent of its literary quality, as the cheap acid-based paper that the stories were printed on began to oxidize and literally burn-up as soon as it rolled off the press. In the introductory essay to a 1950 collection of pulp-detective Philip Marlowe stories (*Trouble Is My Business*), mystery writer Raymond Chandler commented on this when he wrote 'pulp fiction never dreamed of posterity.' Pulp fiction was synonymous with trash fiction, and the nature of much of early pulp SF has been aptly described as 'scientific pornography for the mechanically minded,' and 'writing which drooled over descriptions of technology.'"

When publisher Hugo Gernsback (1884–1967) brought out the first issue of *Amazing Stories* in April 1926, it was the first pulp devoted totally to science fiction. With its masthead motto of "Extravagant Fiction Today—Cold Fact Tomorrow," and with the illustration on the contents page of each issue showing a muscular Jules Verne bursting from his grave in the heroic, up-up-and-away pose made famous years later by Superman, there could be no doubt as to what kind of fiction the reader would find under the dramatic, multi-colored cover art. It was fiction populated with mad scientists, and half-naked woman about to be ravished by alien

---

[15]Maxim Jakubowski, *The Mammoth Book of Pulp Fiction*, Carroll & Graf 1996.

[16]P. J. Nahin, *Holy Sci-Fi!: where science fiction and religion intersect*, Springer 2014.

invaders from outer space; all in all, stuff of interest only to teenage boys and imbecilic adults.[17]

How else, after all, to explain the publication of one tale[18] that was given the following heart-stopping editorial introduction: "Professor Lambert deliberately ventures into a Vibrational Dimension to join his fiancée in its magnetic torture-fields"? In defense of many of the readers of early pulp science fiction, however, not all were attracted by such nonsense. Just 2 months later (June 1931) one reader wrote to the same magazine to complain of masculine heroes saving weeping women from ungodly horrors: "Just why do you permit your Authors to inject messy love affairs into otherwise excellent imaginative fiction? Just stop and think. Our young hero-scientist builds himself a space flyer, steps out into the great void, conquers a thousand and one perils on his voyage and amidst our silent cheers lands on some far distant planet. Then what does he do? He falls in love with a maiden—or it's usually a princess—of the planet to which the Reader has followed him, eagerly awaiting and hoping to share each new thrill attached to his gigantic flight. But after that it becomes merely a hopeless, doddering love affair ending by his returning to Earth with his fair one by his side. Can you grasp that—a one-armed driver of a space-flyer! ... We buy A.S. for the thrill of being changed in size, in time, in dimension ... not to read of love ... I wish ... for plain, cold scientific stories sans the fair sex."

Here's another example, this one of the sort of tale that gave an aroma of the sophomoric to 'golden age' time travel science fiction. It was a story of a young man of the far future, with access to a time machine, who wants to see a dinosaur before he dies. So back he travels, back, back, until he at last finds himself in a "subterranean cave, dark and foul-smelling." At first he is puzzled (did dinosaurs live underground?), but then suddenly he hears a thundering roar and sees a huge black shape in the gloom. There can be no doubt now; it *is* a dinosaur, and he sees its red, gleaming eyes just as it crushes him into a pancake. But that's okay; he saw a dinosaur before he died. Then comes the dénouement. He hadn't really gone back quite as far as the Jurassic period, but only to the twentieth century, where he has been run down by the local express train in a subway tunnel![19]

---

[17]This was particularly thought to be the case for readers of the romance pulps, written for young women in the 1930s and 1940s (a separate and distinct audience from that of the science fiction pulps). As one commentator wrote on that genre, the heroes and heroines of such tales often displayed the "mental equipment of a banana split," with the implication that the same might be said of the readers, themselves. (See Margaret MacMullen, "Pulps and Confessions," *Harper's Monthly Magazine*, June 1937.) I don't think, however, that this particular complaint generally applied to the pulp science fiction readership. I'll have much more to say about Gernsback and early pulp science fiction speculations concerning time travel, in Chap. 4.

[18]T. Curry, "Hell's Dimension," *Astounding Stories*, April 1931.

[19]R. G. Thompson, "The Brontosaurus," *Stirring Science Fiction*, April 1941. In the editors of *Stirring*'s defense, notice the month: maybe this story was *meant* to be a joke. If so, it was an admirable success.

Vibrating into new dimensions was, apparently, a popular idea in 1930s pulp science fiction. This 'super science' gadget operated by vibrating an object faster than light, whereupon the Lorentz-FitzGerald contraction formula (see Chap. 3) predicts an *imaginary* size for the object—which means (so we are told) that the object has entered "another plane of existence." The inventor (the fellow with the gun) is inviting his grim-faced assistant to give the gadget a try. The original caption reads "Get into that vibrator! Get in, I say!"

Illustration for "Into Another Dimension" by Maurice Duclos, *Fantastic Adventures* November 1939 (art by Kenneth J. Reeve), © 1939 by Ziff-Davis Publishing Co., reprinted by arrangement with Forrest J. Ackerman, Holding Agent, 2495 Glendower Ave., Hollywood, CA 90027

Today, however, the need to apologize for science fiction tales about time travel isn't quite so necessary. Now and then, in fact, you'll even find one of the better pulp stories cited in highly mathematical papers on time machines in the *Physical Review D*, one of the most important scholarly physics journals. Even those physicists and philosophers who mostly ignore science fiction—except perhaps to make slightly condescending remarks—would, if honest, admit that their early teenage interest in time travel was sparked by reading a really good science fiction story, and not by working their way through a physics textbook. Yes, when the physics eventually came later, it was very good—but the science fiction came *first*, and it was pretty good, too.[20] It's in a 1937(!) tale, for example, that we find the claim for consistency around a closed loop in time, *decades* ahead of the physicists and philosophers.[21] And when you get to the final section of Chap. 4, I think you'll find it difficult to believe that Everett's many-worlds interpretation of quantum mechanics, dating from the late 1950s (which avoids the standard paradoxes of time travel) wasn't inspired by some youthful reading of science fiction from the 1930s and 1940s.

In a number of places in this book you'll find my comments on how science fiction has occasionally anticipated physicists on the subject of time machines and time travel. This is *not* to be interpreted as some sort of 'gotcha' in favor of science fiction. Far from it. When push comes to shove, physics *always* wins. This situation was specifically addressed by Joe Haldeman, in an afterword to his 2007 novel *The Accidental Time Machine*. There he wrote, about when he started in 1971 to write his earlier, now classic novel *The Forever War*, "I needed a way to get soldiers from star to star within a human lifetime, without doing too much violence to special and general relativity. *I waved my arms around really hard* [my emphasis] and came up with the 'collapsar jump'—at the time, collapsar was an alternate term for 'black hole,' though I was unaware of the latter term [because John Wheeler had invented it only 4 years before, as discussed in Chap. 1 and note 106]." And then Haldeman admitted "It's a truism of science fiction that if you predict enough things, a few of them are going to come true. . . . What I think it actually demonstrates is that *if you wave your arms around hard enough* [my emphasis], sometimes you can fly."

Now, there is one feature common to all books on time travel to the past (which is the central topic treated here, of course) that I would like to clearly state. It's obviously a subject of vast interest to physicists, and yet it offers (as far as I know) *absolutely no hope of suggesting even a single experiment for study.* (As far as I know, nobody is building a time machine in their basement.) A suggestion *has* been made that it may be possible to detect, in the *present*, the effects of the *future* operation of "man-made time machines, which could be of a size traversable by humans," that is, machines with a 1-m spatial extent offering a one second trip into the past.[22] With the best technology available today, however, the calculated effects on the proposed two-particle scattering experiment are orders of magnitude too small to measure.

---

[20]The view expressed by Vladimir Voinovich's time traveler in his 1986 novel *Moscow 2042* (Science fiction . . . is not literature, but tomfoolery like the electronic games that induce mass idiocy.) is, I think, wrong. For an interesting presentation on the role of science fiction in exciting an interest in science among youngsters, see the paper by Frederik Pohl (1919–2013), "Science Fiction: the stepchild of science," *Technology Review*, October 1994, pp. 57–61. In this essay Pohl, a well-known writer of science fiction and editor of *Galaxy Science Fiction* and *If* magazines, writes "Science fiction is [the ultimate protection] against future shock . . . if you read enough of it, nothing will take you entirely by surprise." Not even time travel.

[21]P. S. Miller, "The Sands of Time," *Astounding Stories*, April 1937.

[22]S. Rosenberg, "Testing Causality on Spacetimes with Closed Timelike Curves," *Physical Review D*, March 15, 1998, pp. 3365–3377.

This situation is really unprecedented in the history of science.[23] To cynics, it may seem to be a bit like writing learned papers on the thermodynamics of fire-breathing dragons (which, like other mythological entities—and time machines, too—have yet to be seen)! This one fact has opened the doors—and has kept them open for decades—for philosophers and science fiction writers, who can endlessly debate back and forth on all aspects of time travel to the past with nary a single experimental fact to complicate their lives. For physicists the situation is naturally frustrating, but for philosophers and science fiction writers it's a dream come true. This isn't to say it's *all* basically theological in nature. Both the physicists and the philosophers *have* written many fascinating papers and books and, of course, so have science fiction writers. Mathematical physics *has* been advanced.

Still, despite all of the theoretical work done in the last 30 years, work that has made it reasonable to seriously talk of 'time travel' and 'time machines,' I suspect many would nonetheless agree with these words from more than 75 years ago: "Of all the fantastic ideas that belong to science fiction, the most remarkable—and, perhaps, the most fascinating—is that of time travel . . . Indeed, so fantastic a notion does it seem, and so many apparently obvious absurdities and bewildering para-doxes does it present, that some of the most imaginative students of science refuse to consider it as a practical proposition."[24] For some, time travel is an even more unlikely possibility than (as declared by Robert Lewis Stevenson) is the "welding of ice and iron." Not all physicists and philosophers view the time travel/paradox arguments as convincing, however. Provocative, yes, of course, but many are not yet prepared to write 'signed, sealed, and delivered' at the end.

So, keep reading and I think you'll discover why there *are* those who are not so quick to dismiss the possibility of following the fantastic world line of H. G. Wells' intrepid Time Traveller[25] into the future. And, just maybe, into the distant past, too.

---

[23]Perhaps, however, I am too hasty. More recent theoretical calculations suggest that wormholes connecting our universe with other universes would, after converting into time machines, have characteristic thermal signatures. See P. F. González-Díaz, "Thermal Properties of Time Machines," *Physical Review D*, 2012, pp. 105026-1 to -7 which, however, concludes that a search for such signatures would be "quite difficult [with the] instruments available."

[24]I. O. Evans, "Can We Conquer Time?" *Tales of Wonder*, Summer 1940.

[25]The Time Traveller is never named in Wells' 1895 novel *The Time Machine*. An earlier (1888) attempt at a time machine story, with the awful title *The Chronic Argonauts* (the "chronic" was apparently inspired by the word *chronology*), so embarrassed Wells that he later called it "imitative puerile stuff," "clumsily invented, and loaded with irrelevant sham significance," and "inept," and so he hunted down and destroyed every copy of it that he could find. You can find *The Chronic Argonauts* reprinted in *The Definitive Time Machine* (H. M. Ceduld, editor), Indiana University Press 1987. The hero in that work *was* named: Dr. Moses Nebogipfel. There is one passage in *The Time Machine* that does tantalize; as the Time Traveller explores a museum of "ancient" artifacts in the Palace of Green Porcelain (they are, of course, artifacts of our *future*) he reveals that "yielding to an irresistible impulse, I wrote my name upon the nose of a steatite monster from South America that particularly took my fancy." Thus, the Traveller *has* given his name, but his signature exists only in the future, in a museum of the past that is yet to be built.

For Further Discussion

For time travel to the past to make any sense, the past must in some sense 'still be there.' This is a concept that we'll find later in the book to have significant support in relativistic physics, but for now let's limit ourselves to a purely romantic view. As an example of this, consider this passage by Canadian writer Grant Allen (1848–1899), from the Introduction to his 1895 time travel novel *The British Barbarians*: "I am writing in my study on the heather-clad hill-top. When I raise my eye from my sheet of foolscap, it falls upon miles and miles of broad open moorland. My window looks out upon unsullied nature. Everything around is fresh and pure and wholesome ... But away below in the valley, as night draws on, a lurid glare reddens the north-eastern horizon. It marks the spot where the great wen of London heaves and festers." I personally find it quite tempting to imagine Allen somehow still there in his study of 1895, and of heaving and festering late-Victorian London, too, with H. G. Wells himself in the middle of it, still reading the first rave reviews of *The Time Machine*. In Wells' novel The Time Traveller journeys into the far future, while in Allen's work the protagonist is a twenty-fifth century anthropologist who has traveled back to the past of the late nineteenth century to study the 'British barbarians.' Read Allen's novel (it's available on the Internet, for free, as a Project Gutenberg book) and comment on the significance of its appearance at virtually the same time as Wells' great work. Why do you think Wells' novel is remembered, and Allen's is not?

In the opening paragraph of his paper "The Conundrum of Time Travel" (*Croatian Journal of Philosophy*, No. 37, 2013, pp. 81–92), Anguel Stefanov writes "Needless to say ... the problems concerning time travel are being still tackled by science fiction only, but resolved by science proper neither theoretically, nor practically." Do you think this is correct?

Eventually every genre of writing becomes the target for parody, in which the *form* of the genre serves as the framework for what (it is hoped) is a humorous mockery. The most famous example of this, perhaps, is the annual

(continued)

Edward Bulwer-Lytton contest in writing a take-off on the long-winded opening line of the 1830 novel *Paul Clifford*, by Bulwer-Lytton (1803–1873). That opening line *is* a wonder (a masterpiece of purple prose): "It was a dark and stormy night; the rain fell in torrents—except at occasional intervals, when it was checked by a violent gust of wind which swept up the streets (for it is in London that our scene lies), rattling along the housetops, and fiercely agitating the scanty flame of the lamps that struggled against the darkness." Here's a recent (from the 2015 contest) spoof: "The Contessa's heart was pounding hard and fast, like an out-of-balance clothes washer, which can get that way if you mix jeans with a lot of light things, though the new ones have some sensor thing to counteract that or shut off, but the Contessa's heart didn't have anything like that, so she had to sit down and tell Don Rolando to keep his hands to himself for a while." Science fiction isn't immune to such fun, and a good example of that can be found in the September 14, 2015, issue of *The New Yorker*, which has (on p. 50) "Eight Short Science-Fiction Stories" by Paul Simms. Here's the one I laughed hardest at: "The Gene-Splicers had tinkered with the DNA, producing a race of warriors who craved just two things: the thrill of battle and the taste of their own feet. They hungered for battle. They literally ate their own feet. None survived to reproduce, and within a few short years they were all gone. The Gene-Splicers chalked it up to experience, and decided to try harder the next time." That, and the other seven spoofs by Simms, cut across a wide swath of science fiction, but one theme noticeably absent was that of time travel. Try your hand at writing a *short* (fewer than 500 words) time travel spoof, and be prepared to read it aloud to an audience of your peers.

The tale "Through the Dragon Glass" by Abraham Merritt (1884–1943) appeared in the early pulp magazine *All-Story Weekly* of November 24, 1917. It described the discovery of a passage through an ancient Chinese mirror into an alternate world. One *might* think of this as an early conception of a wormhole, but more likely it may remind you mostly of Lewis Carroll's *Through the Looking Glass*. More interesting for us, in this book, is a story written 75 years ago that describes a gadget connecting two regions of spacetime, with a time shift of a week between the two regions. (See "Time Locker" by Lewis Padgett, in the January 1943 issue of *Astounding Science Fiction*.) The gadget falls into the hands of a crooked lawyer who, not understanding what he has, ends up accidently killing himself. As the story ends, the inventor of the gadget ruefully muses to himself that the lawyer "must have been the only guy who ever reached into the middle of next week—and killed himself!" The gadget is, in everything but name, a wormhole time machine. Speculate on how such a spacetime structure could appear in a *science fiction magazine*(!) decades before there was any discussion of such a possibility in the physics world.

Adjectives used to describe many of the stories in the science fiction pulps included *primitive, trashy, tawdry, silly, absurd, crummy, ludicrous,* and *cheap.* One early pulp magazine actually boasted, of its contents, that they contained "sensational fiction with *no* philosophy." Speculate on how such a low-level 'literary' form could have been so successful in finding an enthusiastic audience for time travel paradox tales, tales that are in fact by their very nature simply *stuffed* with philosophical issues. As an example of the tremendous emotional power a particularly well-written time travel story can deliver, read Isaac Asimov's "The Ugly Little Boy" (*Galaxy Science Fiction,* September 1958). Asimov rated this story as among his most favorite of all the many he wrote. If you can read it without ending in tears, well, ... An excellent modern historical work on the pulps (of all genres, not just science fiction) is by Lee Server, *Danger Is My Business: an illustrated history of the fabulous pulp magazines, 1896–1953,* Chronicle Books 1993.

A literary fascination with time was already 'in the wind' when Wells wrote his *Time Machine,* as with Oscar Wilde's 1890 novel *The Picture of Dorian Gray.* Even decades earlier than that one can find a hint of time travel of a sort in Edgar Allen Poe's 1841 short story "Three Sundays in a Week." And just 4 years later Henry Wadsworth Longfellow wrote his haunting poem "The Old Clock on the Stairs," with these opening words:

Somewhat back from the village street
Stands the old-fashioned country-seat.
Across its antique portico
Tall poplar-trees their shadows throw;
And from its station in the hall
An ancient timepiece says it all,—

    "Forever—never!

(continued)

Never—forever!"

The most interesting of all pre-Wells time travel fiction to appear in a mass-audience publication was, I think, the short story "The Old Folks Party" by Edward Bellamy, printed in the March 1876 issue of *Scribner's Monthly*. In this story a group of teenagers, who belong to a weekly discussion club, agree that at their next meeting they will all come dressed and behaving as they believe they will be dressing and behaving 50 years in the future. Also attending will be the grandmother of one of the young ladies. The meeting of the "old folks" takes place, and it invokes such powerful feelings of mortality that, at last, one of the young men can stand it no more: "Suddenly Henry sprang to his feet and, with the strained, uncertain voice of one waking himself from a nightmare, cried:—'Thank God, thank God, it is only a dream,' and tore off the wig, letting the brown hair fall about his forehead. Instantly all followed his example . . .." The young people then began to laugh with relief at once again being young, until they notice the grandmother is crying. Her granddaughter instantly knows what is wrong and says, "Oh, grandma, we can't take you back with us." Read, compare, and contrast, these works by Wilde, Poe, Longfellow and Bellamy, with the 'scientific' presentation of time travel by Wells.

Acknowledgements

My name is on *Time Machine Tales* as the author, but there are many others without whose help I could not have written it. For this, the revised and updated third edition of the original 1993 book *Time Machines*, I am enormously grateful to the publisher, Springer Science + Business Media, and to my physics editors at Springer, Dr. Sam Harrison and his assistant Ho Ying Fan in New York City. They made an intellectually demanding and inherently lengthy project (one that could easily have turned into a nightmare) a pleasant one. Springer's New York City-based editorial assistant Irene Bruce smoothly handled the administrative details of transforming the book from typescript to print. And to Serguei Krasnikov in Russia, Frank Arntzenius and Roberto Casati in England, Francesco Gonella in Italy, and Geoff Goddu and Archille Varzi in America, I offer grateful thanks for permission to reprint from their publications.

But the story of this book actually begins long before 1993.

Nearly four decades in the past, in 1979, when time travel was still mostly just a nutty idea used by science fiction writers and a few rogue philosophers—proposing to give a serious seminar on time travel at the weekly college *physics* seminar would almost certainly have resulted in getting the bum's rush out the nearest door, or maybe even tossed through the closest window (or, God forbid, *off the roof!*)—I wrote my first time travel tale for *Analog Science Fiction* magazine. Bought by then editor Ben Bova (you can read it in Appendix A), he also bought another time travel tale from me soon after he had become the fiction editor at the newly created *Omni Science Fiction* magazine (you can read it in Appendix B). (Years later, Ben's successor as *Omni*'s fiction editor, Ellen Datlow, bought a third time travel story from me—"The Invitation"—which you can read if you can locate the July 1985 issue.)

Those fictional experiences, and my discovery of Princeton philosopher David Lewis' seminal 1976 essay on the *logical* possibility of time travel, led me to plunge into a deeper study of the *physics* of time travel, to the point where as the 1980s ended (1988, to be precise) I thought a book should be my next writing project. Alas, nobody else shared that thought. I spent the next 3 years looking for a publisher among numerous university presses; as I looked, the pile of rejection letters, like entropy, grew steadily larger.

The editor at one well-known university press, in fact, simply laughed at the idea of a scholarly time travel book when I called him on the telephone (e-mail was then still in its infancy) to ask, after a long period of no response to my written proposal, if he had gotten it. (Perhaps, I told myself, the mail truck carrying my precious document had fallen into a hyper-dimensional spacetime warp: how pathetically desperate is the anxious academic writer!) Later, I published a number of math/physics books with that same press (but now with new editors) and so there were no lasting hard feelings.

That laughing, unresponsive editor wasn't alone in his opinion, I have to admit, and it wasn't until 1991 when Maria Taylor, the publisher at the Press of the American Institute of Physics (AIP Press), decided to take a chance (a *big* chance) and publish the first edition of *Time Machines*. She made that decision in large part because of two extremely supportive reviews of my proposal from the academic physicists Edwin Taylor (MIT, and a former editor of *The American Journal of Physics*) and Gregory Benford (University of California, Irvine, as well as being an award-winning writer of science fiction who occasionally used the time travel theme). After the original *Time Machines* appeared, Ben Bova asked me to use it as a guide to writing a book on time travel for the new Writer's Digest series on science fiction that he was editing, aimed specifically at would-be story writers; that book came out in 1997 as *Time Travel*. Later, Trevor Lipscombe, then editor-in-chief at the Johns Hopkins University Press (and my former math editor at Princeton University Press), reprinted *Time Travel* (with a new Preface) in 2011.

After Springer acquired AIP Press (and all of its books) in 1995, and Maria Taylor herself went over to Springer, she and I collaborated again on bringing out the second edition of *Time Machines* in 1999. And now, 18 years later, here (with a title slightly altered for reasons I give in "Some First Words") is its successor.

To Sam, Ho Ying, Irene, Ben, Ellen, Maria, Edwin, Gregory, and Trevor, thank you for your support. But my greatest debt of all, one I can never even begin to repay, is to my wife of 55 years, Patricia Ann, a woman of infinite tolerance. Who else would put up with someone who plays first-person-shooter video games at midnight (the bigger and the louder the explosions, the better!) and writes books on time machines, all the while claiming not to be crazy?

University of New Hampshire, Exeter, NH, USA   Paul J. Nahin

Introduction

> Over the last few years leading scientific journals have been publishing articles dealing with time travel and time machines. . . . Why? Have physicists decided to set up in competition with science fiction writers and Hollywood producers?
> —John Earman (see note 25 of *Some First Words*)

Writing about time travel is, today, a respectable business. It hasn't always been so. After all, time travel, *prima facie*, appears to violate a fundamental law of nature; every effect has a cause, with the cause occurring before the effect. Time travel to the past, however, seems to allow, indeed to *demand*, backwards causation, with an effect (the time traveler emerging into the past as he exits from his time machine) occurring *before* its cause (the time traveler pushing the start button on his machine's control panel years *later* to start his trip backward through time).

Thus, when H. G. Wells published his breakout masterpiece, *The Time Machine*, in 1895, even those readers who loved it as a *story* (and not all did) were still quick to dismiss it as a *romantic fantasy*. It was, in their view, certainly an emotionally powerful tale of pure imagination, but nothing more. Reviewers of the day used such words as "hocus-pocus" and "bizarre," and called the work a "fanciful and lively dream."[26] Any one of the novels by Wells' contemporary, Jules Verne (even such super-technology ones like the 1865 *From the Earth to the Moon*) would have been ranked *far* above Wells' novella in terms of 'it could actually happen.'

Wells himself always denied that his time machine was anything more than a literary device[27] to get his Time Traveller into the far future. Indeed, in 1934, in the

---

[26]These reviews are reprinted in P. Parrinder, *H. G. Wells: The Critical Heritage*, Routledge & Kegan 1972. A modern reviewer has applied such negative characteristics to the Time Traveller, himself, calling him "a kind of Trickster figure" and "a quack and magician." See Robert J. Begiebing, "The Mythic Hero in H. G. Wells's *The Time Machine*," in *Essays in Literature*, Fall 1984, pp. 201–210.

[27]Wells was not the first to use a machine to enable time adventures, as the Spanish writer Enrique Gaspar (1842–1902) used one in his 1887 story *The Time Ship: A Chrononautical Journey*. It's Wells' tale we remember, however.

preface to *Seven Famous Novels* (published by Knopf), a collection of his novel-length scientific romances (as science fiction had been known before the term *science fiction* came into use), including *The Time Machine*, Wells made his position perfectly clear: "These stories of mine collected here do not pretend to deal with possible things; they are exercises of the imagination ... They are all fantasies; they do not aim to project a serious possibility; they aim indeed only at the same amount of conviction as one gets in a good gripping dream." Wells then went on to say in that same preface that all attempts before at writing fantastic stories depended on magic. But not in his works. "It occurred to me that instead of the usual interview with the devil or a magician, an ingenious use of scientific patter might with advantage be substituted." Wells' great contribution to time traveling story-telling was his introduction of a *machine*; science instead of magic, drugs, dreams, blows on the head, or suspended animation.[28] Not all modern science fiction writers have followed Wells' lead, however.

A science fiction tale by Clifford Simak (1904–1988), for example, the 1978 novel *Mastodonia*, incorporates an alien creature marooned on Earth (because of a spaceship crash centuries earlier) who 'makes time tunnels.' One of the characters in the story, who is attempting to start a time-travel agency using these tunnels, explains why not having a time *machine* is causing her difficulties with prospective clients: "The whole trouble was that I couldn't tell them about some machine—a time-travel machine. If I could have told them we'd developed a machine, they'd have been more able to believe me. We place so much trust in machines; they are magic to us. If I could have outlined some ridiculous theory and spouted some equations at them, they would have been impressed." I think that's off the mark. We trust in machines not because they are magic, but for precisely the opposite reason. They are *not* magic, but rather are *rational*. And to dismiss mathematics is to say that some non-natural—some supernatural—influence is at work.

But is a time *machine* actually possible? Or is the idea of a time machine simply "Nonsense" and "A bilgeful of crap," as a character bluntly puts it in the 1972 novel *The Dancer from Atlantis* by Poul Anderson (1926–2001). Wells, himself, addressed this point in an autobiographical essay (published in the *Cornhill Magazine*) that he wrote in July 1945 (just 13 months before his death) in even blunter words. Writing under the name of "Wilfred B. Batterave," he penned a very funny summary of his life titled "A Complete Exposé of This Notorious Literary Humbug." There he described *The Time Machine* as "[A] tissue of absurdities in which people are supposed to rush to and fro along the 'Time Dimension.' By a few common tricks of the story-teller's trade, Wells gets rid of his Machine before it can be subjected to a proper examination. He cheats like any common spook raiser. Otherwise it is plain commonsense that a man might multiply himself indefinitely,

---

[28]Examples of 'non-machine' time travel stories of the last four types are, respectively, H. G. Wells' "The New Accelerator" (1901), Charles Dickens' *A Christmas Carol* (1843), Mark Twain's *A Connecticut Yankee in King Arthur's Court* (1889), and Edward Bellamy's *Looking Backward, 2000–1887* (1888).

pop a little way into the future and then come back. There would then be two of him. Repeat *da capo* and you have four, and so on, until the whole world would be full of the Time Travelling Individual's vain repetitions of himself. The plain-thinking mind apprehends this in spite of all the Wellsian mumbo-jumbo and is naturally as revolted as I am by the insult to its intelligence." Funny, yes, but still pretty harsh stuff.

As one writer has argued,[29] Wells was, rather than presenting a scientific discovery, simply attempting to refute the nearly suffocating, unjustified (in his mind), smug optimism of the well-to-do of the Late Victorian Age. And so, on his journey to the year A.D. 802,701, the Time Traveller finds the awful decay of humanity in the cannibalistic subjugation of the Eloi by the Morlocks, the end result of class warfare between the working class (Morlocks) and the idle, parasitic upper class (Eloi).

The German social philosopher Karl Marx, if he hadn't already been dead for 12 years in 1895, would surely have nodded in vigorous agreement as he read *The Time Machine*, even as he would have regretted Wells' decision to have the victory of oppressed workers take so long. (What irony that he is buried in London's Highgate Cemetery, the Victorian Valhalla where he has spent the last century and more quite literally mingling with many of the capitalistic ancestors of the Eloi!) What Marx would have thought of *time travel* as a possibility is, however, far less certain.

How things changed in the years that followed *The Time Machine*. There was, at first, admittedly a 'slight' decline in literary merit as the newly developing pulp science fiction magazines picked-up and ran with the time travel genre. Many of the magazine time travel tales of the 1920s, 1930s, and 1940s were, frankly, simply awful. BUT—some were pretty good, too. And some were, in fact, *very* good. From the 1950s on, there have been ever more sophisticated time travel tales from ever more sophisticated writers.

In the academic communities of philosophers and physicists, too, big events occurred. I give the philosophers the edge, in fact, with the 1976 publication of a hugely important paper that opened with these dramatic words: "Time travel, I maintain, is possible. The paradoxes of time travel [to the past] are oddities, not impossibilities. They prove only this much, which few would have doubted: that a possible world where time travel took place would be a most strange world, different in fundamental ways from the world we think is ours."[30] That writer wasn't the first philosopher to write on time travel to the past, but none had expressed themselves in such powerful and unequivocal words in unmistakable support of the concept.

---

[29]R. M. Philmus, *"The Time Machine*; Or, the Fourth Dimension as Prophecy," *Publications of the Modern Language Association*, May 1969, pp. 530–535.

[30]David Lewis, "The Paradoxes of Time Travel," *American Philosophical Quarterly*, April 1976, pp. 145–152. Lewis (1941–2001) was a Princeton University philosophy professor.

Lewis' paper is also notable because it gives what seems to be a clear definition of just what it means to say one has 'traveled in time,' either to the past *or* to the future:

> What is time travel? Inevitably, it involves discrepancy between time and time. Any traveler departs and then arrives at his destination; the time elapsed from departure to arrival (positive, or perhaps zero) is the duration of the journey. But if he is a time traveler, the separation in time between departure and arrival does not equal the duration of the journey.

To understand this, we need to appreciate the distinction between the *personal time* of the time traveler and the *external time* of remote observers of the time traveler. A time traveler's personal time is measured, for example, either by the time kept by his wrist watch or, perhaps, by a burning candle. (This distinction had actually appeared earlier in Horwich's paper—see note 27 in *Some First Words*—published the year before Lewis' paper.)

I say I 'give the edge to the philosophers' because, while the first *physics* time travel paper had appeared decades earlier, its author wasn't really a physicist at all but rather was Einstein's friend, the world-famous mathematical logician Kurt Gödel. Gödel's paper was, in retrospect, a pivotal event in establishing the 'respectability' of *scientific* time travel; it's worthwhile to take some time here to explain this important point. For physicists (and for philosophers and science fiction writers, too) a 'time machine,' one either constructed by intelligent beings or occurring naturally, manipulates (all the while obeying the known laws of physics) finite amounts of matter and energy in a finite region of spacetime.[31] A 'time machine' would be declared to be *plausible* if it could be explained by a rational, scientific theory. Such a rational theory is found in Einstein's general theory of relativity. (His *special* theory of relativity applies in those situations where there is no gravity.)

Until Einstein, the theory of gravity used by scientists was Newton's—a theory that, although amazingly accurate for any situation encountered on Earth, does have observable errors in certain astronomical applications. In addition, Newton's theory is a descriptive one; it makes possible the calculation of gravity effects without offering any explanation for gravity itself. Einstein's theory not only gives the right answers, even in those cases where Newton's theory doesn't, but it also explains gravity. It does that by treating the world as a four-dimensional structure in which all four dimensions (three of space and one of time) are in a certain sense on equal footing. The resulting Einsteinian description of the world is that of a unified spacetime in which time and space are intimately intertwined, whereas Newton's theory keeps time and space separate and distinct.

---

[31]I am going to feel free to use words like *spacetime* without having to first write introductory essays on relativity theory and tensor mathematics, because such words have entered common use. All those Hollywood science fiction movies, even the crummy ones that routinely trash the laws of physics, have at least expanded the general imagination!

As Newton wrote of time, at the start of his 1687 masterpiece *Principia*, a work that revolutionized physics, "Absolute, true, and mathematical time, of itself, and from its own nature, flows equably without relation to anything external, and by another name is called duration." This view of time would be, of course, discarded with the arrival of Einstein and his view of variable time depending on the state of the observer.

Unlike Einstein's view, Newton's view of the nature of time was entangled with theology. As one modern theologian has written, "Newton conceived of absolute time as grounded in God's necessary existence."[32] To quote Newton himself, in the *General Scholium* to the second edition of *Principia* (1713) he added words that didn't appear in the original: "God is a living, intelligent, and powerful Being; and, from his other perfections, [it follows] that he is supreme, or most perfect. He is eternal and infinite, omnipotent and omniscient; that is, his duration reaches from eternity to eternity; his presence from infinity to infinity; he governs all things, and knows all things that are or can be done. He is not eternity and infinity, but eternal and infinite; he is not duration or space, but he endures and is present. He endures forever, and is everywhere present; and, by existing always and everywhere, he constitutes duration and space. Since every particle of space is *always*, and every indivisible moment of duration is *everywhere*, certainly the Maker and Lord of all things cannot be *never* and *nowhere*."

Okay, I'll be honest—I really am not at all sure just what that means! Newton added these words to the *Principia* in response to criticism (from the influential philosopher George Berkeley (1685–1753)) that his original statements about absolute time were "pernicious and absurd notions," notions that were in fact atheistic in conception. That was a most serious charge in Newton's day, and he was trying (I think) to find some cover from those critics who spent more hours of the day thinking about God than of physics. Much more honest (in my opinion) are the witticisms 'time is just one damn thing after another' and 'time is what keeps everything from all happening at once.'[33] More funny than useful, yes, of course, but at least they're funny.

Newton's theological view of time is simply irrelevant to the modern physicist (although perhaps of more interest to the philosopher-historian) but in many cases it is of *central interest* to the science fiction writer. For example, Newton's religious

---

[32]William Lane Craig, "God and the Beginning of Time," *International Philosophical Quarterly*, March 2001, pp. 17–31, which discusses the question 'Why didn't God create the world sooner?' One irreverent answer is 'He was busy creating Hell for all those who ask that question,' but a more scholarly analysis can be found in Brian Leftow, "Why Didn't God Create the World Sooner?" *Religious Studies*, June 1991, pp. 157–172.

[33]This last 'definition' first (as far as I know) appeared in the work of the science fiction writer Ray Cummings (1887–1957), in his 1921 story "The Time Story," published in *Argosy-All-Story* magazine. He repeated the phrase in his 1929 novel *The Man Who Mastered Time*, and then again in the 1946 novel *The Shadow Girl*. ("This same Space; the spread of this lawn ... what would it be in another 100 years? Or a 1000? This little space, from the Beginning to the End so crowded with events and only Time to hold them apart!")

mindset and its (perhaps!) connection with time travel is treated in my short story "Newton's Gift," originally published in *Omni Magazine* (January 1979) and reprinted in Appendix B at the end of this book. Wells' Time Traveller's view of time is more Newtonian than it is Einsteinian—and perhaps that's not such a big surprise, considering that Einstein was only 16 years old when *The Time Machine* was published.

From the first (1905) it has been known that Einstein's special theory allows time travel into the future via the well-known mechanism of *time dilation*. (The faster a rocket ship travels relative to Earth, the slower is the tick-tock of a wrist watch worn by a rocketeer, compared to that of an identical watch back on Earth.)[34] To return from the future, however, to travel back into the past to the instant after the traveler began his journey, had been thought to be impossible. It was Gödel's discovery that showed the general theory, which has passed every experimental test it has been subjected to (most recently, the September 2015 detection, from two massive colliding black holes, of gravitational waves—'ripples in spacetime'—generated more than a billion years ago in an effect predicted by the general theory a century ago), does allow time travel to the past *under certain conditions*. It is this availability of a *theory* that distinguishes time travel speculations from the outlandish fantasy speculations with which it is often unjustly lumped—speculations that *are* in the province of quacks (such as ESP, astrology, and mind over matter a' la spoon bending).

In his general theory, Einstein showed how spacetime can be either 'flat' (in the no-gravity, special relativity case of what is called a *Minkowski spacetime*[35]) or 'curved' (those situations with gravity), and he did that not by verbal hand waving, but rather by writing mathematical equations that obey all the known laws of physics: his famous gravitational field, nonlinear differential tensor equations. These complicated equations are notoriously difficult to solve in general, but in certain, special cases they *have* been solved. Those solutions describe how matter and energy and spacetime interact. As the popular saying puts it, "Curved spacetime tells matter how to move, and energy and matter tell spacetime how to curve." In that sense, gravity *is* curved spacetime.

In 1949 Gödel found one such special solution to the field equations that describes the movement of mass-energy not only through space but also *backward in time* along trajectories in spacetime that are called *closed time-like lines* or *curves*

---

[34]One pulp magazine science fiction story (F. J. Bridge, "Via the Time Accelerator," *Amazing Stories*, January 1931) got this right when its time traveler explains how his time machine works with these words: "Time as we know it is not universally absolute. The rate of its passage depends to a great extent upon the velocity of its observer with regard to some certain reference system. A moving clock will run slower with respect to a selected coordinate system than a stationary one." (Recall my earlier comments on the personal time of a time traveler.)

[35]Named after Hermann Minkowski (1864–1909), Einstein's mathematics professor in Zurich who gave the now well-known spacetime diagram interpretation of special relativity which, when originally presented by Einstein, was in the form of pure mathematics.

(called CTLs or CTCs, respectively).[36] These trajectories are such that if a human traveled along one, *always at a speed less than that of light* (that's what *time-like* means), he would see everything around him happening in normal causal order from moment to moment (for example, the second hand on his wrist watch would tick clockwise into the local future), but eventually the CTL/CTC closes back on itself and the traveler finds himself in his own past.

On the scale of the Solar System, general relativity has causality built into itself, but on much larger scales things can be a good deal more complicated. On a very large, astronomical scale, in fact, curved spacetime can result in violations of causality, with effects occurring before their causes. That is what the physics and the mathematics of Gödel's solution imply. That is what is meant by saying there is a scientific, rational basis for discussing time travel to the past. It is particularly important to note that travel along one of the closed time-like world lines discovered by Gödel requires a *machine*, some kind of accelerating rocket ship. That's because none of Gödel's CTLs/CTCs are what is called a *geodesic*. That is, none are *free-fall* world lines.[37] This machine does not, however, generate CTLs/CTCs where none existed before (CTL/CTC *creation* requires what physicists call a *strong* time machine) but rather simply makes use of the CTLs/CTCs that are inherent in Gödel's spacetime. A Gödelian rocket ship then is an example of a *weak* time machine.

I mentioned earlier that "certain, special cases" of Einstein's gravitational field equations result in CTLs/CTCs. What was the "special case" that Gödel solved? His solution of the field equations is for a rotating, infinite, static universe composed of a perfect fluid at constant pressure. In such a universe Gödel found that naturally occurring CTLs/CTCs pass through every point in spacetime; that is, time travel in Gödel's universe is *not* the result of a machine *manipulating* mass and energy on a *local* scale (the classic science fictional description of a time machine); rather, in Gödel's spacetime time travel is a naturally occurring phenomenon! The observable

---

[36]Kurt Gödel, "An Example of a New Type of Cosmological Solutions of Einstein's Field Equations of Gravitation," *Reviews of Modern Physics*, July 1949, pp. 447–450. A CTL/CTC is a special type of *world line*; the trajectory through spacetime of every particle in the universe is a world line that extends from each particle's past to its future. Our everyday experiences are with world lines that never cross or come close to themselves (which would put a particle at or near the same spacetime point more than once). That lack of experience with CTLs/CTCs that self-intersect is what makes time travel to the past so difficult for humans to grasp. For a discussion of *how* Gödel did what he did, see Wolfgang Rindler, "Gödel, Einstein, Mach, Gamow, and Lanczos: Gödel's Remarkable Excursion into Cosmology," *American Journal of Physics*, June 2009, pp. 498–510.

[37]It was discovered in 1969, however, that this isn't strictly true *if* one allows for a test particle (our 'time traveler') to be electrically charged. Then, naturally present electromagnetic forces acting on the particle could be sufficient to propel the particle along a Gödelian CTL/CTC. That is, no rocket would be required. See U. K. De, "Paths in Universes Having Closed Time-Like Lines," *Journal of Physics A*, July 1969, pp. 427–432. There are other solutions to Einstein's equations that do allow time travel on free-fall geodesics: see, for example, I. D. Soares, "Inhomogeneous Rotating Universes with Closed Timelike Geodesics of Matter," *Journal of Mathematical Physics*, March 1980, pp. 521–525.

universe is, however, *non*-rotating and expanding (astronomers see red-shifts in the spectrums of distant stars) and so, although Gödel's spacetime satisfies the general relativity field equations, its time travel property does not hold in the spacetime in which we live. (This may account for why the initial reaction in the physics/ philosophical communities, to Gödel's discovery that time travel is *not* nonsense according to general relativity, was mostly indifference.) The failure to observe time travel in our universe may (somewhat surprisingly, I think) still have possible implications for us, however, as one philosopher has cleverly argued.[38] He points out that naturally occurring Gödelian time travel would endow the universe with properties particularly useful for the survival of intelligence (presumably that includes humans) against extinction from a multitude of cosmic disasters. So, for those who argue that the universe we live in was *made* for us (the advocates of various proofs of God's existence that have Him as Designer), we have an obvious question: why did He (apparently) skip incorporating time travel?

In an invited essay that appeared the same year as his time travel physics paper, Gödel specifically addressed the seemingly paradoxical aspect of what he had discovered: "By making a round trip on a rocket ship in a sufficiently wide course, it is possible in these [rotating] worlds to travel into any region of the past, present, and future, and back again, exactly as it is possible in other worlds to travel to distant parts of space. This state of affairs *seems* [my emphasis] to imply an absurdity. For it enables one, e.g., to travel into the near past of those places where he has himself lived. There he would find a person who would be himself at some earlier period of life.[39] Now he could do something to this person which, by his memory, he knows has not happened to him."

Gödel's nerve then failed him, and he defended the possibility of the paradox of a time traveler meeting himself in the past with what I think an astonishingly unconvincing argument (particularly so for a logician) based primarily on *engineering* limitations: "This and similar contradictions, however, in order to prove the impossibility of the worlds under consideration, presupposes the actual feasibility of the journey into one's own past. But the velocities which would be necessary in order to complete the voyage in a reasonable time are far beyond everything that

---

[38] Alasdair M. Richmond, "Gödelian Time-Travel and Anthropic Cosmology," *Ratio*, June 2004, pp. 176–190. Not all physicists think Gödel's result is actually time travel. At least two think it is all simply the result of mathematical hijinks, and that time machines must remain "an aspect of science fiction fantasy": see F. I. Cooperstock and S. Tieu, "Closed Timelike Curves and Time Travel: Dispelling the Myth," *Foundations of Physics*, September 2005, pp. 1497–1509. This skepticism towards Gödel actually started much earlier, when two physicists (one a Nobel physics laureate) incorrectly claimed Gödel had simply gotten his math wrong: see S. Chandrasekhar and J. P. Wright, "The Geodesics in Gödel's Universe," *Proceedings of the National Academy of Sciences*, March 1961, pp. 341–347. It was those two physicists who had erred, however, as was pointed out by the philosopher Howard Stein, in his "On the Paradoxical Time Structures of Gödel," *Philosophy of Science*, December 1970, pp. 589–601.

[39] You'll recall that this is *precisely* the situation that Wells mentions in his "Notorious Literary Humbug" essay. If only he had lived just three more years, to see what he thought to be an absurdity actually appear in the serious writings of a brilliant mathematician!

can be expected ever to become a practical possibility. Therefore it cannot be excluded a priori, on the ground of the argument given, that the space-time structure of the real world is of the type described."[40] That is, Gödel was trying to head off critics of his rotating universe model who might point to the time travel result as proof that the model had to be flawed.

In a footnote Gödel says that the time traveler would have to move at least as fast as nearly 71 % of the speed of light, and that if his rocket ship could "transform matter completely into energy" then the weight of the fuel would be greater than that of the rocket by a factor of 10^{22} divided by the square of the duration of the trip (in rocket years). A trip to the past in Gödel's universe would require a time machine that looked like Dr. Who's telephone booth attached to a fuel tank the size of several hundred *trillion* ocean liners. These are formidable numbers,[41] but they require no violation of physical laws, and that's what really counts if time travel is to be disproved. Gödel's use of engineering limitations for explaining away backwards time travel is actually worse than simply being wrong, because the puzzle is not in practicality but rather in showing, assuming that general relativity is correct, how correct mathematical physics can lead to what seems to be a paradoxical conclusion. (And see note 12 again, for another reason the 'fuel argument' really has no force at all against the possibility of time travel in Gödelian spacetime.)

So, what did the great man himself, Einstein, think of all this? In the same publication as Gödel's essay, he *cautiously* replied as follows: "Kurt Gödel's essay constitutes, in my opinion, an important contribution to the general theory of relativity, especially to the analysis of the concept of time. The problem here involved disturbed me already at the time of the building up of the general theory of relativity, without my having succeeded in clarifying it ... the distinction 'earlier-later' is abandoned for world-points which lie far apart in a cosmological sense, and those paradoxes, regarding the *direction* of the causal connection arise, of which Mr. Gödel has spoken ... It will be interesting to weigh whether these are not to be excluded on physical grounds."

Despite the mathematical physics of Gödel, showing the possibility of time travel to the past, many philosophers are not quite so sure. As one expressed his concerns, "No science-fiction staple poses more philosophical difficulties than time travel, but there is still no consensus as to whether time-travel fictions exhibit logical, metaphysical, or physical impossibility."[42] The best-known and possibly

---

[40]Kurt Gödel, "A Remark About the Relationship Between Relativity Theory and Idealistic Philosophy," in *Albert Einstein: Philosopher-Scientist*: volume 7 of *The Library of Living Philosophers* (P. A. Schilpp, editor), Open Court 1949.

[41]For the analysis of a rocket powered by matter/anti-matter, a known physical process that satisfies Gödel's energy requirement for time travel, see E. Purcell, "Radioastronomy and Communication Through Space," in *Interstellar Communication* (A. G. W. Cameron, editor), W. A. Benjamin 1963.

[42]Alasdair Richmond, "Time-Travel Fictions and Philosophy," *American Philosophical Quarterly*, October 2001, pp. 305–318.

oldest of the paradoxical situations that seem to be part-and-parcel of time travel is the so-called *grandfather paradox*,[43] expressed this way by philosopher David Lewis in his pioneering 1976 paper (see note 5):

> Consider Tim. He detests his grandfather, whose success in the munitions trade built the family fortune that paid for Tim's time machine. Tim would like nothing so much as to kill Grandfather, but alas he is too late. Grandfather died in his bed in 1957, while Tim was a young boy. But when Tim has built his time machine and traveled to 1920, suddenly he realizes that he is not too late after all. He buys a rifle, ... and there [Tim] lurks, one winter day in 1921, rifle loaded, hate in his heart, as Grandfather walks closer, closer ...

So, there's the puzzle. Tim can obviously achieve his goal—he has a loaded gun, he's an excellent shot, a clueless granddad is coming ever closer—but if Tim actually does kill grandfather, years *before* Tim was (will be) born, then how can Tim *be* born? And if he is not born, then how can Tim ('now' not in existence) travel back through time to kill grandfather? What a confusing mess, right? So, the only possible conclusion to all this is that the starting premise, that time travel makes sense, must actually be nonsense. Right?

Well, maybe, but then what of Gödel with his time traveling rocket ship? That's hard-as-diamond, unshakeable mathematical physics, for heaven's sake. We can't just ignore *that*! Lewis offers a way out of this conundrum, and when we get to the book's discussions on paradoxes (that's plural because, believe it or not, there are other paradoxes even *more* perplexing than that of killing granddad in the distant past) we'll return to his solution.

Ever since Lewis wrote his paper, philosophers have been particularly fascinated by the grandfather paradox and have shown themselves to be at least as inventive as the science fiction writers in discussing it, or variations on it.[44] Here, for example, is a twist on that paradox that I think particularly clever, one that avoids the murderous spirit of the tale told by Lewis and Horwich:

---

[43]The origin of this paradox is probably lost in time (the irony of that is *so* appropriate!), but I have traced it at least as far back as to the science fiction pulp magazine *Science Wonder Stories* which published, in its December 1929 issue, an editorial essay titled "The Question of Time Traveling." It challenged readers to think about the following scenario: "Suppose I can travel back into time, let me say 200 years; and I visit the homestead of my great great great grandfather, and am able to take part in the life of his time. I am thus enabled to shoot him, while he is still a young man and as yet unmarried. From this it will be noted that I could have prevented my own birth ...."

[44]Even before Lewis' paper, Paul Horwich had reduced the grandfather paradox to *autoinfanticide*—a time traveler tries to kill his younger *self*—in "On Some Alleged Paradoxes of Time Travel," *The Journal of Philosophy*, August 14, 1975, pp. 432–444. But not all philosophers share this fascination. Earman (see the opening quote), for example, dismisses *all* of the science fiction paradoxes that are so beloved by fans of the genre as "while always good for a chuckle," they are just "crude and unilluminating means of approaching some delicate and deep issues about the nature of physical possibility." I think Earman is fundamentally correct, although I wouldn't go so far as to characterize the paradoxes as mere "chuckles." They are, after all, the source of much of the intellectual motivation prompting the exploration of the physics of time travel. An excellent example of this is found in the paper by the Russian physicist S. V. Krasnikov, "Time Travel Paradox," *Physical Review D*, February 14, 2002, pp. 064013-1 to 064013-8. The physics of the grandfather paradox is of *great* interest in this paper.

Sarah has just completed building her time machine. She decides to test the machine on herself tomorrow morning at which time she intends to travel back one day. In the meantime, she goes home, puts some salve on the burn she received that day, and goes to bed. In the morning, Sarah, with coffee in hand, sits down to read the morning paper. She opens the paper to the following headline: 'Famous physicist found dead.' On the front page is a picture of her body, salve burn clearly visible on her arm, inside her pristine time machine. Underneath is the caption. 'Nobel-prize winning physicist found dead yesterday in mysterious device that materialized near city hall.' Extremely shaken, Sarah returns to the lab and destroys the time machine.[45]

Can any sense be made of this? We'll come back to this question later in Chap. 5, when we discuss the possibility (or not) of time being *multi*-dimensional.

Now, to conclude this Introduction, let me end with two amusing, connected short stories (in epistle form) that nicely describe the issues we'll take up in the rest of this book. The rejection letter for the denial of a research grant to fund the construction of a time machine has just been received . . .

That Useless Time Machine[46]

Dear Review Committee:
It is not our practice to raise complaints against a negative review report. We believe in peer refereeing and we respect it, whatever its content and consequences. However, in the case of our latest grant application (project named 'The Time Machine') we find it necessary to express our astonishment at the motivations with which our request for funding was turned down. Your main objection appears to be that our project is 'philosophically interesting' but 'practically useless', by which you mean that the project 'has no potential for applications.' We do not quite think that the main criterion for judging the scientific value of a project should be its practical usefulness, but never mind that. Let us agree that usefulness is a relevant criterion, especially when large amounts of money are involved. Why should that be a reason to turn down our project? Quite frankly, we cannot think of a project with better application potential than ours. Some examples:

- Cultural tourism: one could send herds of history fans back in time to witness the crucial episodes of the French Revolution, or to watch the Egyptians build the pyramids, or to videotape Socrates' lectures.
- Exotic safaris: we have already received several applications for dinosaur hunting expeditions (they got extinct anyway).

---

[45]G. C. Goddu, "Time Travel and Changing the Past: (Or How to Kill Yourself and Live to Tell the Tale)," *Ratio*, March 2003, pp. 16–32.

[46]Story by Roberto Casati (Senior researcher at CNRS, Paris) and Achille C. Varzi (Professor of Philosophy at Columbia University). Originally published in *Philosophy*, October 2001, pp. 581–583, and reproduced here by kind permission of the authors.

- Error detection: we could take a closer look at our past mistakes and learn how to avoid them in the future.
- Historic documentaries: think of the huge saving in set design, costumes, special effects, etc. (How much did *Gladiator* cost?)

And so on and so forth. Honestly, can you think of a project with better prospects for useful and thrilling applications?

Sincerely Yours,
The 'Time Machine' Research Group

Dear 'Time Machine' Research Group:
Thank you for your letter. We agree that it would be interesting to exploit a time machine for the uses that you suggest. It would also be remarkable if we could use it to prevent all sorts of unpleasant events that happened in the past. It would be remarkable, for instance, to be able to go back to November 22, 1963, and prevent Lee Harvey Oswald from killing John Kennedy, or to go back to April 14, 1912, and steer the Titanic around the iceberg. It would be excellent indeed to be able to do such things. However, suppose your project were to be successful. Suppose you *will* manage to build a time machine. Then why *didn't* you do any of those things? Why is it that our past history is still full of such sad events? Either this means that your project is doomed to fail and you will never manage to build a time machine; *or* it means that the project will succeed but that you are not going to use your time machine for these good purposes. In the first case, logic shows it would be pointless to support your project. In the second case, ethics dictates that it would be wrongdoing. Either way, you must concede that the reasons against your project are overwhelming.

Cordially Yours,
The Review Committee

Dear Review Committee:
Certainly you have noticed that our suggestions for practical applications of the time machine did not include any uses that could result in an alteration of the natural course of history. As a matter of fact, we believe that no such alteration is logically possible. According to our project, it is logically possible to *visit* the past but not to *modify* the past. No time traveler can undo what has been done or do what has not been done. So the logic is safe. This does not mean that the time traveler will be ineffectual during her stay in the past, of course; it simply means that what she is going to do is something that she has already done. An accurate catalogue of all the past events would include an account of the arrival of the Time Machine from out of nothing as well as an account of all the actions and reactions that followed. And ethics is safe, too. For, if indeed we managed to go back to Dallas, we could not stop Oswald from doing what he did. Nobody would be able to stop Oswald because nobody was able to stop him (and nobody was able to stop Oswald because nobody will ever be able to do so, even if they came from the future). Alas, the past is full of sad events but there is nothing that we can do about that.

Respectfully Yours,
The 'Time Machine' Research Group

Dear 'Time Machine' Research Group:
We appreciate the distinction between changing the past (impossible) and affecting the past (possible). However, this simply reinforces our initial impression: your project has no practical value. If in order to travel to the past one has to have been there already, and if one can only do what has already been done, then *á quoi bon l'effort*? Why should we invest in a 'Time Machine' at all? We are afraid that our decision is now final.

Yours with best wishes,
The Committee
Well, all seems to be certainly lost with *that*. But, wait, perhaps not. Maybe, with just one more *really good* appeal, The Committee's rejection can be reversed! If *you* were on the Review Committee, and had just read the following letter, how would *you* vote?

A Useful Time Machine[47]

Dear Review Committee:
We regret your continued decision to reject our proposal. Even though you have told us your decision is now final, we humbly ask your indulgence for one last appeal. We believe you have misinterpreted a crucial part of our proposal.

You maintain that our 'Time Machine' project 'has no potential for applications' and has 'no practical value.' You ultimately base this claim on the fact that "If in order to travel to the past one has to have been there already, and if one can only do what has already been done, then *á quoi bon l'effort*? Why should we invest in a 'Time Machine' at all?" Your argument however is a misinterpretation of our own comments that 'According to our project it is logically possible to visit the past but not to modify the past ... This does not mean that the time traveler will be ineffectual during her stay in the past, of course; it simply means that what she is going to do is something that she has already done.' We regret the awkward and easily misleading locution of the last sentence, but such are the perils of talking about time travel. Regardless, please consider our clarification.

Certainly if we were proposing that the time traveler *be* 5 years old again, we would be proposing something not worth the effort—our proposed time traveler has already turned five and cannot do so again. But we are not proposing that the time traveler do things that have already occurred in her own personal past, but rather in

---

[47]Story by Geoff Goddu (Professor of Philosophy at the University of Richmond, Virginia). Originally published in *Philosophy*, April 2002, pp. 281–282, and reproduced here by kind permission of the author.

her personal future. The time traveler has not yet, from her personal temporal perspective, travelled back to, say, the library at Alexandria in 100 BCE. When she does travel back to 100 BCE to obtain scans of the books in the library before its destruction, she will be older than she is now. When she returns she will be still older (and we hope wiser, i.e., in possession of valuable information to which neither you nor we currently have access).

But is it true that as of 2002 AD [the year this letter was written] the time traveler has already visited Alexandria in 100 BCE? It could well be. But whether or not it is depends upon whether it is *also* true that our project will be successfully funded and completed. Because time travel into the past involves reverse causation, certain past events, such as the time traveler visiting 100 BCE, will be dependent upon certain future events, such as the successful funding and completion of our project. Hence, if it is not true that our project is both funded and completed, then it is not true that our time traveler has of 2002 already visited 100 BCE.

But suppose we were to learn now, before the funding and completion took place, that our time traveler had indeed been present at the library in Alexandria in 100 BCE. Would this imply that there was no reason to expend the effort to fund our project? After all, if the travel has 'already' happened, why bother funding the project? Firstly, such an argument does not imply that a 'Time Machine' would have no practical application, but rather expresses the futile hope that one could in fact get the practical benefits (if time travel is successful, we obtain the desired information) without expending the effort at all. Secondly, the hope is futile, for if we learn right now that our time traveler had been present at the library in 100 BCE, we would then know, assuming no other possible funding source, that you *will* expend the effort to fund our project. To deny this last is to make the impossible suggestion that even though your support is truly a causal antecedent of the successful trip, there is now no need for you to actually expend the effort to provide funding.

Hence, the effort is far from pointless, for the project will only succeed through your and our efforts. And success will generate, not only all the practical applications we outlined in our first letter, but, in addition, a host of information gathering applications such as more accurate historical research, lost item location identification, legal testimony verification, etc. Even if, as we (and you) acknowledged, no one could now prevent Oswald from killing Kennedy, wouldn't it be worth verifying that Oswald was the lone killer of Kennedy? Also, the information gathering need not be restricted to the past. For example, information concerning the prices of various stocks 10 years from now would be extremely valuable to a suitably cautious and prudent investor. Surely you cannot object to our information gathering in the future on the grounds that 'it will already have been done.' And just think, the information we obtain could be what allows you to obtain at very low prices those stocks that in the future will be extremely valuable and allows your esteemed committee to dramatically increase your support of worthy scientific endeavors.

Again, we ask you to reconsider your original decision.
Respectfully yours,
The 'Time Machine' Research Group

A time machine inventor makes an experimental test of the grandfather paradox! (Illustration from "Thompson's Time Traveling Theory" by Mortimer Weisinger), *Amazing Stories* March 1944 (art by Malcolm Smith). Reprinted by arrangement with Forrest J. Ackerman, Holding Agent, 2495 Glendower Ave., Hollywood, CA 90027.

Not everybody likes time machines as a science fiction gadget, not even otherwise enthusiastic devotees of the genre. For example, in a Letter-to-the-Editor published in the December 1931 issue of Astounding Stories, *one seventeen-year-old fan had this to say: "There is only one kind of Science Fiction story I dislike, and that is the so-called time-traveling. It doesn't seem logical to me. For example, supposing a man had a grudge against his grandfather, who is now dead. He could hop in his machine and go back to the year that his grandfather was a young man and murder him. And if he did this how could the revenger be born? I think the whole thing is the 'bunk.'" As this book will demonstrate, this young reader was not alone in that opinion. As this book will also demonstrate, in the last few decades that view has been rapidly evolving.*

For Further Discussion

Read again the penultimate sentence in the last letter from The 'Time Machine' Research Group, and then think about how you would respond to the following questions.

(1) Would *you* invest in a stock market if you knew somebody else had a time machine giving *them* advance information on stock performance?

(2) How might the existence of a time machine influence the future of the stock market, in general? For an early science fiction look at these questions, see Lee Laurence, "History in Reverse," *Amazing Stories*, October 1939.

One writer has speculated that Wells' model for the Time Traveller was the American inventor Thomas Edison. (See Martin T. Willis, "Edison as Time Traveler: H. G. Wells' Inspiration for His First Scientific Character," *Science Fiction Studies*, July 1999, pp. 284–294.) As Wells worked his way from *The Chronic Argonauts*, through revisions, to the final *Time Machine*, the story's hero evolved from Dr. Nebogipfel to the Philosophical Inventor to the Time Traveller. The one individual who could have inspired all of these various hero types was, according to Willis, Edison, a world-famous Victorian-age celebrity whose story was well known to Wells. If Wells had today's scientific personalities available as potential inspirations, who do you think he would use? How might that choice affect the story and structure of *The New Time Machine*?

The idea of *personal time*, used by the philosopher David Lewis (note 5) to consistently interpret time travel stories, has been used in a quite different way (although time travel gets a few words, too) by the philosopher Roy Sorenson. In his paper "The Cheated God: Death and Personal Time," *Analysis*, April 2005, pp. 119–125, Sorenson asks you to imagine an immortal god. For some reason this god runs afoul of a demon, who curses the god in a curious way. (The 'telling of a story' is a common technique in philosophical papers and, while foreign to what readers of physics papers are used to seeing, is not without some charm. Just be sure to always keep in mind that its

(continued)

primary use is as an *attention-grabbing* device, but as far as having any other merit, well, that's often another story.) The curse is such that the life span of the once immortal god is reduced to that of a normal human life span and yet, perhaps surprisingly, the god will still never die. As Sorenson writes, "[The god] will live forever. But [the god] will not have a better life than a mortal. The demon has harmed [the god] as gravely as death harms mortals." How, you might wonder, is this to be done? As Sorenson explains, "[The god] lives half of its now mortal span, followed by a trillion years of nothingness, then a quarter of its mortal span followed by a trillion years of nothingness, then an eighth of its mortal span followed by a trillion years of nothingness and so on ad infinitum." Sorenson's argument is simply an exotic form of the high school summation of the geometric series $\frac{1}{2} + \frac{1}{4} + \frac{1}{8} + \cdots = 1$, where there are an *infinite* number of terms to the left of the equality. (Each term represents a period of time during which the god is conscious, and each + represents a trillion years.) Sorenson picked a trillion years of nothingness between consecutive periods of consciousness for (I suggest) dramatic reasons, but suppose instead that he had picked 1 μs for the period of nothingness. Discuss what effect this would have (if any) on the life of the god. Consider two cases:

(a) There is no minimum time duration for consciousness, and
(b) There *is* a minimum time duration such that, for any shorter duration, a consciousness remains 'unaware' even though it is *not* in a state of nothingness.

After working all night making some final calculations, a physicist carefully solders a final resistor into the control module of the world's first time machine and then steps into the gadget that is a sure bet to win the next Nobel Prize in physics. As she does, she notices that it is precisely 8:10 in the morning, as indicated on both her wrist watch and the clock on the lab wall. After settling into a plush leather seat she pushes the time machine's power button, the machine glows with a flickering blue-red halo and hums with a mighty throb for a while and then, at precisely 8:15 by her wrist watch, she steps out of the machine and back into her lab. She notices the clock on the wall now reads 8:05. That is, she took 5 min of personal time (8:10 to 8:15) to travel 5 min of external time into the past (8:10 to 8:05). On the one hand she certainly seems to be a time traveler, in that she exits the machine before she enters it. (Ignore the issue of there being *two identical* physicists in the lab from 8:05 to 8:10!) On the other hand, the elapsed personal and external

(continued)

times are *equal*. Does this suggest a need to modify or expand David Lewis' definition of a time traveler? As you ponder this question, you might want to read the following four papers: (1) Paul R. Daniels, "Lewisian Time Travel in a Relativistic Setting," *Metaphysica*, October 2014, pp. 329–345, (2) Douglas Kutach, "Time Travel and Time Machines," in *A Companion to the Philosophy of Time* (H. Dyke and A. Bardon, editors), Wiley-Blackwell 2013, pp. 301–314, and (3) Frank Arntzenius, "Time Travel: Double Your Fun," *Philosophy Compass*, November 2006, pp. 599–616. A bit more demanding (but worth the effort) is the long chapter "Time Travel and Time Machines" by Chris Smeenk and Christian Wüthrich, in *The Oxford Handbook of Philosophy of Time* (C. Callender, editor), Oxford 2011, pp. 577–630 (see page 580, in particular).

> The idea that information is physical has given rise to a series of discoveries which indicate that physics has much to say about fundamentals of computer science.

The above quotation is the opening sentence to a most interesting paper by the physicist Dave Bacon, "Quantum Computational Complexity in the Presence of Closed Timelike Curves," *Physical Review A* (70), 2004. (When he wrote, Bacon was at Caltech, but he is now a software engineer at Google.) The title of Bacon's paper, translated into blunt English, is "It Would Be Really Neat If We Could Merge a Time Machine With a Computer." That is, to further quote from Bacon's paper, "One could [efficiently] solve a hard problem by trying out a solution to the problem, sending one's computer back in time, attempting a different solution to the problem, sending one's computer back in time, etc., until a solution to the problem has been found." There then follows a pretty sophisticated analysis on the self-consistent time evolution of a quantum system, ending with Bacon's frank admission that "we would not be honest if we did not end this paper with the caveat that this work is at best a creature of eager speculation … Practical considerations are humorous at best." Read Bacon's paper and discuss what he means by "a hard problem." (There *is* a technical term used by computer scientist for such problems: NP-complete.)

The occasional theological commentary in this book may strike some as a bit odd for a topic treated with heavy doses of deep mathematics in the physics literature but, as you'll see on the following pages, theology is an unescapable dimension to any informed discussion of time travel. A literary connection

(continued)

between time travel and theology has, in fact, existed for a long time. As pointed out in Paul Alkon's *Origins of Futuristic Fiction* (University of Georgia Press 2010), "The first time-traveler in English literature is a guardian angel who returns with state documents from 1998 to the year 1728 in Samuel Madden's *Memoirs of the Twentieth Century*" (published in 1733, nearly three centuries ago). Madden was an Irish-Anglican clergyman whose book was satire rather than science fiction, but its time traveling aspect was a first. As Professor Alkon also writes, "Madden [was] the first to write a narrative that purports to be a document from the future. He deserves recognition as the first to toy with the rich idea of time-travel in the form of an artifact sent backward from the future to be discovered in the present." Your assignment: read and discuss Alkon's book.

You'll recall that Gödel cast his view of time travel in the form of a self-encounter in the past. In Frederik Pohl's "Let the Ants Try," we find a science fiction tale that appeared essentially simultaneously with Gödel's paper (*Planet Stories*, Winter 1949), in which a time traveler journeys back forty million years. Upon stepping out of his time machine, he hears a "raucous animal cry" from somewhere in the nearby jungle. Later, after other adventures in time, he returns to near the same point in spacetime. After stepping out of his time machine, he sees himself in the distance—the earlier version of himself during the first trip. Then, suddenly, the time traveler meets a violent death: "As his panicky lungs filled with air for the last time, he knew what animal had screamed in the depth of the Coal Measure forest." In fact, self-encounters had appeared in science fiction *years* before Gödel's paper. In the 1942 story "Minus Sign" (*Astounding Science Fiction*, November) by Jack Williamson, for example, a spaceship battles with itself while traveling backward in time. How do you think a scientist like Gödel would have liked these two stories? (Who knows, maybe he *did* read them!) If you could travel back in time to 1949 to ask him if such tales had been an inspiration, do you think he would be intrigued, amused, or instead would he be insulted?

Contents

About the Author

Paul J. Nahin was born in California, and did all of his schooling there, including a PhD in electrical engineering from UC Irvine (1972). He is now emeritus professor of electrical engineering at the University of New Hampshire. Prof. Nahin has published a couple of dozen short science fiction stories in *Analog*, *Omni*, and *Twilight Zone* magazines, and has written 18 books on mathematics and physics. He has given invited talks on mathematics at many prestigious institutions, and has appeared on National Public Radio's "Science Friday" show (discussing time travel) as well as on New Hampshire Public Radio's "The Front Porch" show (discussing imaginary numbers). He advised Boston's WGBH Public Television's "Nova" program on the script for their time travel episode.

Chapter 1
A Broad Look at Time Travel

"Hold infinity in the palm of your hand and eternity in an hour."

—William Blake, writing in "Auguries of Innocence" (1863), with words that could quite well describe what it would be like to time travel

"I need a place to hide, that's why I believe in yesterday."

—The Beatles (*Yesterday*, 1965)

1.1 Time Travel in the Fantasy and Science Fiction Literature

"Woodn't it be grate to go back in tyme and correct your mistakes? Wouldn't it be great to go back in time and correct your mistakes?"

—motto of *Time Twisters* comics

To travel in time.

Could there possibly be a more exciting, more romantic, more wonderful adventure than that? I don't think so, and in this opening section I want to just briefly discuss how fascinating many writers (and their readers) have found the concept of time travel, and to point out that the fascination began *long before* mathematical physicists discovered time travel lurking in Einstein's general theory of relativity.[1]

Before the arrival of humans on the surface of the Moon in 1969, the only other 'fantastic voyage' that could compare with time travel was traveling through outer space. During the seventeenth and eighteenth centuries, in fact, such voyages were the center of a genre of fiction (now called *science fiction*) called the "imaginary voyage" or "extraordinary voyage." Marjorie Hope Nicolson's[2] 1948 book *Voyages*

---

[1] An excellent, book-length *literary* treatment of time travel is by David Wittenberg, *Time Travel: the popular philosophy of narrative*, Fordham University Press 2013.

[2] Marjorie Hope Nicolson (1894–1981) was a literary scholar of the first rank at both Smith College and Columbia University.

© Springer International Publishing AG 2017
P.J. Nahin, *Time Machine Tales*, Science and Fiction,
DOI 10.1007/978-3-319-48864-6_1

to the Moon carefully documents just how popular that form of literature was—and still is. Since 1969 the first such voyages have become history, of course, and time travel has replaced space travel as the modern "imaginary voyage."

It seems a safe bet that that, given a random selection of middle-aged adults, the vast majority of them would respond enthusiastically if asked whether time travel interests them. This fascination with time travel has actually been 'scientifically' documented. In one intriguing study,[3] several hundred men and women were asked to consider the possibility of spending an hour, a day, and a year back in both their personal past (since birth) and their historical past (before birth). They were further told that it would cost $10,000 to purchase such time travel services. Their response indicated that 10 % would be willing to spend that much money for an hour in the historical past, 22 % for a day, and 36 % for a year. As might be expected, the numbers rose as the cost dropped and, if such trips were free the interest was almost universal. As one writer put it, "Time travel [is] the ultimate fantasy, the scientific addition to the human quest for immortality."[4] And as a philosopher observed, "[T]he popular appeal of time travel . . . is no doubt due to a nostalgia for the past, which is almost an omnipresent aspect of the human condition."[5]

Fiction writers have, for *centuries*, recognized the fantasy appeal of time travel. The common fairy tale theme of 'The Three Wishes,' in which the recipient ends up using the final wish to undo the unforeseen consequences of the first two, is the precursor to all modern change-the-past time travel stories. Indeed, the means of time travel in the Norwegian poet Johan Wessel's 1781 play *Anno 7603 is* a fairy. Some of the best modern science fiction stories have played with the fantasy appeal of time travel by having gifts arrive by accident from the future: the moral of such tales is generally that unearned gifts usually bring grief.[6] The editor's introduction to a time travel story involving the Civil War referred to the fantasy aspect of time travel—with a reference to another age-old adult fantasy—when he wrote "time travel stories about the Civil War have one thing in common with pornography; they serve to titillate an impulse [in the case of time travel stories, the impulse to change history] and to frustrate [history]."[7] This is the motivation for the time traveler in Stephen King's 2011 novel *11/22/63*, who uses what appears to be a naturally occurring wormhole (connecting a Maine diner to 1958) for his attempts at preventing the assassination of John F. Kennedy.

---

[3]T. J. Cottle, "Fantasies of Temporal Recovery and Knowledge of the Future," in *Perceiving Time*, John Wiley 1976.

[4]T. Paul, "The Worm Ouroboros: Time Travel, Imagination, and Entropy," *Extrapolation*, Fall 1983, pp. 272–279.

[5]J. W. Smith, "Time Travel and Backward Causation," *Cogito* 1985, pp. 57–67.

[6]Among the many such tales, five particularly good ones are "Something for Nothing" by Robert Sheckley, "Mimsy Were the Borogoves" by Lewis Padgett (pseudonym of the married couple Henry Kuttner and C. L. Moore), "Child's Play" by William Tenn (pseudonym of Philip Klass), "The Little Black Bag" by Cyril Kornbluth, and "Thing of Beauty" by Damon Knight. All can be found in various anthologies.

[7]*The Fantastic Civil War* (F. McSherry, Jr., editor), Baen 1991.

A character in the 1985 novel *The Bird of Time* by George Effinger nicely captures the fantasy appeal of time travel with the declaration "The past ... is the home of romance." On a less poetic level, time travel and the movies and stories about it fascinate most people because they turn our everyday world view upside down and inside out.[8] Such movies and stories make people *think*. It is therefore not surprising that time travel movies have been popular for decades, from the pioneering *Berkeley Square* in 1933, to the classic 1960 filming of *The Time Machine*, to *Back to the Future* in 1985 (*the* top film that year in a *Boxoffice* magazine poll), to the flawless 1989 *Bill & Ted's Excellent Adventure*, to the clever *Terminator* action films, to the ingenious 2012 *Looper*, to the commercially successful 2014 *Interstellar*. Each of these films, and others, too, will be discussed later in the book.

When we discuss time travel, we should really be careful to distinguish between two quite different versions: to the future, and to the past. There is no dispute, today, about the first. As two *severe critics* of the possibility of time travel to the past wrote decades ago, "After 1900, special relativity made scientific discussion of time machines possible."[9] What they were referring to is the fact that, by traveling in a rocket ship fast enough (but never, unlike Superman, faster than the speed of light), and far enough, one could leave Earth, loop out on a vast journey perhaps halfway across the universe, and then return hundreds, thousands, even millions of years in the future. You could theoretically (ignoring all the *engineering* difficulties) do this, in fact, with the apparent passage of your 'personal time,' (as measured by your wrist watch or the beating of your heart) as brief as you'd like. (Physicists call 'personal time' *proper time*, and I'll return to this in Chap. 3.) This astonishing conclusion from special relativity, that time travel to the future makes physical sense, literally put a lot of Victorian-era trained physicists into shock.

A quite sophisticated use of this idea appeared early in science fiction, in the tale of a space traveler who returns from a high-speed trip out to the blue supergiant star Rigel in the constellation Orion.[10] The 900 or so light years of the round trip had taken just 6 months of ship or personal time (proper time), but a thousand years of back-home time. The traveler returns to Earth to find all he had left behind long dead and returned to dust: "Sometimes I waken from a dream in which they are all so near ... all my old companions ... and for a moment I cannot realize how far away they are. Beyond years and years."

Another story[11] of a trip into the future that delivers an equally powerful emotional impact, this time via a Wellsian-type time machine (more on what that

---

[8]Somewhat more pompous (but no less correct) was this observation by an academic: "The time-travel [film] romance is an attempt to reenchant the world, to regain a sense of belongingness, to reinstate the magical, autocentric Universe of the child and the primitive." See W. Wachhorst, "Time Travel Romance on Film," *Extrapolation*, Winter 1984, pp. 340–359.

[9]S. Deser and R. Jackiw, "Time Travel?" *Comments on Nuclear and Particle Physics*, September 1992, pp. 337–354.

[10]R. H. Wilson, "Out Around Rigel," *Astounding Stories*, December 1931.

[11]W. Tucker, *The Year of the Quiet Sun*, Gregg 1979.

means later in this chapter) rather than by rocket travel, tells of a time traveler trapped in post-nuclear war times, where there is no energy available to power his machine for the return trip. As the story ends, he finds the woman he had loved and left behind in the past. She is now the elderly widow of another man, having married his rival because the time traveler (just like Wells' Time Traveller) never returned.

A trip into the future does not *have* to be serious or sad. A nice example of that is the story of a spaceship crew that sets off for the Alpha Centauri triple star system, more than four light years distant.[12] They survive the trip, which requires 500 years of both personal *and* external time (this story is the *only one* I'll mention in this book that uses *a preserving drug* rather than physics). What causes me to include it in a book on *time travel* (stories in which proper and external time are one-in-the-same are simply *not* about *time travel*) is that, long before they arrive at the end of their journey, the secret of faster-than-light (FTL) travel is discovered back on Earth and so they arrive at their destination to find a human reception committee! As you'll see when we get to Chap. 3, knowledge of FTL travel is equivalent to knowing the secret of travel into the past, and so the crew is sent back in time, to Earth, to just one year after they left—and they listen to their own radio communications arriving from deep space.

The *real* adventure in time travel, as suggested by "Far Centaurus," would be to go backward in time, to visit the past. The editor of the science fiction pulp magazine *Thrilling Wonder Stories* used the powerful emotional hook of changing the past in a 1950 blurb announcing a time travel story coming in the next issue: "What's the biggest mistake you ever made? Don't worry about it. You may have pulled some awful boners in your time, but there's a sure-fire remedy for them all. It's simple. Just look up at that old time-clock on the wall—and turn it back to the moment just preceding your terrible blunder. Then make your corrections—and set your time-clock back to the present. You may be starting a new chain of error, but why fret? You can go back in time again..." Or, as the promotional text on the video package of the 1986 movie *Peggy Sue Got Married* says, "to do it again" is "the golden opportunity almost everyone has longed for at least once."

Writing less romantically, a philosopher declared that a "major source of interest in the time travel question is our general fascination with the exotic and the child-like frustration we sometimes feel at being confined to the present. We wish that the benefits of moving through space could be supplemented with the benefits which would accrue from movements through time."[13] Robert Silverberg, a science fiction writer who has used the time travel theme often and effectively, expressed this sentiment quite clearly when he wrote "Suppose you had a machine that would enable you to fix everything that's wrong in the world ... The machine can do anything ... it gives you a way of slipping backward and forward in time ... Call this machine whatever you want. Call it Everybody's Fantasy Actualizer. Call it a

[12]A. E. Van Vogt, "Far Centaurus," *Astounding Science Fiction*, January 1944.

[13]R. A. Sorenson, "Time Travel, Parahistory and Hume," *Philosophy*, April 1987, pp. 227–236.

Time Machine Mark Nine."[14] He gives a masterful demonstration of what he means by that in one of his own stories, a tale[15] set in a year when time machines actually exist. Even so, the characters use their imaginations to explore their fantasy worlds and wishes—wishes that could (if they *really* wanted to) be realized with a real time machine. Time travel fiction is, you see, the ultimate escapist literature!

The adventure promised by time travel to the past doesn't necessarily mean *pleasant* adventure, and science fiction has used that idea to great effect. The unstated horror of a trip backward in time, if you think just a bit about it, is that it would bring the dead past, filled with all its dead occupants, alive again, literally resurrected from dank and moldering graves. The top of Mount Everest, the bottom of the Marianas Trench, the sands of Mars—none of these exotic places can even be mentioned in the same breath with the *past*. The capture of the mystery and, yes, the sheer terror of the past, is in this opening line to a 1950s tale: "When Dr. Flitter came into the room, it seemed as though the past and its dead people came in with him, clinging to him like stale surgery smells, like the cold sweat of ancient autopsies."[16] In another equally macabre story, we read of a time traveler in the past anticipating a meeting with a long-dead lover as he "shivered with a renewal of horror . . . She ought to be grateful to him for having raised her from the dead, even briefly."[17]

Another tale, slightly less gruesome, tells us of a character who delights in pointing out all the bad aspects of living in the past.[18] Tell him *when* in time, and he quickly ticks off the disadvantages of being *then*. To live in ancient Greece would let you rub shoulders with Aristotle, sure, but you already know what he said and you'd soon regret the lack of modern plumbing. The year of the American Revolution might let you exchange greetings with George Washington, but you'd also have to put up with cholera in Philadelphia, malaria in New York, and the fact that if you needed an operation there would be no anesthesia anywhere. The Victorian Age appeals to modern romantics but, before you go back, you'd better have your eyes and teeth checked. The time traveling historian in one novel[19] takes these medical warnings to heart and has her appendix prophylactically removed, and is further advised to have her nose cauterized against all the awful stinks of her destination, the fourteenth century.

On the other hand, poor health care isn't *all* there is with time travel to the past. One time traveler from the future, for example, makes a very good living in the past

---

[14]R. Silverberg, "Ms. Found in an Abandoned Time Machine," *Beyond the Safe Zone*, Warner 1986.

[15]R. Silverberg, "Many Mansions," *Beyond the Safe Zone*, Warner 1986.

[16]R. Bretnor, "The Past and Its Dead People," *Magazine of Fantasy and Science Fiction*, September 1956.

[17]L. Marlow, *The Devil in Crystal*, Faber and Faber 1944.

[18]A. Bester, "Hobson's Choice," *Magazine of Fantasy and Science Fiction*, August 1952.

[19]C. Willis, *Doomsday Book*, Bantam 1992.

by winning bets on yet-to-happen events whose outcomes he knows.[20] Time travel to the past allows a Parisian curio shop in the present to offer remarkably authentic looking newspapers from 1804 whose only 'flaw' is that they appear to be fresh off the press—which of course they are![21] And the failed professor in one romantic story[22] finds the Paris of 1482 infinitely better than the Paris of 1961.

A popular fictional appearance of time travel is the use of the past as a hiding place, as a sanctuary for those wishing to escape the troubles of modern times.[23] An interesting twist on this idea was presented in one tale[24] in which the past is used for later military gain in the present. In this story the Earth of a thousand years in the future is ruled by a dictator, and the oppressed masses are unable to arm themselves for revolt. So, back into the past travels an agent to arrange for the construction of weapons, which are then stockpiled in hidden caverns where they can be retrieved for use ten centuries later. The past is used in this story as both a sanctuary *and* a repository from which to make war in the future, and so we have a time travel fantasy for both doves *and* hawks in the same tale! This military use of the past is passive; other writers have more aggressively used time travel to the past for military gain as, for example, mining uranium deposits before they have had time to reduce themselves to lead via radioactive decay,[25] or in drilling for Middle East oil in the past to deprive adversaries of it in the present.[26]

Time travel to the past would, perhaps, interest criminals, too. As the science fiction writer Larry Niven wrote, "If one could travel in time, what wish could not be answered? All the treasures of the past would fall to one man with a submachinegun. Cleopatra and Helen of Troy might share his bed, if bribed with a trunkful of modern cosmetics."[27] Or, as the tragically flawed inventor of the first time machine dreamed, before using time travel to commit what he thought would be the perfect locked-room murder, "The Great Harrison Partridge would have untold wealth. He could pension off his sister Agatha and never have to see her

---

[20]C. Sprague, "Time Track," *Startling Stories*, January 1951.

[21]M. Leinster, *Time Tunnel*, Pyramid Books 1964.

[22]U. K. Le Guin, "April in Paris," *Fantastic Stories*, September 1962.

[23]There are *many* excellent examples of such tales, a few of which are Clifford Simak, "Over the River & Through the Woods"; Ray Bradbury, "The Fox and the Forest"; Jack Finney, "Such Interesting Neighbors"; J. B. Priestly, "Mr. Strenberry's Tale"; James Gunn, "The Reason Is With Us"; and H. B. Piper, "Flight from Tomorrow." All can be found in various anthologies.

[24]R. F. Young, "Not to be Opened—," *Astounding Science Fiction*, January 1950.

[25]C. Simak, "Project Mastodon," *Galaxy Science Fiction*, March 1955.

[26]P. Anderson, "Wildcat," *Magazine of Fantasy and Science Fiction*, November 1958, and W. Jeschke, *The Last Day of Creation*, St. Martin's Press 1982.

[27]L. Niven, "The Theory and Practice of Time Travel," in *All the Myriad Ways*, Ballantine 1971.

again. He would have untold prestige and glamour, despite his fat and baldness, and the beautiful and aloof Faith Preston would fall into his arms like a ripe plum."[28]

Instead of viewing the past as an aid to crime, some writers have used it as the perfect dumping ground for criminals, as a highly convenient place to remove them from society.[29] After all, there can be no breakout from the prison of the past—at least not without a time machine. What might happen to criminal recidivists in a world that has mastered time travel is nicely explained in one story as follows: "If you cannot live among people, then off to the reptiles—one hundred or one hundred twenty million years before the present. There you wouldn't freeze in a tropical pre-glacial climate, and you could nourish yourself on plants. But there is no one to talk with, boredom, and in the end you offer yourself up as an afternoon snack to a tyrannosaurus."[30] With an interesting twist on this is the tale[31] of a physics professor who *helps* criminals disappear into the past to escape relentless police pursuit.

Museum curators, too, would seem to be obvious clients for time machine companies, as would collectors of extinct species who work for zoos. An example of the first case is in a novel[32] about a time travel business called *Time Researchers*, with the corporate mottos 'We Sift the Sands of Time.' It works as a futuristic version of Indiana Jones, as finders of lost historical artifacts for customers who can pay the substantial charges. A typical mission is to make original sound recordings of one of Lincoln's unreported speeches. We find the same idea in a short story[33] about a business called *Genealogy, Inc.*, with the corporate motto

"An Ancestor for Everybody." It uses a 'time scanner' to provide its clients with a list of distinguished predecessors. And in another novel we read of the Historical Corps, whose time travel agents are "writing the definitive history of mankind.".[34]

Another use for time travel to the past, one of the most unusual I have seen, was suggested in a philosophical article[35] which considers an age-old question that has

---

[28]A. Boucher, "Elsewhen," *Astounding Science Fiction*, January 1943. Partridge's dream is shattered, however, because he overlooks a few details about time travel, ones that he wouldn't have missed if he could have read this book. We'll come back to this classic story, which merges a time machine with murder, later in the book.

[29]This is a popular science fiction scenario, and three of the best stories playing with it are P. Anderson, "My Object All Sublime," *Galaxy Science Fiction*, June 1961; I. Watson, "In the Upper Cretaceous with the Summerfire Brigade," in *Stalin's Teardrops*, Victor Gollancz 1991; and R. Silverberg, "Hawksbill Station," *Galaxy Science Fiction*, August 1967.

[30]S. Gansovsky, "Vincent Van Gogh," in *Aliens, Travelers, and Other Strangers*, Macmillan 1984.

[31]J. Finney, "The Face in the Photo," in *About Time*, Simon and Schuster 1986.

[32]W. Tucker, *The Lincoln Hunters*, Rinehart 1958. See also A. Bitov, "Pushkin's Photograph," in *The New Soviet Fiction*, Abbeville Press 1989.

[33]M. Shaara, "Man of Distinction," *Galaxy Science Fiction*, October 1956.

[34]L. A. Frankowski, *The Cross-Time Engineer*, Del Rey 1986. The same idea is in the 1991 film comedy *The Spirit of '76*, in which time-traveling historians from 2176 visit the past in an attempt to reconstruct the lost records of the founding of America.

[35]J. C. Graves and J. E. Roper, "Measuring Measuring Rods," *Philosophy of Science*, January 1965, pp. 39–56.

long bedeviled schoolboys: "If everything in the Universe doubled in size overnight while we slept, could we tell what had happened when we woke up next morning?" The usual answer to this puzzle (called the *Universal nocturnal expansion* by philosophers) is *no*, but the authors of the article suggest that 'all' we need do is take a yardstick back to yesterday and compare it with itself! Great idea, for sure, but it had appeared years earlier in a science fiction story.[36]

More ingenious uses for the past are discussed in the story[37] of a time travel business called *Time Associates*. One use comes in the form of a request from a United States senator who wants to send the disadvantaged of today back into the remote past, where they could have a fresh start on a virgin Earth. Yet another use comes from a religious fringe group that wants to purchase exclusive rights to the time of Jesus—not to visit, but to prevent *anyone* from visiting. The group fears that any such visitors would "learn the truth," which might contradict the very legends that form the heritage of Christianity. And, in what may be the most ingenious idea of all, *Time Associates* itself does not do business in the present, but rather 150,000 years in the past, in a 'new' country called Mastodonia. The corporate lawyer, you see, has determined that such an arrangement legally means the company is a *foreign* company doing business outside the United States, and so it is not liable for taxes to the IRS!

The tourist trade is a booming business in science fiction, with dinosaur hunting at the top of the list. There are many such tales,[38] including the cerebral stories in L. Sprague de Camp's short-story collection *Rivers of Time*, starting with the classic "A Gun for Dinosaur." The earliest (that I know of) fictional use of time travel to the past for hunting was not for dinosaur hunting, however, but rather for saber-toothed tigers, wooly mammoths, and cave bear.[39] Historical tours to the great events of the past are also an entertaining use of time travel.[40] Even mundane events may one day be on the 'to do' lists of time travelers to the past. In one, for example, curious crowds from the very far future show up, nightly inside the home a twentieth century family, much as tourists today visit Monticello.[41]

Perhaps the most direct use of the past's unique resource, itself, is realized in science fiction by Hollywood. In one tale,[42] after purchasing the motion picture

---

[36]H. M. Sycamore, "Success Story," *Magazine of Fantasy and Science Fiction*, July 1959.

[37]C. Simak, *Mastodonia*, Ballantine 1978.

[38]Two are B. W. Aldiss, "Poor Little Warrior," in *The Science Fictional Dinosaur*, Avon 1982, and a famous one by Ray Bradbury, "A Sound of Thunder," in *The Stories of Ray Bradbury*, Alfred A. Knopf 1980.

[39]C. Simak, "The Loot of Time," *Thrilling Wonder Stories*, December 1938, in which a time machine inventor raises money for his research by transporting hunters back 70,000 years to the Old Stone Age.

[40]Three such tales are R. Silberberg, *Up the Line*, Ballantine 1969 and "When We Went to See the End of the World," in *Beyond the Safe Zone*, Warner 1986; and G. Kilworth, "Let's Go to Golgotha!," in *The Songbirds of Pain*, Victor Gollancz 1985.

[41]B. Tucker "The Tourist Trade," in *Tomorrow the Stars*, Doubleday 1952.

[42]L. Laurence, "History in Reverse," *Amazing Stories*, October 1939.

rights to H. G. Wells' *Outline of History*, the head of a movie studio uses a time machine to send his ace cameraman into the past to get live action footage. Prehistoric animals, the ice age, Cheops building his pyramid, the destruction of Pompeii by the eruption of Vesuvius, the Battle of Hastings, Columbus, all the *originals* of these historical events appear in the final film. Years later this idea was developed even further in a very funny novel,[43] in which a movie director uses the eleventh century as a realistic setting for a picture. Realism isn't always the result, however, as portrayed in another tale; the films produced by a gadget that can 'look' into the past are failures because they don't look "authentic enough" to Hollywood moguls![44] (Fig. 1.1).

Fig. 1.1   The inventor of a time machine demonstrates it by sending the family cat on a trip. In the story the inventor, himself, travels back to 1901, where he accidently kills his grandfather in an early pulp magazine, non-paradoxical version of the famous riddle. Illustration for Raymond A. Palmer's "The Time Tragedy" (*Wonder Stories*, December 1934) by Frank R. Paul, ©1934 by Continent Publications Inc.; reprinted by permission of the Ackerman Science Fiction Agency, 2495 Glendower Ave., Hollywood, CA 90027 for the Estate

---

[43]H. Harrison, *The Technicolor Time Machine*, Doubleday 1967.

[44]A. Derleth, "An Eye for History," in *Harrigan's File*, Arkham House 1975.

1.2   Where Are All the Time Travelers?

"If it [time travel] *could* be done, someone will eventually learn how. If that happens, history would be littered with tourists. They'd be *everywhere*. They'd be on the *Santa Maria*, they'd be at Appomattox with [cameras], they'd be waiting outside the tomb, for God's sake, on Easter morning."[45]

The question the title of this section asks is an echo of the one the physicist Enrico Fermi (1901–1954) asked in the 1950s, about the possibility of interstellar space travel and of alien intelligent life in the universe—if such travel is possible and 'they' exist, then *where are they*? Why haven't we at least received radio signals from them? For many, the apparent lack of time travelers among us is similar evidence for the impossibility of time travel. As one famous science fiction writer put it, "The most convincing argument against time travel is the remarkable scarcity of time travelers. However unpleasant our age may appear to the future, surely one would expect scholars and students to visit us, if such a thing were possible at all. Though they might try to disguise themselves, accidents would be bound to happen—just as they would if we went back to Imperial Rome with cameras and tape recorders concealed under our nylon togas. Time traveling could never be kept secret for very long."[46]

Clarke's idea is that, from the moment after the first time machine was constructed, through all the rest of civilization, there would be numerous historians, to say nothing of weekend sightseers, who would want to visit every important historical event in recorded history. They might each come from a different time in the future, but all would arrive (according to Clarke) at destinations crowded with temporal colleagues, crowds for which there is no historical evidence!

Long before Clarke the science fiction writer Robert Silverberg had already used the same idea in his 1969 novel *Up the Line*, where it's called the *cumulative audience paradox*. That paradox claims that as time travelers to the past continue to visit certain historically interesting dates and places, there will be an ever-increasing number of people present. As it is presented in the novel, "Taken to its ultimate, the *cumulative audience paradox* yields us the picture of an audience of billions of time-travelers piled up in the past to witness the Crucifixion, filling all the Holy Land and spreading out into Turkey, in Arabia, even to India and Iran . . . Yet at the original occurrence of [that event] *no such hordes were present!*" And later in the same work, we read "A time is coming [when we] will throng

---

[45]A skeptic's reaction to the idea of time travel in J. McDevitt's story "Time's Arrow" in *The Fantastic Civil War* (see note 7).

[46]Arthur C. Clarke, "About Time," in *Profiles of the Future*, Warner 1985. A story that I recall once having read (but cannot now remember either the author or the title) wonderfully illustrates Clarke's point. A time traveler in disguise at Golgotha for the Crucifixion has a camera hidden beneath his robe to avoid attracting attention. All goes well until he notices odd, clicking noises coming from all those standing near him. It is then he realizes the entire crowd is *nothing but* time travelers, from all through the ages, all with hidden cameras!

the past to the choking point. We will fill our yesterdays with ourselves and crowd out our own ancestors."

Philosophers are well aware of Silverberg's and Clarke's conundrum and, indeed, it can be found in the philosophical literature before Clarke wrote. In one paper, for example, we read "Actually I know of only one argument against the possibility of time travel that seems to carry any weight at all. This is the fact that it does not appear ever to have happened. That is, it might be argued that there will be no time trips from [2100] to [2017] because we were here in [2017] and saw no time travelers. But this argument is far from conclusive."[47] At most, in other words, the absence of temporal visitors amongst us is an objection to the *actuality* of time travel, and not to the *possibility* of time travel.

This same philosopher then mentions some ways around this concern, including one which he called a "pettifogging physical limitation on time travel: perhaps the energy expenditure varies as the fourth power of the time traversed, making only very short trips feasible, and its discovery lies too far in the future for its effects to have yet been felt." Another science fiction writer, as famous as Clarke, used that idea in his 1957 novel *The Door Into Summer* when Robert Heinlein has one character comment "Now if there was some way to photograph the Crucifixion ... but there isn't. Not possible ... there isn't that much power on the globe. There's an inverse-square law tied up in [time travel]." Or, perhaps, time travel *is* possible but it's so extraordinarily dangerous that it's impossible to get anyone to do it. In one provocative tale[48] that takes this idea to the extreme, we read that there is only *one* time traveler, ever, from the future—indeed, from just 18 min (!) in the future—and his first (and last) experiment destroys the Earth.

Clarke presented some other possible science fiction rebuttals to the puzzle of 'where are the time travelers?' As he wrote, "Some science fiction writers have tried to get around this [question] by suggesting that Time is a spiral; though we may not be able to move along it, we can perhaps hop from coil to coil, visiting so many millions of years apart that there is no danger of embarrassing collisions between cultures. Big game hunters from the future may have wiped out the dinosaurs, but the age of *Homo sapiens* may lie in a blind region which they cannot reach."

The idea of time as a spiral was quite popular in early science fiction. Typical is one tale[49] in which the time traveler suddenly finds himself not in 1933 but in 2189. His situation is 'explained' to him thus: "[The] time stream is curved helically in some higher dimension. In your case, a still further distortion brought two points of the coil into contact, and a sort of short circuit threw you into the higher curve."

---

[47]G. Fulmer, "Understanding Time Travel," *Southwestern Journal of Philosophy*, Spring 1980, pp. 151–156. The modern view of this 'paradox' is *not* that it describes a situation so absurd that time travel must be impossible, but rather that *all* the time travelers who *were* (will be?) at the Crucifixion *are* in the historically recorded crowd (see note 46 again). The Crucifixion happened just *once*, not over and over. I'll return to this point later in the book.

[48]D. Plachta, "The Man from When," *Worlds of If Science Fiction*, July 1966.

[49]R. H. Wilson, "A Flight Into Time," *Wonder Stories*, February 1931.

A few years later (in 1937) we find another story[50] with the same spiral-time concept; with a sixty-million year pitch to the time helix, there is no danger of a grandfather paradox. That same year spiral time was the central 'scientific' theme in the stage production "I Have Been Here Before" by the English playwright J. B. Priestley.

The very next year (1938) the young Isaac Asimov used the idea in his first attempt at professional writing, despite a life-long unhappiness with the concept of time travel. Titled "Cosmic Corkscrew," it was initially rejected, indeed it was *never* published, and eventually lost. Though, perhaps, not for long, since one enterprising modern writer has used it as the basis for his own time travel tale.[51] In it a traveler from the near future travels back to 1938 to retrieve "Cosmic Corkscrew" before Asimov loses it (perhaps that's *why* it was lost!) Even when writing introductions to other writers' time travel tales, Asimov would often insert personal comments on his opinion of the concept. For example, in one volume of an anthology series he edited, *The Great Science Fiction Stories* (of 1954), he wrote "To my way of thinking it is precisely because time travel involves such fascinating paradoxes that we can conclude, even in the absence of other evidence, that time travel is impossible." And in *The Great Science Fiction Stories* (of 1961) he bluntly declared "I think scientists who think up methods of time travel are probably all wrong."

Spiral time is a close cousin to circular time. One story[52] dealing with circular time has a time traveler who finds, after a trip one hundred years into the future, that he can't get all the way back to his own time because the required energy rises exponentially with increasing penetration into the *past*. Still, it's very cheap in energy to go forward in time in this tale, and that's what the traveler does, in search of help from the future's advanced technology. He never finds what he needs, however, and so goes forward right into the collapse of the universe and through a new Big Crunch that forms an identical new cycle of time. He thereby returns home to just before he left.[53] This eternal recycling of identical, circular time is so terrifying that the traveler decides to suppress what he has learned about how to time travel—and so maybe *that's* why there are no apparent time travelers. Science fiction writer Larry Niven has argued, however (see note 27), that while this may be a conceptually valid (?) way to travel into the past, he also warns that "Removing your time machine from the reaction of the Big Bang/Crunch could change the final configuration of matter, giving an entirely different . . . history." (I strongly suspect that Niven wrote that with a big smile on his face!)

---

[50]P. S. Miller, "The Sands of Time," *Astounding Stories*, April 1937.

[51]M. A. Burstein, "Cosmic Corkscrew," *Analog*, June 1998.

[52]P. Anderson, "Flight to Forever," *Super Science Stories*, November 1950.

[53]Turning this idea on its head is the approach of the 1978 novel *The Way Back* (DAW) by A. B. Chandler. Its characters return from the past to their own time by traveling even further backward, right through the Big Bang and into the *previous* (and identical) cycle of time.

The famed English physicist Stephen Hawking is so taken with the question of 'where are all the time travelers?' that he has elevated their apparent absence ("we have not been invaded by hordes of tourists from the future"[54]) to the status of being a demonstration of the impossibility of time travel to the past. His so-called *Chronology Protection Conjecture*, Hawking likes to say, "makes the universe safe for historians": that is, there is nothing to worry about (if you're concerned that time travelers could change the past) because time travel is simply impossible. You'll see later in the book that there are other possible ways to insure the safety of history *without* denying the possibility of time travel, and Hawking himself has backed away just a bit from the Conjecture, saying now that he was simply looking for a humorous line.

While Hawking's endorsement of it has made the Conjecture famous, he wasn't the first to state it. Two years earlier it had appeared, in of all places, a *financial* publication: "[If] time travel was possible, someone from the future would eventually either discover a time tunnel or build a time machine and come visit us."[55] And even before that, an economist presented a 'proof' for concluding that "time travelers do not and cannot exist."[56] He argued that if time travelers from the future were actually amongst us (our 'now' is their 'past') then, by virtue of their knowledge of things to come (our 'future') they would make financial deals so numerous and extensive that interest rates would be driven to zero. Interest rates are *not* zero, however, and thus no such time travel hanky-panky has occurred.

These sorts of financial arguments aren't like to convince many physicists or philosophers of the Conjecture's merit. At most we can only conclude from them that time travelers from the future have *not* influenced financial affairs, which doesn't mean they aren't here. In any case, the Conjecture was actually stated more than 20 years before Hawking by Larry Niven (see note 27), who declared what is called *Niven's Law*: "If the universe of discourse permits the possibility of time travel, and of changing the past, then no time machine will be invented in that universe." And Hawking's concern over time travelers meddling with the past was anticipated in science fiction, too, by at least half a century; in a 1950 tale, for example, we learn of a Master Historian, and the graduate students in his course on 'Experimental History' in the forty-sixth century, trying to correct a problem created by a previous tampering with the past![57]

Not all physicists and philosophers feel intellectually comfortable with the Conjecture, as it seems (to them) a too quick surrender: 'Time travel is a problem

---

[54]S. W. Hawking, "Chronology Protection Conjecture," *Physical Review D*, July 15, 1992, pp. 603–611. See also J. F. Woodward, "Making the Universe Safe for Historians: Time Travel and the Laws of Physics," *Foundations of Physics Letters*, February 1995, pp. 1–39.

[55]J. Queenan, "Time Warp: Or, Investing in the Future Is a Bust," *Barron's*, January 8, 1990, p. 46.

[56]M. R. Reinganum, "Is Time Travel Impossible? A Financial Proof," *Journal of Portfolio Management*, Fall 1986, pp. 10–12.

[57]L. Jones, "Sunday is Three Thousand Miles Away," *Thrilling Wonder Stories*, June 1950. A more recent, two-*novel* treatment of historians tinkering with history is by Connie Willis (*Blackout* and *All Clear*, both published in 2010).

so hard to do let's simply *define* it to be non-existent and then we won't have to worry about it anymore.' To *really* show that time travel is impossible, however, one needs to demonstrate how it would violate one or more of the laws of physics. Hawking, of course, understands this and has stated that, as one who is no fan of time machines and time travel, he believes there is new physics yet to be discovered that *will* forbid would-be time travelers from roaming up and down the centuries. Finding that new physics is the lure the study of time machines has for him. As he correctly writes in his autobiography,[58] "Even if it turns out that time travel is impossible, it is important that we understand why it is impossible."

One mathematical physicist who agrees with Hawking on the matter of the unlikely possibility of making a time machine is the New Zealand theoretician Matt Visser. Noting that while quantum field theory, and the general theory of relativity, are each amazingly good theories in many applications within their respective realms, they are not so good in spacetime regions at the so-called *Planck scale* (that is, when the density of mass-energy reaches the fantastic level of 10^{94} grams/cm^3 and beyond) where chronology violations (that is, time travel) seem to be spawned. As Visser has observed,[59] this situation won't change until 'we wander into the guts of quantum gravity,' the unification that will merge gravity with the quantum to give a theory that *always* works. Without quantum gravity, physics will continue to be "infested" (Visser's word) with "sick" (Visser's word) spacetimes that allow time travel. Visser believes that the discovery of the theory of quantum gravity can be 'guided' by *building* causality into it,[60] and the result will finally consign time machines to where (in his mind) they belong, the dust-bin of crackpot physics.

Well, perhaps so, but we don't have a quantum theory of gravity yet, and probably won't for some time to come, and so the puzzling questions about time travel remain. To end this section on a slightly gloomy note, an idea appeared in science fiction,[61] when Hawking was still a teenager, offering a possible rebuttal to the Conjecture. It opens with one of the inventors of the first time machine just returning from a trip to the past of 1938. Still, despite this success, the inventors are puzzled by what they call 'the problem': "But if *we* have time traveled, then obviously men of the future have time traveled. They will be able—*are* able to come back. [So] where are they?" They finally conclude that there can only be two

---

[58]Stephen Hawking, *My Brief History*, Bantam 2013, p. 113.

[59]Matt Visser, "The Quantum Physics of Chronology Protection," in *The Future of Theoretical Physics and Cosmology*, Cambridge 2003. This paper was Visser's contribution to the celebration of Hawking's 60th birthday, held in January 2002.

[60]This may seem like something new, but it really isn't. General relativity has causality built into it on a *local* level (where it belongs); a failure of causality (that is, time travel—see J. Sharkey, "The Trouble With Hyperspace," *Fantastic* April 1965) occurs in general relativity only when one studies *global* regions of certain spacetimes. Forcing a physical theory to have a *prescribed global behavior* would be to undo all of physics since the development of *local field theories*, along with all their amazing successes in explaining nature.

[61]M. Shaara, "Time Payment," *Magazine of Fantasy and Science Fiction*, June 1954.

possible answers. Either there is nobody in the future, or time travel is so dangerous (is that why the future might be empty—humanity misused time travel and killed itself off?) that all who invent it will suppress it. And that's what they decide *they* must do.

1.3 Skepticism About Tales of Time Travel

"May it not be that our inability to leap into the fiftieth century, A.D., seems impossible to us, merely because of certain prejudices we entertain or certain facts and tricks of which we are still hopelessly ignorant? Assuredly, this is not a foolish query. Its answer, whatever that may be, carries immeasurable consequences for metaphysics."

—a scholar wonders[62]

A thought-provoking possibility for explaining the scarcity of certified time travelers is the central thesis of a fascinating paper in the philosophical literature. The author of that paper argues (note 13) that nobody would believe a time traveler even if he willingly confessed and revealed his knowledge of the future, or even gave the details of his time machine. He goes on to make the astonishing assertion that even the *time traveler himself* would have doubts! This perhaps shocking suggestion deserves some elaboration, especially because it invokes the authority of the patron saint of skeptics for support, the Scot David Hume (1711–1776). The crucial point to keep in mind is explicitly stated in the argument: "The key question will not be 'Is time travel possible?' We shall instead ask whether it is possible to justify a belief in a report of time travel." This gets to the real heart of Clarke's puzzle from the previous section.

Much of the resistance to the idea of time travel lies in sheer skepticism. For many, time travel (to the past, in particular) is simply too much out of the ordinary to be taken seriously. For many, time travel would literally be miraculous. Hume's great work, *An Enquiry Concerning Human Understanding*,[63] contains a section on how a rational person should react to a claim that a miracle has occurred. Hume proclaimed that a miracle *by definition* violates scientific law and that, because such laws are rooted in "firm and unalterable experience," any violation of one or more of these laws immediately provides a refutation of the report of a miracle. In Hume's own words:

"Nothing is esteemed a miracle, if it ever happened in the common course of nature. It is no miracle that a man, seemingly in good health, should die on a sudden; because such a kind of death, though more unusual than any other, has yet been frequently observed to happen. But it is a miracle, that a dead man should come to life; because that has never been

---

[62]W. B. Pitkin, "Time and Pure Activity," *Journal of Philosophy, Psychology and Scientific Methods*, August 27, 1914, pp. 521–526. Pitkin's essay was a critique of time travel as presented in Wells' *The Time Machine*, which Pitkin called "one of the wildest flights of literary fancy."

[63]Making its first appearance in 1748, *Enquiry* has been reprinted numerous times since. I used the 1963 edition published by Open Court.

observed in any age or country ... When anyone tells me, that he saw a dead man come to life, I immediately consider with myself, whether it be more probable, that this person should either deceive or be deceived, or that the fact, which he relates, should really have happened. I weigh the one miracle against the other; and according to the superiority, which I discover, I pronounce my decision, *and always reject the greater miracle* [my emphasis]."[64]

It is a strict interpretation of Hume that Sorenson (note 13) has adopted in claiming that a time traveler would have no success (among rational persons) with tales of 'different times.' As he explains, "Clearly the time traveler cannot persuade a reasonable person by baldly asserting 'I am a time traveler.' The improbability of his claim places a heavy burden of proof on him. But perhaps he could shoulder the burden by means of artifacts, predictions, and demonstrations." Sorenson dismisses all of these possibilities, however, by reminding us of the slightly sleazy history of parapsychology and ESP, both of which run counter to known scientific laws, but which have still duped "many a respected scientist." Any artifact, prediction, or demonstration of time travel, argues Sorenson, is more likely to be the result of deception and fraud than of actual time travel: "Should the time traveler take observers for a spin in his time machine, the skeptics will have us compare their adventures with séances." The rational reaction to such a spin around the centuries, according to Sorenson's presentation, would be like that of a magician who cannot figure out how a colleague has just done his newest act: 'Nice trick! How did you do it?'[65]

The time traveling tourist stranded in the past in one story is used to getting a skeptical reaction because he can provide his questioners no technical explanation for his situation. "How the hell should I know? I'm just a tourist. It has something to do with chronons [see the Glossary]. Temporal Uncertainty Principle. Conservation of coincidence. I'm no engineer."[66] Somewhat more successful (perhaps) is a time traveler born in 2003 who turns up in 1975. After he tries to convince an interrogator of how that can be, he apparently succeeds. As the time traveler later tells a new friend in the past of 1975, "What amazed me ... was that he really believed me in the end." But the friend doesn't buy that, replying "He did? I think he just

---

[64]What Hume is alluding to here should be plain; as expressed in P. Heath, "The Incredulous Hume," *American Philosophical Quarterly*, April 1976, pp. 159–163, Hume was "an exposer of bad arguments in rational theology." For Hume, second-hand (or even more remote) tales of the return of a man from the dead—the claim that literally kept Christianity alive after Christ's execution—were suspect.

[65]This skeptical reaction was nicely captured in the story "E for Effort" by T. L. Sherred (*Astounding Science Fiction*, May 1947). As one character laments, "I've watched scribes indite the books that burnt at Alexandria; who would buy, or who would believe me, if I copied one. ... What sort of padded cell would I get if I showed up with a photograph of Washington or Caesar? Or Christ?" The padded cell was indeed the fate of the time traveler in "The Ambassador from the 21st Century" (*Startling Stories*, March 1953) by H. J. Shay, the story of a man who journeyed from A.D. 2007 back to 1952 to warn of a future war; he was committed to a mental institution to receive help for his "illusion."

[66]J. Haldeman, "No Future In It," *Omni*, April 1979.

pretended. A scientist isn't likely to believe a thing that is against all logic."[67] If the reception committee is a crowd of conservative, cautious Humeans, it would seem that a time traveler is almost certainly doomed. Early science fiction time travelers from 2030, for example, were warned about receiving a skeptical response as follows (the editorial introduction to this tale[68] called it "a curious study of psychology"): "Our wisest men advised against [our trip to the past]. They said we could hope to be received only as imposters and fakirs, that . . . we would find only twentieth-century barbarians, suspicious, ill-tempered, likely to do us bodily harm."

For many, such skeptical reactions to self-proclaiming time travelers seems dogmatic in the extreme—the response of people with no imagination, no spirit, and heads full of cement. Humean skepticism requires, so it would seem, the rejection of anything and everything that is profoundly surprising, leaving the world a place of utter predictability and boredom. As one science fiction writer put it, "When the miraculous occurs, only dull, workaday mentalities are unable to accept it."[69] Sorenson answers this harsh criticism as follows: "Humeans respond [to Sheckley] by distinguishing between surprises. Most surprises in science do not violate accepted scientific laws. The strange wildlife in Australia was not excluded by biology, X-rays were not precluded by physics."

Sorenson does well, however, to avoid mentioning such profound surprises as, for example, the spectrum of black-body radiation and, later, the photoelectric effect, which were not in the domain of known classical science at the beginning of the twentieth century. Those puzzling, surprising, *totally mystifying* effects required new science—the discovery of the quantum concept by Max Planck. (Explaining the photoelectric effect, *not relativity*, is what won Einstein his Nobel prize.) A strict Victorian-age Humean, as described by Sorenson, would have wrongly rejected the experimental reports of all quantum phenomena and would also (perhaps just as wrongly) have rejected all reports of time travel.

A strict Humean definition (as described by Sorenson) that a miracle has occurred requires a violation of one or more of the *known* [my emphasis] scientific laws of nature.[70] As one modern philosopher defines a miracle, it is any event that "can be explained *only* [my emphasis] by reference to the intervention of a supernatural force."[71] Time travel, by that interpretation, is *not* a miracle because general relativity, not God, is all that is required. C. S. Lewis (1898–1963), late

---

[67]G. Gor, "The Garden," in *Russian Science Fiction* (R. Magidoff, editor), New York University Press 1969.

[68]P. Bolton, "The Time Hoaxers," *Amazing Stories*, August 1931.

[69]Robert Sheckley, "Something for Nothing," in *Citizen in Space*, Ballantine 1955.

[70]The word *known* is important. As a character in one early science fiction story puts it, "These things [four-dimensional object] sound like miracles; but, after all, what are miracles but phenomena which, *on account of our ignorance* [my emphasis], we cannot explain?" See B. Olsen, "The Four-Dimensional Roller-Press," *Amazing Stories*, June 1927.

[71]D. M. Ahern, "Miracles and Physical Impossibility," *Canadian Journal of Philosophy*, March 1977, pp. 71–79.

professor of Medieval and Renaissance Literature at Cambridge University, however, absolutely rejected Hume's view on how a rational person should react to certain surprising events. Lewis, one of the most thoughtful modern writers on Christian theology, had no patience with skeptics (or, as he called them, materialists).[72]

Professor Lewis graphically illustrated the dug-in position of the extreme skeptic as follows: "If the end of the world appeared in all the literal trappings of the Apocalypse; if the modern materialist saw with his own eyes the heavens rolled up and the great white throne appearing, if he had the sensation of being himself hurled into the Lake of Fire, he would continue forever, in that lake itself, to regard his experience as an illusion and to find the explanation of it in psychoanalysis, or cerebral pathology."[73] If the end of the world would receive such a skeptical response, then a mere time traveler would surely have no hope at all of being believed.

Lewis would certainly have rejected Sorenson's most astonishing assertion: "So far I have concentrated on the time travel question from the perspective of the time traveler's audience. What about the time traveler himself? Can he at least know he is a time traveler?" Sorenson argues that a time traveler, if authentic, should be able to convince his audience, and that if he can't (and he *cannot* if they are true Humean skeptics), then the time traveler must entertain doubts, too! It doesn't matter (says Sorenson) that the time traveler has memories of his adventures, and it doesn't matter that he knows in his heart that he speaks the truth. Using words that echo Lewis' sarcasm, Sorenson quickly dismisses the importance of the time traveler's self-knowledge, declaring such memories to be merely the symptoms of some deep psychosis, and the traveler's introspective sincerity to be a product of gross self-deception.

Sorenson specifically mentions the traditional Humean response to astonishing reports when he cites earlier writers on time travel in the philosophical literature. In one of those analyses, for example, we find an argument *for* the reasonableness of a rational belief in time travel ("I have been amused and irritated by the spate of articles proving that time travel is a 'conceptual impossibility'") by claiming such proofs must be faulty because there is a mathematically consistent explanation for such a belief.[74] (This author was referring to spacetime diagrams, which we'll get to in Chap. 3.) This paper received a very sharp rebuttal from another philosopher who convincingly used fundamental physics to show a simple use of spacetime diagrams in a special relativity setting does *not* support time travel to the past.[75] (I'll return to

---

[72]In Lewis' eerie, unfinished story "The Dark Tower," a tale of the 'chronoscope,' a gadget that "does to time what the telescope does to space," the persistent skeptic in the story is a Scot, surely created in the image of Hume. See C. S. Lewis, *The Dark Tower and Other Stories*, Harcourt 1977.

[73]C. S. Lewis, *The Grand Miracle*, Ballantine 1986.

[74]H. Putnam, "It Ain't Necessarily So," *Journal of Philosophy*, October 1962, pp. 658–671.

[75]J. Earman, "On Going Backward in Time," *Philosophy of Science*, September 1967, pp. 211–222.

this point in Chap. 3.) Even later a Humean-style rebuttal came from yet another philosopher, who showed how to explain the time travel phenomenon that Putnam (note 74) described *without* invoking time travel.[76] This isn't to say that Weingard doesn't invoke some pretty astonishing gadgetry (and more) himself, like matter transmitters and anti-matter humans. (You'll see how anti-matter ties-in with time travel a bit later in the book.) A resurrected Hume would surely applaud these rebuttal analyses (although he might also doubt his own fresh existence).

Hardly anybody is happy with Weingard's approach for avoiding time travel (including, I suspect, even Weingard). His 'explanations' seem, just like a time machine, to be incredible and, as Arthur Conan Doyle's Professor Challenger says in one tale not staring Sherlock Holmes, "You cannot explain one incredible thing by quoting another incredible thing."[77] An interesting science fiction exposition illustrating Professor Challenger's Humean philosophy occurs when a copy of *The New York Times* for December 1 shows up for some subscribers a week early, on November 22. It seems the only explanation is either that the paper really is from the future (due to some sort of fluke of the fourth dimension), or that it is a hoax. The first-person narrator of this 1973 tale[78] provides us with his reason for believing the former: "I don't find either notion easy to believe but I can accept the fourth-dimensional hocus-pocus more readily than I can the idea of a hoax." Hume couldn't have said it better.

It should be clearly understood that Hume was not arguing for disbelief in absolutely anything surprising, but rather for rational analysis. Historically, the context of Hume's times was that of what he took to be non-rational arguments for a belief in God, particularly those 'proofs' so beloved by theologians based on Design (Heath [note 64] calls such 'proofs' "philosophical museum pieces"). As Heath writes, "Hume ... makes no attempt to deny the supposed facts; he simply argues that they are consistent with other explanations and other analogies of a less ambitious kind. There is no right to attribute to the causes of such phenomena abilities more extensive than are needed to produce the observed effects."

As a matter of fact, even Hume could be convinced of quite strange matters, and I think Sorenson does interpret the philosopher a little too narrowly. In his essay concerning Hume's position on holding a belief in God, Heath wonders whether there is "empirical evidence [imaginable] which would persuade any reasonable mind of the real existence of an infinite God." Heath answers his own question as follows: "If the stars and galaxies were to shift overnight in the firmament, rearranging themselves so as to spell out, in various languages, such slogans as I AM THAT I AM, or GOD IS LOVE—well, the fastidious might consider that it

---

[76]R. Weingard, "On Travelling Backward in Time," *Synthese*, July–August 1972, pp. 117–132.

[77]Arthur Conan Doyle, "The Disintegration Machine," *The Strand Magazine*, January 1929. Professor Challenger is nothing like Wells' thoughtful Time Traveller; in the original 1912 Challenger novel *The Lost World*, he was described as a "primitive cave-man in a lounge suit."

[78]R. Silverberg, "What We Learned from This Morning's Newspaper," in *Beyond the Safe Zone*, Warner 1986.

was all very vulgar, but would anyone lose much time in admitting that this settled the matter? ... Confronted with such a demonstration, the hard-line Humean [but not Hume, himself, I think] could continue, of course, to argue that, for all its colossal scale, the performance is still finite, and so cannot be evidence of more than the finite, though immense power that is needed to achieve it."

Skepticism about 'time travel' was around long before the specific idea of a 'time machine' was conceived. For example, the eleventh-century Persian poet-philosopher Omar Khayyam was blunt in his evaluation of the likelihood of reliving the past. As he so beautifully wrote in one of the quatrains of the *Rubaiyat*,

The Moving Finger writes; and having writ,
Moves on: nor all your Piety nor Wit
Shall lure it back to cancel half a Line,
Nor all your Tears wash out a Word of it.

Quite a bit later the English poet Thomas Heywood, in his 1607 play *A Woman Killed with Kindness*, had one of his characters express a similar thought:

God, O God, that it were possible
To undo things done, to call back yesterday;
That Time could turn up his swift sandy glass
To untell the days, and to redeem these hours.
Or that the Sun
Could, rising from the west, draw his coach backward,
Take from the account of Time so many minutes,
Till he had all these seasons called again,

. . . . .

But O! I talk of things impossible,
And cast beyond the moon. . .

When Gödel's discovery of time travel in his rotating universe was announced, the skeptics were easy to find. One philosopher[79] wrote of it "This property [of time travel] must be judged an absurdity by anyone committed to the ordinary modes of speech." And another[80] was only slightly less charitable: Gödel's solution was a "bizarre conception" and a "mere mathematical curiosity." Science fiction wasn't immune to skepticism, either, even though you might have expected that to be the one place where the high drama of time travel would be welcomed. Four *years* after Gödel's paper appeared we find one respected anthologist writing,[81] as part of his introduction to a story, "In this tale we meet our first Mad Scientist. Just as in reality the thoroughly cracked pots used to be found inventing perpetual-motion machines, so in science fiction we find the lunatic fringe more often than not trying to perfect

---

[79]J. D. North, *The Measure of the Universe*, Oxford University Press 1965.

[80]C. T. K. Chari, "Time Reversal, Information Theory, and 'World-Geometry'," *Journal of Philosophy*, September 1960, pp. 579–583.

[81]Groff Conklin, editor of *Science-Fiction Adventures in Dimension*, Vanguard 1952.

time-travel mechanisms." And that same year the founding editor of *Galaxy Science Fiction Magazine* declared "Time travel requires a suspension of disbelief that is almost unbelievable ... Scientifically, time travel can't stand inspection.".[82]

Years later matters had not much changed. For example, in his marvelous 1985 book *The Past is a Foreign Country*, David Lowenthal repeatedly refers to time travel as "fantasy," and to science fiction stories about time travel as "unbridled by common sense." (Lowenthal is a professor of *geography*, not physics.) Science fiction writers were still often not much more enthusiastic about time travel. The well-known science fiction writer and critic Alexei Panshin, for example, agrees with Lowenthal, at one point, long after Gödel, writing "Time travel is a philosophical concept, not a scientific one. It is, in fact, as has often been pointed out, scientific nonsense."[83]

Skepticism does have its uses, however. Modern science fiction writers have often used it as a dramatic means of building conflict and tension in their time travel stories. A skeptical reception is extreme, for example, for a soldier-in-time who has fought in numerous wars, from the ancient past to a billion years in the future.[84] He finds that nobody believes him when he speaks openly of his temporal adventures during a visit to a present-day bar. Everybody merely thinks it is all a hilarious gag. This is in great contrast to one 1870s story[85] in which suspicion of a stranger plays a central role, but which finds its offered explanation in something entirely different from time travel. It tells of a man who suddenly appears in the midst of a Union military camp during the American Civil War.

This man quickly displays strange lapses in his background, as well as possessing knowledge of many different things well beyond anything that could be called common. The details of the story are not important for us but, if it were published in a modern science fiction magazine, this man would almost surely be identified in most readers' minds as a time traveler. In 1875, however, the author's narrator found his punch line in "his firm conviction that the quiet, gentle, well-behaved, modest gentleman, so singularly gifted ... is, in plain terms, the devil!" Time travel certainly never entered the author's thoughts or, if it did, he lost his nerve at the idea of using it in this pre-Wells story. You'll recall from the opening of the Introduction that it was this 'use of the devil to explain mysterious happenings' that Wells wanted to move away from, and that was the motivation for his introduction of a time *machine*.

Hollywood has at least gotten the skeptical part of the psychology of time travel right (later discussions in this book will focus on how film makers have been less successful with the physics). When, for example, the time traveling villain in the

---

[82]H. L. Gold, editor of *The Galaxy Reader of Science Fiction*, Crown 1952.

[83]In his introduction to Robert Heinlein's classic time travel tale (to be discussed later) "All You Zombies—," in *The Mirror of Infinity*, Canfield 1970.

[84]F. Leiber, "The Oldest Soldier," *Fantastic*, May 1960.

[85]G. C. Eggleston, "Who Is Russell?" *American Homes*, March 1875.

1989 movie *Time Trackers* is confronted in the medieval past, he simply laughs-off a threat to reveal his true identity. "Go ahead," he says in effect, 'the only thing your talk of time machines from the future will accomplish is for people to think you are crazy!'

What would Arthur C. Clarke have thought of all this skepticism being directed toward those who claim to have a time machine? His thoughts about the difficulty time travelers would have in maintaining low profiles were what started the previous section, after all. My guess is that he would have had little patience with extreme incredulity. The surprise of being confronted by a time traveler would soon have turned to awe and pleasure IF—and I emphasize the IF—Clarke had been taken for a spin around the centuries in the stranger's machine. He would surely have ended-up quoting his own famous 'third law' to explain the wonder of it all: "Any sufficiently advanced technology is indistinguishable from magic."

Near the end of his paper, Heath writes what I think is the perfect rebuttal to anyone who would refuse to admit to time travel, even after taking a quick trip backward a few tens of millions of years to the late-Mesozoic era to hunt *Tyrannosaurus rex*, and even after seeing instant photographs of the dead monster with the skeptic's *own foot* on the great creature's head, or of his *own boots* dripping a bloody puddle of unholy size on the floor of the time machine. Writing about the Humean-unconvinced, even when faced with a rearranged firmament, Heath observes "But this now seems a cavil, designed only to prove that even omnipotence is powerless against the extremer forms of skeptical intransigence." Where God would fail to convince, a simple time traveler could hardly hope to do better!

1.4   Troubles with (some) Time Machines

"If you don't stop this senseless theorizing upon something that's an obvious impossibility, you'll find yourself working alone! Your ridiculous ideas sound like the ravings of a madman. Anyone with average intelligence realizes that the mere thought of traveling through time is absurd."[86]

If the previous section seemed just a bit gloomy concerning time travel, there is a very big reason for that. The sentiment expressed in the above opening quote to this section was a common one among philosophers long before physicists began to seriously think on the topic. While there *are* issues with time travel to the *future*, they are of an *engineering* nature, centered on how to build a big enough rocket ship with enough fuel to make the high speed, looping trip out into space and back again described in the opening section of this chapter. 'Mere' engineering problems are of no concern to physicists and philosophers. What *does* concern them are the far deeper puzzles of time travel to the *past*, the puzzles presented by what appear to be

---

[86]A science fiction physicist receives harsh criticism from a colleague in L. A. Eshbach's "Out of the Past," *Tales of Wonder*, Autumn 1938.

logical paradoxes. Before we get into the paradoxes, however, we need to first clear our minds of two common, popular notions of just what a time machine *is*. Both are false notions (one is due to H. G. Wells) which are, today, *rejected* by physicists (and most philosophers, too).[87]

As you'll see later in the book, all the theoretical time machines that have appeared in the modern physics literature involve spatial displacement. That is, they require *movement*. (On this point, the speedy DeLorean time car in the *Back to the Future* films has it right.) Wells' time machine, however, did not move; it always remained in the Time Traveller's laboratory (or at least on the spot where the laboratory would have been) unless he pushed it about after a trip in time. This, alas, results in a particularly troublesome problem: a Wellsian time machine heading into the past would run into itself!

Consider: There sits my time machine as I prepare for the first time journey ever, a trip back to the late-Mesozoic era to hunt dinosaur. I load my Continental. 600 super-high-power rifle with Nitro Express cartridges the size of bananas, kiss my wife good-bye, and climb in. I pull the lever. Now, Wellsian-type time machines don't *jump over* time but rather travel *through* time (see the Time Traveller's own description of how things looked to him, a description faithfully and spectacularly reproduced in the 1960 film). Therefore, the time machine will instantly collide with itself at the micro-moment *before* I pull the lever!

The resulting destruction obviously introduces a nice paradox: Given that this happens before I pull the lever, how did I manage to pull it? Many of the early science fiction writers were not totally oblivious to this collision problem and, in order to avoid materializing inside of an object in the future or the past, it was common to combine the time machine with an airplane.[88] Even that though might not be enough, as one writer thought a Wellsian time traveler would get "a severe case of the bends" if his body materialized in air![89] Of course, one might argue that Wells' machine does actually move because it is attached to the Earth, which is certainly moving, but it is not clear why this should result in the time machine arriving in the temporal past of the Earth, rather than in some past region of space (almost surely a vacuum).[90]

---

[87]Both of these notions still routinely appears in science fiction, however, because they are 'just too neat' to let 'mere physics' get in the way of a good tale. I use one, without apology, in my own story "Newton's Gift" in Appendix B.

[88]Three such tales are M. J. Breur, "The Time Valve," *Wonder Stories*, July 1930; F. J. Bridge, "Via the Time Accelerator," *Amazing Stories*, January 1931; E. Binder, "The Time Cheaters," *Thrilling Wonder Stories*, March 1940.

[89]J. Lafleur, "Time as a Fourth Dimension," *Journal of Philosophy*, March 1940, pp. 169–178, and "Marvelous Voyages—H. G. Wells' *The Time Machine*," *Popular Astronomy*, October 1943.

[90]Philosophers seem to be becoming more aware of the collison problem (which they have dubbed "the double occupancy problem"), and at least three papers published since the second edition of *Time Machines* discuss it: W. Grey, "Troubles With Time Travel," *Philosophy*, January 1999, pp. 55–70; P. Dowe, "The Case for Time Travel," *Philosophy*, July 2000, pp. 441–451; R. Le Poidevin, "The Cheshire Cat Problem and Other Spatial Obstacles to Backward Time Travel," *The Monist*, July 2005, pp. 336–352. Physicists don't concern themselves with the collison problem simply because they *aren't interested* in *Wellsian* time machines; I'll explain why I say this by the end of this section.

The general problem of 'where the past is' was nicely illustrated by the physicist Gregory Benford in his 1980 novel *Timescape*. In that story the world of 1998 is on the verge of total ecological collapse, and an attempt is made to change the past by aiming a backward-in-time message via faster-than-light tachyons (these hypothetical particles are discussed in Chap. 5) at the pivotal year 1963. When the principal scientist involved in this effort is explaining the process to a potential financial backer, he is asked, "Hold on. Aim for *what*? Where *is* 1963?" The scientist replies, "Quite far away, as it works out. Since 1963, the Earth's been going around the Sun, while the Sun itself is revolving around the hub of the galaxy, and so on. Add that up, and you find 1963 is pretty distant." An understanding of the question 'Where is the past?' actually goes quite a bit further back in science fiction. For example, after looking through a TV-like gadget to view the past, one character in a 1940s story complains, "You said you'd find Captain Kidd's treasure, but all I can see is fog and static." He is told that's because "It's too far back—1698 or thereabouts. The Earth was billions of miles from here then, and there are too many cosmic rays between."[91]

But let's suppose we ignore this concern about where things are for a time traveler, as do most science fiction stories. Still another problem with a true Wellsian-type time machine is that because it travels *through* time, the machine must *always* appear to be located in the same place. For example, to travel from Ford's Theater today to Ford's Theater on the evening of Good Friday, April 14, 1865, in a misguided attempt to save Lincoln from Booth's bullet (*why* this would be misguided will be discussed at length later in the book), a Wellsian-type time machine would have to occupy every instant of the intervening century and more. For observers outside the machine, the machine would appear to have been sitting in the same place all those years. There is an amusing illustration of a failure to understand this point by the scriptwriters of the 1989 film *Time Trackers*, who have time travelers 'hide' their Wellsian machine from accidental discovery by 'parking' it 5 s in the future!

Wells was well-aware of the "does a time traveling object disappear or not?" issue, and tried to have it both ways in *The Time Machine* by invoking what he had the Time Traveller call "diluted presentation." As we are told in the novel, the reason why we cannot see the model time machine he sends on its way into the future as a demonstration is that "the spoke of a wheel spinning, or a bullet flying through the air" is invisible because if those objects are "traveling through time fifty times or a hundred times faster than we are . . . the impression [they create] will of course be only one-fiftieth or one-hundredth." Similarly for the model. This explanation breaks down when one remembers that, even if you cannot see the spoke or bullet, they are still there and you can get in their way—Wells, unfortunately, has one of his characters stick a hand into the space where the model time machine was last seen.

---

[91]M. Jameson, "Dead End," *Thrilling Wonder Stories*, March 1941. The "cosmic rays" are presumably the cause of the interference.

This objection to Wellsian time travel was raised soon after the 1895 publication of the novel, and then again in 1914 by Pitkin (note 62), who noted that violent disaster awaited once the time journey ended. Wells, it is only fair to note, seemingly anticipated Pitkin when he had the Time Traveller say "So long as I travelled at a high velocity *through time* [my emphasis] ... I was, so to speak, attenuated—was slipping like a vapor through the interstices of intervening substances!" What this is getting at is that for the Time Traveller to stop 'inside' anything (Pitkin's example was the pile of bricks the Time Traveller's laboratory is certain one day to become) would, as Wells had his hero say, cause "a profound chemical[92] reaction—possibly a far-reaching explosion—[that would] blow myself and my apparatus out of all possible dimensions." Just why this spectacular event doesn't occur when the time machine simply stops *in air*, never mind inside Pitkin's pile of bricks, is never addressed.

In any case, it seems clear from all of this that Wells' machine travels *through time*, just as the Time Traveller claims. But Wells, himself, raises doubt when he describes the observed effects of a departing time machine. At the beginning of the novel, when the Time Traveller sends his model machine into the future, we read "There was a breath of wind, and the lamp flame jumped. One of the candles on the mantel was blown out ... and it [the model time machine] was gone—vanished!" And, at the end, when the Time Traveller makes his final exit, the narrator of the tale just misses the departure but tells us "A gust of air whirled around me as I opened the door, and from within came the sound of broken glass falling on the floor. The Time Traveller was not there ... Save for a subsiding stir of dust, the further end of the laboratory was empty. A pane of the skylight had, apparently, just been blown in." Both of these descriptions read as *implosions*, air rushing in to fill a spatial void, as though the time machines had *jumped* in time. Is there an inconsistency here? Well, perhaps not, if one accepts the curious idea of "slipping like a vapor" for an operational Wellsian-type time machine.

One famous science fiction story[93] nicely illustrates these points. The inventor of the first time machine demonstrates it to colleagues by sending a brass cube 5 min into the future. After being placed in the machine, the cube vanishes and then, 5 min later, reappears. Did the cube travel *through* time, or was its journey 'instantaneous,' so to speak? If *through* time, the cube was present at every instant after the start of its trip—so why did it vanish? The cube gets to each instant before the observers do, but why this should produce the visual effect of disappearing is unclear. The description in the story implies the cube traveled 5 min into the future without existing at any of the in-between instants, and so the story's time machine certainly was *not* Wellsian.

An immediate implication of the immobility of a Wellsian time machine is that if you are being chased by an angry mob somewhen in time (perhaps because you unwittingly violated a sensitive social taboo), then hopping into your Wellsian-type

---

[92]Actually *nuclear*, but don't forget when Wells wrote his novel.

[93]F. Brown, "Experiment," *Galaxy Science Fiction*, February 1954.

time machine isn't going to help because the machine just sits there. The mob could simply take its deliberate time in first building a roaring fire and then pushing the machine (and you) into it. As one author (with a wonderfully appropriate name!) expressed this, "You might as well try to escape by taking a nap."[94]

Pulp science fiction, always alert to a good story gimmick, used this character-istic of Wellsian time machines in one clever tale[95] in which a criminal attempts to hide his crimes by sending the bodies of his victims into the far future. His mistake is to use a Wellsian time machine in which he escapes into the future. The police, however, having learned of his foul deeds, simply build a cage around the machine and arrest him when he exits 23 years later!

If a Wellsian time machine that moves *through* time suffers from a fatal collision problem, then how about that other favorite of science fiction, a time machine that *jumps* in time? (Recall the final departure of Wells' Time Traveller.) That certainly would avoid the self-colliding problem. When you pull the lever inside the machine you simply disappear from 'now' and (from your point of view) then instantly pop into existence 'then.' The problem with this sort of time machine is that a Time Traveller who uses it will have a *discontinuous* world line, with the break occurring at the moment his time machine 'jumps.' In the modern physicist's view of time travel, however, based on general relativity, a Time Traveller's world line should always be continuous. That's because general relativity is a smooth, local field theory described by differential equations, resulting in *continuous* CTLs/CTCs.

Imagine, for example, that a 'jumping' time machine inventor starts building his gadget at time $t = A$ and expects to finish building it at time $t = B > A$. At time $t = C < B$, however, he runs into a problem. Fortunately, just at that moment a fully-functional time machine suddenly appears in the lab, and from it emerges a slightly older version of the inventor. The older version has the solution to the problem and, after telling the younger version the answer, gets back into the operational time machine and jumps off to ... somewhen. The younger version then completes his machine at time $t = B$, gets into it, jumps back to time $t = C$, and ....[96]

In the past, philosophers have gotten themselves all tangled-up in debates over *personal identity*, that is, which version is *the* inventor, the younger or the older? Can they both be the same person, even though the older version has a world line (*starting* at $t = C$) that is separate and distinct from the world line of the younger (that *stops* at $t = B$)? One philosopher (note 74) left physics behind and pursued this question into the following *legal* question concerning our two (?) inventors: if the older version commits a crime and then vanishes in his time machine before the police can apprehend him, can the younger version be punished even though *he*

---

[94]M. Cook, "Tips for Time Travel," in *Philosophers Look at Science Fiction*, Nelson-Hall 1982. One modern story that gets Cook's point right is by I. Watson, "The Very Slow Time Machine," in *The Best Science Fiction of the Year* (T. Carr, editor), Ballantine 1979.

[95]M. Jameson, "Murder in the Time World," *Amazing Stories*, August 1940.

[96]This little story I've just told you involves what is called a *bootstrap paradox* (just where did that solution come from, that is, who thought it up?) and it is one of the real puzzles of time travel. I'll say *lots* more about such curious doings later in the book.

hasn't yet committed the crime? While certainly 'interesting,' this really is a non-issue for the modern physicist who is concerned only with the physical possibility (or not) of time travel to the past.

Well, okay, you might now say, if neither a Wellsian time machine or a 'jumping' time machine will do, then just what *are* physicists studying in their papers on time travel? The short answer here (in Chap. 3 I'll say more) is that physicists don't view time machines as super-tech gadgets covered with wires, meters, dials, and levers, humming away beneath a seated Time Traveller as gigawatts of power throb through massive copper/crystal rods, with the whole business surrounded by a pulsating red-blue glow. Hollywood absolutely loves that sort of thing, but it's simply all wrong. For modern physicists, a time machine is *a region of spacetime with special topological structure*. Then, to time travel, a Time Traveller moves through that region of spacetime (in a rocket, perhaps) along an appropriate path. To 'make a time machine' therefore, in modern terms, means to (somehow) manipulate finite amounts of matter/energy in such a way as to alter the topology of a finite region of spacetime from one that has no CTLs/CTCs to one that does.[97] The most famous example of such a spacetime topology alteration (or *warp*) is the creation of a *wormhole*. A wormhole is a *topological* artifact of a spacetime; wormholes were popularized in Carl Sagan's 1985 novel *Contact* (under the guidance of physicist Kip Thorne) and are now common in science fiction.[98] As mentioned at the start of this chapter, for example, even Stephen King uses one in his 2011 *mainstream* novel *11/22/63*.

I'll return to the 'topology of spacetime' in Chap. 3 but, just so we don't leave it here as a mysterious phrase, here's a simple illustration of a topology change. Imagine a long, flat, narrow, two-dimensional strip of paper. The strip has the following topology features of interest to us here: (1) it has a beginning (its left end) and an ending (its right end), and (2) it has two sides (the top surface) and the flip-side surface. Now, imagine that we take the right end of the strip, give it a half-twist of 180^0 through our three-dimensional space, and then finally we glue that twisted end to the left end of the strip. The half-twist and gluing (our *warp*) has changed both of the topological properties of the strip. That's because the strip now has *no end* (you can travel forever along the strip, always going 'forward' and never reaching a point where can't go forward some more), and the strip now has just *one* side. You can convince yourself that it is one-sided by coloring the strip with a

---

[97]There is a hint of this in one prescient science fiction story, in which the inventor of a time machine, when asked about how it works, replies "An electromagnetic *warping* [my emphasis] of the spacetime continuum." See N. Schachner, "When the Future Dies," *Astounding Science Fiction*, June 1939.

[98]A wormhole is featured in the 2014 film *Interstellar*, whose Executive Producer and technical advisor was Thorne. The film's special effects are relativistically correct (not the typical Hollywood 'fantasy physics'), and you can read how that was achieved in Oliver James, Eugénie von Tunzelmann, Paul Franklin and Kip S. Thorne, "Gravitational lensing by spinning black holes in astrophysics, and in the movie *Interstellar*," *Classical and Quantum Gravity*, February 2015. See also (same authors) "Visualizing *Interstellar*'s Wormhole," *American Journal of Physics*, June 2015, pp. 486–499.

crayon. During the coloring, *do not lift the crayon from the strip*. When you can color no more, you'll find that every last bit of the strip has been colored. You can't do that with the original strip without lifting the crayon and turning the strip over because the original strip was two-sided. Many readers will recognize that what we've done is make a *Möbius strip*, named after the German astronomer and mathematician August Möbius (1790–1868) who described it in 1858.

Here's another astonishing property our half-twist warp has introduced. Cut the Möbius strip lengthwise with a scissors; most people believe you will then get two strips, each the length of the original strip but each only half as wide. Actually, you get one strip with a *full* 360° twist, which means the result is back to having two sides. (To see this, make a Möbius strip, cut it, and *then* apply the crayon.) And if you cut this new strip lengthwise once more, you get two separate loops, linked together. Try it and see, but be very careful. As the late science fiction writer Cyril Kornbluth (1923–1958) warned, there may be horrific potential dangers in unschooled experimentation with topology warps:

A burleycue dancer, a pip
Named Virginia, could peel in a zip;
　　But she read science fiction
　　And died of constriction
Attempting a Möbius strip.[99]

To end this section, I should point out that a change in the topology of a spacetime is *not* a necessary requirement for that spacetime to support time travel to the past. Gödel's rotating spacetime, for example, has a remarkably simple topology and, as you'll recall, it's literally *stuffed* with CTLs/CTCs, to the point that time travel to the past in Gödelian spacetime would be an everyday occurrence. You might think a world that presents time travel as a *fundamentally allowed* physical phenomenon, as does Gödel's spacetime, would be irresistible to science fiction writers. (So far as I know, however, no one has written a time travel story using the rotating universe idea.[100]) In Chap. 6, in fact, I'll show you just how easy it would be to time travel in Gödel's spacetime, using a rocketship as the means to move through that spacetime. Of course, *our* universe is not Gödelian, so the 'time travel to the past' question is not so easily answered for the spacetime we appear to actually inhabit.

---

[99]As you can see from this, science fiction writers have had fun with the Möbius strip. Two early examples *not* involving time travel are N. Bond, "The Geometrics of Johnny Day," *Astounding Science Fiction*, July 1941, and W. H. Upson, "A. Botts and the Möbius Strip," *The Saturday Evening Post*, December 1945. The use of the Möbius strip for time travel occurs, for example, in M. Clifton's "Star, Bright," *Galaxy Science Fiction*, July 1952.

[100]If he had lived, perhaps the well-known science fiction writer James Blish (1921–1975) would have written such a tale. In David Ketterer's biography of Blish (*Imprisoned in a Tesseract*, Kent State University Press 1987), there is this comment from a 1970 letter written by Blish: "I am especially intrigued by the spinning-universe form of time travel, especially since ... nobody has touched it ... But I should really stop mentioning the spinning-universe in public, or somebody will nobble onto it before I can get into it!"

1.5   Quantum Gravity, Singularities, Black Holes, and Time Travel

"A [spacetime] singularity is where God is dividing by zero."

—Anonymous

"A theory that involves singularities and involves them unavoidably, moreover, carries within itself the seeds of its own destruction."

—Peter Bergmann (1915–2002), Einstein's research assistant at the Institute for Advanced Study, Princeton

A fundamental objection to general relativity's suggestion of the possibility of time travel to the past is that, in a very deep sense, general relativity is known to be incomplete. That is, it is incompatible with quantum mechanics, which is the physics of the very, *very* small—the physics of atomic-size objects and *smaller*. We touched on this at the end of Sect. 1.2, and here we'll take a longer look at the issue of merging quantum mechanics with general relativity.

In quantum mechanics, the discrete nature of the atomic world appears in such phenomena as the photoelectric effect, in which light acts like individual particles (photons) rather than as continuous waves. Einstein's general relativity works beautifully on a cosmological scale but, like Maxwell's theory of electromagnetism, and unlike quantum mechanics, it fails when applied deep in the interior of the atom. Quantum theory, however, seems to work *everywhere*. As one physicist put it, "As far as we can tell, there is no experiment that quantum theory does not explain, at least in principle … Though physicists have steered quantum theory into regions far distant from the atomic realm where it was born, there is no sign that it is ever going to break down."[101]

One of the central concepts in relativity is the *world line*, which is the complete story of a particle in spacetime. A world line assigns a definite location to the particle at each instant of time. This is a classical, pre-quantum concept, however, and today physicists use the probabilistic ideas of quantum mechanics to describe the location and momentum of a particle once they get down to the atomic scale of matter. Quantum theory is a discrete theory in which the values of physical entities vary discontinuously (in 'quantum jumps'), whereas in classical theories the values of physical entities are continuous. The difference between the two types of theories is something like the difference between sand and water. Mixing the two theories— the classically smooth, continuous general relativity and the discrete quantum mechanics—to get something called *quantum gravity*, is the Holy Grail of physicists today, and nobody has more than an obscure idea of how to do it.

---

[101]N. Hebert, *Faster Than Light: Superluminal Loopholes in Physics*, New American Library 1988.

Just one of the more curious results of the fusing of quantum mechanics with general relativity may be *quantum time*.[102] That is, in quantum gravity the smallest increment of time that has physical meaning—sometimes called the *chronon*, a term first used in a non-time travel science fiction story[103]—may have a non-zero value. As we'll see later in the book, much of the controversy over the possibility of time machines hinges on what is called the *quantum gravity cut-off*. This is the end-result of destructive spacetime stresses that tend to grow toward infinity whenever a time machine spacetime topology attempts to form. This process goes under the general name of the *back reaction*, and is conceptually similar to a rubber band growing ever more taut as it is stretched, an effect that resists more stretching (and, of course, if stretched too far the rubber band breaks).

The cut-off of those stresses, at some *finite* value, is imagined to occur when the terminal phase of the growth would take place in less than the minimum possible time interval. The cut-off happens because, it is thought, nothing can actually occur in less than the minimum time. The debate is over just what that minimum duration is, and over whether the cut-off would occur before the stresses could reach *finite* values large enough to destroy the putative time machine topology. If the cut-off occurs before the back reaction stresses climb to the critical value, then the time machine survives. Otherwise, not.

To see how this 'works,' consider the two fundamental physical constants associated with classical gravity, the *gravitational constant*[104] G and the speed of light c, and the fundamental physical constant associated with quantum mechanics (*Planck's constant*) ℏ. Now, if you play around with combinations of these constants it is easy to show that the following expressions have the units of length, time, and mass, called the *Planck length* (l_P), the *Planck time* (t_P), and the *Planck mass* (m_P), respectively:

$$l_P = \sqrt{\frac{\hbar G}{c^3}} \approx 1.6 \times 10^{-33} \text{ cm,}$$

$$t_P = \sqrt{\frac{\hbar G}{c^5}} \approx 5.3 \times 10^{-44} \text{ s,}$$

$$m_P = \sqrt{\frac{\hbar c}{G}} \approx 22 \times 10^{-6} \text{ g.}$$

The extremely tiny values of l_P and t_P (the chronon), in particular, indicate (*roughly*) where it is expected that the smooth, continuous spacetime of general

---

[102]H. Kragh and B. Carazza, "From Time Atoms to Space-Time Quantization: the idea of discrete time, ca 1925–1936," *Studies in History and Philosophy of Science*, June 1994, pp. 437–462.

[103]S. Weinbaum, "The Ideal," *Wonder Stories*, September 1935.

[104]This is the constant in Newton's famous inverse-square law for gravity; the attractive force F between two point masses m_1 and m_2, distance r apart, is $F = G\frac{m_1 m_2}{r^2}$.

relativity will itself become quantized, and so will have to give way to a quantum theory of gravity.

The associated value of the mass-energy density when this transition is imagined to occur is enormous; the so-called *Planck density* is the Planck mass divided by the cube of the Planck length and has the value of about 10^{94} grams/cm^3. This is where physicists expect classical and quantum gravity to part company. Can such an enormous mass-energy density actually occur?[105]

Yes, and more, in what physicists call *singularities*.

This was all still very speculative until about 50 years ago, but today the search for how to connect general relativity and quantum mechanics is serious business. That search is related to time travel studies via a fantastic sequence of discoveries in relativistic physics, made during the last 80 years, beginning in 1931 with the work of the young Indian astrophysicist Subrahmanyan Chandrasekhar (1910–1995). He combined quantum mechanics and *special* relativity to show that a non-rotating star above a certain mass (about 1.4 times the mass of the Sun) cannot evolve into a white dwarf, which had until then been thought the eventual fate of all stars. Stars more massive than 1.4 Solar masses (but not *too* massive) would, instead, become *neutron* stars. But what then happens to stars that are too massive for even *that* bizarre eventuality?

General relativity predicts that a sufficiently massive star—greater than about four times the mass of the Sun—will, when its fuel is nearly exhausted and its nuclear fires are beginning to fade, experience a truly spectacular event called *total gravitational collapse*. When its fuel-starved, weakened radiation pressure is no longer able to keep a massive, aged star inflated against the collapsing force of its own gravity, the star will suddenly implode and crush itself into what is called a *black hole*, a dramatic term coined in 1967 by the Princeton physicist John Wheeler (1911–2008) in an address before the American Association for the Advancement of Science. A black hole is an object with a gravitational field so strong that even light cannot escape—that's why it's black!—at whose center is something called a *singularity*. This is all well-known lore in the physics world.[106]

Indeed, cataclysmic views of the collapse of matter are actually quite old. In Lucretius' first-century B.C. *The Nature of the Universe*, for example, we find the following imagery on what it would be like if matter itself collapsed: "The ground will fall away from our feet, its particles dissolved amid the mingled wreckage of

---

[105]By comparison, the density of a neutron star is on the order of a 'mere' 10^{16} grams/cm^3.

[106]Perhaps not so well-known, however, is that science fiction was there long before Wheeler. In one classic tale (M. Leinster, "Sidewise in Time," *Astounding Stories*, June 1934) a scientist explains at the end, "We know that gravity warps space ... We can calculate the mass necessary to warp space so that it will completely close in completely ... We know, for example, that if two gigantic star masses of certain mass were to combine ... they would simply vanish. But they would not cease to exist. They would merely cease to exist in our space and time." And then, as another character sums it up, "Like crawling into a hole and pulling the hole in after you." The explicit use of the complete term *black hole* for a region of weird spacetime also appeared in science fiction before Wheeler (P. Worth, "Typewriter from the Future," *Amazing Stories*, February 1950).

heaven and earth. The whole world will vanish into the abyss, and in the twinkling of an eye no remnant will be left but empty space and invisible atoms. At whatever point you allow matter to fall short, this will be the gateway to perdition." These words were actually inspired by earthquakes, not black holes and their singularities, but could a modern expert in general relativity and singularities have said it any better? But, what *is* a singularity?

As one theoretical physicist has dramatically written, "once gravity runs out of control, spacetime smashes itself out of existence at a singularity."[107] Or to quote Hawking, "A singularity is a place where the classical concepts of space and time break down as do all the known laws of physics."[108] One particular view of a singularity is that it is a place in spacetime that has infinite density and a gravitational field that is infinitely strong. The curvature of spacetime (more on curvature in Chap. 3) at this sort of singularity, sometimes called a *crushing* singularity, is also infinite. This is the sort of singularity believed to be at the center of non-rotating black holes. Historically, however, the occurrence of infinities in physical theories has been thought the red flag signaling that the theories have simply been extended too far, and their calculated results are nonsense.

Perhaps, then, singularities occur only in unrealistic physical applications of general relativity, and so it is only *perfectly* spherical collapsing stars that can end-up (on paper) as a black hole singularity. For a while physicists tried to establish that, but they were forced to abandon the attempt when it was shown that singularities are *unavoidable* and not just the result of idealistic assumptions.[109] This result worried many, and so the concern that general relativity was failing with its prediction of black holes and their singularities continued. In the case of a crushing singularity, perhaps all that meant is that once the collapsing star had fallen into a region even smaller than an electron, general relativity is no longer valid and the singularity is simply the 'math gone wild.' Einstein, himself, held that view. In his book *The Meaning of Relativity* (based on lectures he gave at Princeton in 1921), he wrote (concerning the use of the general theory to study the origin of the universe as a "big bang," which was a crushing singularity), "For large densities of field and matter, the field equations [of general relativity] and even the field variables which enter into them will have no real significance. One may not therefore assume the validity of the equations for very high density of field and matter, and one may not conclude that the 'beginning of the expansion' must mean a singularity in the mathematical sense."

Well, what does general relativity say about the singularity at the center of a black hole? To start, the theory says that, at a distance directly proportional to the

---

[107]P. C. W. Davies, *The Edge of Infinity*, Simon & Schuster 1981.

[108]S. Hawking, "Breakdown of Predictability in Gravitational Collapse," *Physical Review D*, November 15, 1976, pp. 2460–2473.

[109]S. W. Hawking and R. Penrose, "The Singularities of Gravitational Collapse and Cosmology," *Proceedings of the Royal Society A*, January 27, 1970, pp. 529–548. Ironically, one of the 'realistic' assumptions made in this paper, which appears to *force* singularities to exist, is that time travel is *impossible*!

mass of the collapsed object, a so-called *event horizon* will form. The event horizon is a surface in spacetime through which anything can fall into the hole, but through which nothing, not even a photon of light, can escape outward. The singularity at the black hole's center is therefore not visible to a remote observer (the singularity is said to be "clothed," and so not "naked"[110]). For all observers beyond the event horizon, the only visible properties of the hole are its mass (via its gravitational effects), its angular momentum (its spin rate), and its electric charge, and these properties are independent of the details of the pre-collapsed object (other than the requirement that electric charge and angular momentum are conserved).

There are actually several fundamentally different types of black holes. If the collapsed star forms a non-rotating, spherically symmetric, uncharged[111] object, then the result is called a *Schwarzschild* black hole, after the German astronomer Karl Schwarzschild (1873–1916) who found the first exact solutions to Einstein's general relativity field equations just months after Einstein published them.[112] Soon after that the Finn Gunnar Nordström and the German Heinrich Reissner independently found the solution to the field equations for the slightly more realistic non-rotating, *charged* black hole.[113] This is only slightly more realistic since it is highly unlikely a black hole wouldn't be spinning, as all observed stars are spinning and angular momentum is conserved during gravitational collapse. Another slightly more realistic solution, that of a *rotating*, uncharged black hole, was found by the New Zealand mathematician Roy Kerr in 1963, and this solution had a twist to it that at last explains why I am telling you all this—the singularity at the center of a Kerr black hole is not the *point* singularity of a non-rotating black hole but rather is a *ring* singularity. That is, there is a hole in the Kerr singularity through which matter can travel, without being destroyed, a hole that seems to act as a portal into

---

[110]A naked singularity, with no event horizon behind which to hide, would be particularly bothersome to physicists who don't like the idea of the breakdown of physics being on full display. What they think they'd then see would be completely unpredictable. Whether such a situation can actually exist is still open to debate, but there *are* both analytical solutions and computer simulations (incorporating realistic equations of state on the pressure response of matter as it is compressed) that seem to allow it (as in the gravitational collapse of an *infinitely* long, non-rotating cylinder that appears to result in an axial, thread-like, naked singularity).

[111]The word *charge* means either electrical *or* magnetic charge, although from a practical point *charge* probably does mean just electrical, as the theoretically possible magnetic monopole has yet to be observed and, in any case, it is thought that black holes will not have a significant electrical charge.

[112]Even *Einstein* hadn't yet solved them, and he apparently thought they were too complicated to be solved; when he saw Schwarzschild's result, he was so impressed that Einstein wrote to say "I had not expected that the exact solution to the problem could be formulated. Your analytical treatment of the problem appears to me splendid."

[113]Two years later, the University of Pittsburgh physicist Ezra Newman finally solved the field equations for the realistic, general case of a rotating *and* charged black hole.

other spacetime regions that may include past or *future* regions of spacetime. In other words, the ring singularity seems to be the entrance to a time machine.[114]

A discussion of singularities in general relativity is especially complicated for at least two reasons. First, there is more than one type, with *crushing* being just the (perhaps) most 'obvious.' Another type has no infinite curvature associated with it, but rather is a point in spacetime beyond which the worldline of a freely falling mass cannot be extended. Such a point is called a *geodesically incomplete* singularity, and it represents either an end to space or to time (in either case, that point is on the boundary or edge of spacetime). There are other types, as well—I've mentioned the naked singularity already—and the appearance of *any* of them is distinctly unsettling (recall Bergman's opening quote) to physicists. One that may be the most unsettling of all, however, is the *thunderbolt* singularity. This singularity propagates to infinity at the speed of light! As its discoverers dramatically put it, "It is not a naked singularity because you do not see it coming until it hits you and wipes you out."[115]

The other reason for a discussion of general relativity singularities being complicated is that they simply are not like the singularities of earlier theories. For example, in electromagnetic field theory spacetime is the *given background reference*; that is, a singularity in that theory is a point in spacetime where the *electromagnetic field* is undefined. In gravitational field theory, however, it is

---

[114]You can find discussions on how this is imagined to work in two papers by R. Weingard: "General Relativity and the Conceivability of Time Travel," *Philosophy of Science*, June 1979, pp. 328–332, and "Some Philosophical Aspects of Black Holes," *Synthese*, September 1979, pp. 191–219. See also M. Calvani et al., "Time Machine and Geodesic Motion in Kerr Metric," *General Relativity and Gravitation*, February 1978, pp. 155–163. I won't pursue black hole time machines in this book, as it is not what modern physicists consider a plausible means of time travel (*How are you going to gain access to a black hole?!!!*) For how one science fiction writer *did* use the idea, however, see L. Niven, "Singularities Make Me Nervous," in *Stellar* 1 (J.-L. del Rey, editor), Ballantine 1974. Black holes *are* bizarre objects—nearly as bizarre as time travel—and it seems risky to try to understand one in terms of the other (recall Professor Challenger's observation!).

[115]S. W. Hawking and J. M. Stewart, "Naked and Thunderbolt Singularities in Black Hole Evaporation," *Nuclear Physics B*, July 1993, pp. 393–415. As bizarre as is the thunderbolt, it was anticipated in science fiction by more than half a century. In the story "The Tides of Time" by R. M. Williams (*Thrilling Wonder Stories*, April 1940), the universe is collapsing at faster than the speed of light. Human scientists learn this when fleeing aliens stop their faster-than-light space ships to warn them. One of the human characters then looks out into the night sky and, in words that sound like those of Hawking and Stewart, "There would be no warning, for the rolling tide was traveling faster than light . . . It would come faster than the flicker of an eye. No one would see it come. One instant the world you knew would be around you. The next instant, there would be nothing. You would not even have time to know what had happened. Death, faster than the lightning flash!" This story may have been inspired by a tale published decades earlier, by the Canadian writer Frank Lillie Pollock (1876–1957). In his "Finis" (*The Argosy*, June 1906), written long before the concept of a super-nova, the light of a huge, distant star finally arrives to cook Earth into oblivion.

spacetime itself that is undefined, and there is no background 'something' in which spacetime is embedded to serve as a reference.[116]

One early suggestion on how to avoid the problem of the crushing singularity of the non-rotating black hole (which is, as mentioned earlier, not a realistic model for the gravitational collapse of a rotating star) is that the collapse may stop short of the singularity. That is, the collapsing body might instead rebound. This 'bounce' would occur after the star was inside its event horizon, so an external observer would not see the later expansion, an expansion imagined to be through the event horizon *but into a different region of spacetime.*[117] When Novikov's work was generalized the following year, the authors clearly had a hard time believing this dramatic imagery, despite their own mathematics, concluding with "It then appears necessary to believe in the existence of other [regions of the universe, including the past and the future] which will accommodate the re-expansion. This seems at least as fantastic as the alternative of [a point singularity].".[118]

In 1974 Hawking announced an astonishing partial connection of quantum mechanics with general relativity's black holes. He showed that, contrary to the usual image of black holes as being one-way trap doors to . . .?, black holes actually *must* radiate energy.[119] His analysis, which stunned physicists by its beautifully simple arguments, invokes the famous *uncertainty principle,* one of the corner stones of quantum mechanics. Hawking himself found the result "greatly surprising." He also cautioned (in his 1975 paper) that the following picturesque imagery is "heuristic only and should not be taken too literally," but it has now been in physics for over 40 years and appears to be here to stay.

The uncertainty principle states that there are certain pairs of variables associated with particles, variables that cannot be precisely measured at the same time. Time and energy form such a pair because a non-zero time interval is required to measure a particle's energy, and the product of the uncertainty in both the time interval (Δt) and the energy (ΔE) must be at least as large as a certain non-zero constant. That is, if $\hbar$ is Planck's constant, then $\Delta E \Delta t \sim \hbar$. This allows the process of virtual particle creation, the appearance of particle/anti-particle pairs just outside the event horizon of a black hole. The uncertainty in the energy is what gives the combined mass of the particles in a pair; this uncertainty in the energy is the quantum fluctuation energy of the intense gravity field of the hole. The only

[116]See, for example, R. Geroch, "What Is a Singularity in General Relativity?" *Annals of Physics,* July 1968, pp. 526–540.

[117]See, for example, I. Novikov, "Change of Relativistic Collapse Into Anticollapse and Kinematics of a Charged Sphere," *JETP Letters,* March 1, 1966, pp. 142–144, and V. P. Frolov, *et al.,* "Through a Black Hole Into a New Universe?" *Physics Letters,* January 12, 1989, pp. 272–276. Igor Novikov is a Russian physicist at the University of Copenhagen, and he will appear later in the book when we get to the paradoxes of time travel to the past.

[118]V. De La Cruze and W. Israel, "Gravitational Bounce," *Nuovo Cimento A,* October 1, 1967, pp. 744–760.

[119]S. Hawking, "Black Hole Explosions?" *Nature,* March 1, 1974, pp. 30–31, and "Particle Creation by Black Holes," *Communications in Mathematical Physics,* 1975, pp. 199–220.

constraint is that the energy be returned to the field, via mutual annihilation of the matter/anti-matter pair within the time uncertainty dictated by the uncertainty principle.[120]

As Hawking showed, this time interval, although incredibly short, is still long enough for the two virtual particles to separate before annihilation, one falling into the hole and the other escaping. This would happen, for example, if the particle/anti-particle pair is an electron/positron pair, and so a negatively/positively charged black hole would tend to attract the positron/electron and repel the other particle (either way, driving the charge of the hole towards zero). (Hawking then later suggested[121] that the particle *entering* the hole could be thought of as an *emitted* particle traveling backward in time, an idea that can be traced back decades, to John Wheeler—I'll return to this idea in just a moment.) By this incredible quantum process, then, the black holes of general relativity slowly *evaporate* (!) as they glow with what is now called *Hawking radiation*. That is, black holes appear to be hot bodies. But *hot* is relative, as a black hole with the mass of the Sun would have a temperature of just sixty *nano*-degrees Kelvin above absolute zero, and it would take 10^{66} years (a stupendously enormous time compared to the age of the universe) to completely evaporate.

Indeed, one physicist had already mused that the *entire universe* might have been created by a similar quantum process, *out of nothing*, a so-called *vacuum fluctuation*.[122] The explanation for why the universe doesn't then disappear—and very quickly, too, because the energy for all the mass in the newly created universe is quite large (that is, ΔE is *really big* and so Δt must be *really small*)—is that the *negative* gravitational potential energy of all that newly created matter would cancel the positive mass-energy, and so ΔE is actually *quite small* and so Δt is then *quite large*. To perhaps show he wasn't quite convinced by all that, himself, Tryon whimsically wrote "I offer the modest proposal that our Universe is simply one of those things which happen from time to time."

As a final comment on the suggestion by Hawking of a connection between virtual particles at a black hole event horizon and backwards time travel, the idea

---

[120]The uncertainty principle has long been used in time travel science fiction. In one story, for example, a character is transported from 1950 to 2634 by a scientist of the future. Once there, this character decides he'd like to remain permanently in the 27th century. He is told that he can't because he is like an atom excited into an elevated energy state and, just as quantum mechanics says that eventually an electron in such a state will drop back down into a lower energy state, so do the "laws of time travel" require that he drop back to his normal time. How long can he remain in future, he is told, "depends on the mass [energy] of his body and the number of years the mass [energy] is displaced." That is simply the uncertainty principle. See W. Bade, "Ambition," *Galaxy Science Fiction*, October 1951.

[121]S. Hawking, "The Quantum Mechanics of Black Holes," *Scientific American*, January 1977.

[122]E. P. Tryon, "Is the Universe a Vacuum Fluctuation?" *Nature*, December 14, 1973, pp. 396–397.

originated (as I said before) with Wheeler, in 1941.[123] In an astonishing coincidence, even as Wheeler was telling his student Richard Feynman about this, a science fiction writer was also identifying anti-matter with backward time traveling 'normal' matter.[124] Later, the Polish science fiction writer Stanislaw Lem (1921–2006) took this idea, combined it with the quantum concept of energy fluctuation, and came up with one of his typically outrageous (and typically hilarious) ideas: shooting a single positron out of an accelerator back to the very beginning of time. His story character called this fantastic machine the "Chronocannon" and claimed that's what started the universe.[125]

Soon after Lem, a philosopher used a variant of this idea, in which the Big Bang creation of the universe was caused by a time traveler from the future who saw a need—his own existence—to generate the Big Bang. This leads to philosophical speculations on the cosmological implications of God as a time traveler.[126] Two recent physicists have taken this one step further by suggesting that the universe, via time travel, may have caused itself! As they put it, "the laws of physics may allow the Universe to be its own mother."[127]

The modern hope is that quantum mechanics (as in quantum gravity) will save physics from the horror of general relativity's singularities. This was the view of John Wheeler and, as the man who named black holes, his view is important to consider. General relativity is a classical, *smooth* theory that is fundamentally continuous, while 'our' universe appears to be a quantum one. So, perhaps, general relativity's prediction of singularities may be just an artifact without physical reality in the 'real world.' Wheeler's position was based on the quantum fluctuations of gravity fields, which are related to the uncertainties inherent in our knowledge of the values of physical entities. Such fluctuations are vanishingly small in systems of everyday size, but they increase dramatically at very tiny distances that are twenty orders of magnitude smaller than the nucleus of an atom. In the microscopic region of spacetime that the matter forming a black hole is falling into, these fluctuations might conceivably result in effects that preclude the formation of a singularity. Agreeing were two physicists who asserted that, even without a detailed knowledge of quantum gravity, quantum effects "would smash the idealized interior geometry"

[123]See Richard Feynman's Nobel lecture, reproduced in *Science*, August 12, 1966, pp. 699–708, where he recounts Wheeler's 'proof' for why every electron in the universe has *exactly* the same charge ('there is only *one* electron, weaving its way back-and-forth in time, with positrons being the electron when traveling backward-in-time').

[124]Will Stewart, "Minus Sign," *Astounding Science Fiction*, November 1942. 'Will Stewart' was a pen-name for John Stewart Williamson (1908–2006).

[125]S. Lem, "The Eighteenth Voyage," in *The Star Diaries*, Seabury Press 1976.

[126]G. Fulmer, "Cosmological Implications of Time Travel," in *The Intersection of Science Fiction and Philosophy* (R. E. Meyers, editor), Greenwood Press 1983. Isaac Asimov used a similar idea in his story "The Instability," *The London Observer*, January 1, 1989.

[127]J. R. Gott and L.-X. Li, "Can the Universe Create Itself?" *Physical Review D*, 1998, 023,501.

[that is, the ring singularity] of a rotating, charged black hole, thereby eliminating any possibility of using such a hole for time travel.[128]

And finally, to generalize beyond black holes to the hoped-for pay-off of the coming of quantum gravity in banishing singularities altogether, one recent study has examined how non-crushing singularities (that is, ones of the geodesically incomplete type) *are* apparently "healed" (the authors' term) by quantum effects.[129] With the eventual development of quantum gravity, perhaps all the singularities of general relativity will vanish while leaving the CTCs/CTLs intact, thereby removing a form of doubt in the theory's apparent support for time travel to the past. It may be a long time coming, however: as the University of Sydney philosopher of science Dean Rickles recently (2014) wrote in his book *A Brief History of String Theory*, "quantum gravity is in many ways . . . a revolution still waiting to happen."

1.6   Tipler's Time Machine

"In short, general relativity suggests that if we construct a sufficiently large rotating cylinder, we create a time machine."[130]

The time traveling property of the ring singularity in a rotating black hole once made it a favorite of science fiction writers, as in Joe Haldeman's classic 1974 novel *The Forever War* (in which the term used is not *black hole*, but *collapsar*, which is a nicely descriptive word in its own right). A major difficulty with this approach, however, as I mentioned in the previous section (note 114), is that of 'getting one's hands on' (so to speak) a black hole! So, is there any other 'time machine' that is consistent with general relativity? Yes, there is.

In 1974 a young physics graduate student at the University of Maryland, Frank Tipler, caused a bit of a stir when he published what seemed to be quite specific construction details for a time machine. Indeed, the final sentence (the above quotation) of his paper couldn't be clearer. Nobody had ever before made such a statement in a respectable physics journal and, best of all, there were no apparent spacetime singularities involved. However, a close look at Tipler's analysis does turn up some difficulties.

What Tipler had actually done was to show that if one had an *infinitely long*, *very dense* cylinder rotating with a surface speed of at least half the speed of light (the rotation speed is such that the outward centrifugal forces are balanced by the inward gravitational attraction of the cylinder), then this allowed the formation of closed

---

[128]N. D. Birrell and P. C. W. Davies, "On Falling Through a Black Hole Into Another Universe," *Nature*, March 2, 1978, pp. 35–37.

[129]T. M. Helliwell and D. A. Konkowski, "Quantum Singularities in Spherically Symmetric, Conformally Static Spacetimes," *Physical Review D*, May 13, 2013, 10404.

[130]F. J. Tipler, "Rotating Cylinders and the Possibility of Global Causality Violation,' *Physical Review D*, April 15, 1974, pp. 2203–2206.

timelike curves around the cylinder. This means that by orbiting the surface of such a fantastic cylinder, one could travel through time into the past—but not to earlier than the moment of the creation of the cylinder.

This last point is a very important one, as it does avoid one particularly odd paradox (called a *bootstrap*): a traveler going backwards in time to tell the inventor of a time machine (perhaps an earlier version of the time traveler himself) how to build the time machine. You can find this idea in early science fiction,[131] and a minor variant of it was amusingly illustrated in the 1985 film *Star Trek IV: The Voyage Home* (when you next watch the movie, ask yourself who actually invented "transparent aluminum"?) Bootstrap paradoxes are quite mysterious and still befuddle physicists and philosophers. Science fiction writers, on the other hand, love bootstraps as great story gimmicks.

Tipler's cylinder would also enable a time traveler to return to her original time, to go "back to the future," by orbiting the cylinder in the reverse direction (but no further into the future than when the cylinder ceases to exist). Later in the book I'll show you a simple illustration—based on a similar one in Tipler's PhD dissertation, published in 1976—of how the cylinder works as a time machine. No one, in fact, disputes any of this. It *is* true. On paper.

But Tipler did *not* prove that a time traveling property holds for cylinders of even very long but *finite* length, which are the only kind we could actually build from a finite amount of matter; he merely suggested that such might be the case. This suggestion does seem reasonable, because if the time traveler orbits at the midpoint of the cylinder, near the surface, then the gravitational end-effects of sufficiently remote ends of the cylinder would, you'd think, become negligible. Similar mathematical approximations are routinely made, for example, when calculating the electrical effects of charged cylinders of finite length. But as one physicist has warned, "Extrapolation from cylindrical symmetry to reality is very dangerous, since spacetime is not even asymptotically flat around an infinite cylinder."[132] The issue of whether a spinning, finite-length cylinder can create closed, timelike curves is still open: to quote another physicist, "[In] some respects an infinite cylinder may be a model for a long finite one, and the possibility cannot be dismissed that a time machine might be associated with a long, but finite rotating system."[133]

[131] See, for example, C. Cloukey, "Paradox," *Amazing Stories Quarterly*, Summer 1929. Later in the book I'll discuss even earlier literary occurrences of bootstraps (that is, of *information* on closed loops in time).

[132] K. S. Thorne, "Nonspherical Gravitational Collapse: Does It Produce Black Holes?" *Comments on Astrophysics and Space Physics*, September–October 1970, pp. 191–196. What "asymptotically flat" means will be discussed in Chap. 3.

[133] W. B. Bonner, "The Rigidly Rotating Relativistic Dust Cylinder," *Journal of Physics A*, June 1980, pp. 2121–2132. Tipler was not the first to study rotating cylinders in the context of general relativity. Such cylinders had been around for decades, going back to 1932. A good reference is M. A. Mashkour, "An Exterior Solution of the Einstein Field Equations for a Rotating Infinite Cylinder," *International Journal of Theoretical Physics*, October 1976, pp. 717–721. The first-analyzed configuration of matter that generates closed timelike lines, solved in all its general relativistic detail, was the infinite rotating cylinder studied by W. J. van Stockum,

There is, however, another potential problem besides the length of the cylinder. There is a strong likelihood that a Tipler cylinder under construction would collapse under its own internal gravitational pressure before it could be made nearly long enough to be even 'approximately infinite.' That is, such a finite-length cylinder might crush itself along its long axis into a pancake-shaped blob, something like what happens to a long cylinder of jello stood on-end. An ordinary can of jellied cranberry sauce will also sometimes display this curious behavior.

The required rotational speed raises yet another concern, as well. We are not talking about cylinders the diameter of a pencil, or even of a large water pipe. Recall that for a given surface speed, the larger the diameter the less the centrifugal acceleration at the surface. It is easy to calculate that even a huge cylinder 10 kilometers in radius, with a surface speed of half the speed of light, would have a surface acceleration *hundreds of billions* of times the acceleration of Earth's surface gravity. No known form of ordinary matter could spin that fast and not explosively disintegrate; Tipler has estimated that the required density for a time machine cylinder would be 40 to 80 orders of magnitude above that of nuclear matter. (In a masterful understatement, Tipler calls this astonishing stuff "unknown material.") Made from such incredibly superdense stuff, even a finite cylinder would still be as massive as the Sun but many trillions of times smaller. Showing no lack of imagination, Tipler has suggested the possibility of speeding up the rotation of an existing star as an alternative approach to that of building a cylinder.[134] That, of course, would be project for a far-future society, with a *very* advanced technology.

All of these concerns were discouraging to Tipler (who could blame him?), and his pessimism about the actual likelihood of achieving time travel via one of his cylinders is shown by the words he used to open his 1977 paper (note 134): "Any attempt to evolve a time machine] from [normal] matter will cause singularities to form in spacetime. Thus, if by the word 'manufacture' we mean 'construct using only ordinary materials *everywhere*,' then the theorems of this paper will conclusively demonstrate that a [time machine] cannot be manufactured." But not all physicists agreed.

"The Gravitational Field of a Distribution of Particles Rotating About an Axis of Symmetry," *Proceedings of the Royal Society of Edinburgh*, 1939, pp. 135–154. This is particularly interesting because, while Van Stockum didn't spot the presence of closed timelike lines in his solution, his cylinder is made entirely from *ordinary* matter.

[134]F. Tipler, "Singularities and Causality Violation," *Annals of Physics*, September 1977, pp. 1–36. See also his earlier paper "Causality Violation in Asymptotically Flat Space-Times," *Physical Review Letters*, October 1976, pp. 879–882, where he wrote "There are many solutions to the Einstein equations [of general relativity] which possess causal anomalies in the form of closed timelike lines (CTL). It is of interest to discover if our Universe could have such lines. In particular, if the Universe does not at present contain such lines, is it possible for human beings to manipulate matter so as to create them? [That is, to construct a time machine.] I shall show in this paper that it is *not* [Tipler's emphasis] possible to manufacture a CTL-containing region without the formation of naked singularities, *provided normal matter is used in the construction attempt* [my emphasis]."

Years after Tipler wrote, one physicist replied[135] with *two* pointed observations. First, Tipler's theorems apply only to singularities of the incomplete kind, not to the more convincingly fatal crushing (or curvature) type. Second, to quote Ori at length, "The standard interpretation of Tipler's theorems is to say that the appearance of a singularity in a given [spacetime] model indicates that this model is unrealistic and cannot be physically realized. Even for future-generation engineers it will probably be impossible to use 'singular matter' for the construction of their time machine. However, the theory of black holes provides an obvious counterexample to this interpretation. For, by applying this interpretation to the black hole singularity theorems one could conclude that black holes can never form." Yet black holes with several times the mass of the Sun *have* been detected in orbit about certain stars, and at least one supermassive black hole (with a mass equal to more than three *billion* Suns) has been detected at the core of galaxy M87. Indeed, it is now believed that the center of *every* sufficiently massive galaxy in the universe is home to a black hole (the one at the center of our own galaxy, the Milky Way, has a mass about three million times that of the Sun).

Even less concerned about singularities interfering with time travel were two other physicists who wrote[136] "It would seem that a successful attempt to manufacture [a time machine] within a finite region of space will be accompanied by the creation of a singularity ... This does not immediately imply, however, that with a sufficiently advanced technology one could not make a time machine. *There is no reason to suspect spacetime singularities could not in principle be created through deliberate human action* [my emphasis]."

These optimistic views were, of course, welcome news for science fiction writers, who had been using Tipler cylinders almost from when Tipler first wrote of them. Indeed, Larry Niven liked them well enough to 'lift' the very title of Tipler's paper (note 130) for the title of a short time travel story for inclusion in his 1979 collection *Convergent Series*. Just one year after Tipler's paper appeared, Poul Anderson featured the cylinders in his 1978 novel *The Avatar*, where they are called "T-machines": one can imagine the "T" stands for Time or Tipler or even both. Anderson's story describes the cylinders as having been scattered about the universe by ancient, altruistic aliens called "the Others," for the use by any who come across them and who have the wits to decipher *how* to use them. Anderson recognized the obvious problems with Tipler cylinder construction, and so has one of his characters say of T-machines, "I have no doubt whatsoever that [they are] the product of a technology further advanced from ours than ours is from the Stone Age."

---

[135]A. Ori, "Must Time-Machine Construction Violate the Weak Energy Condition?" *Physical Review Letters*, October 1993, pp. 2517–2520. The weak energy condition is the seemingly 'obvious' requirement that the observed local mass-energy density should never be negative. Quantum mechanics predicts (and it has been experimentally confirmed) that there *are* exceptions.

[136]M. P. Headrick and J. R. Gott, "(2+1)-Dimensional Spacetimes Containing Closed Timelike Curves," *Physical Review D*, December 15, 1994, pp. 7244–7259. The '(2+1)' refers to a toy spacetime with just two spatial dimensions and one time dimension.

Even before (actually *long* before) Tipler's paper, science fiction had foreshadowed his physics. Oliver Saari (1918–2000), for example, had incorporated both superdense matter and the rule of 'no time travel before the creation of a time machine' in a story written *40 years* earlier.[137] Saari's fictional time machine works by warping spacetime via a plate of superdense matter. (An even earlier tale[138] had also used superdense matter, but it was badly flawed by its hocus-pocus invoking of 'rays' emitted by the newly discovered element of *tempium*.) The 'no time travel before the creation of a time machine' rule is the basis for an obvious response to Hawking's Chronology Protection Conjecture, discussed earlier in this chapter, and it was so used by one physicist to rebut the Conjecture: as he wrote,[139]

(1) time machines, if possible, must have the property of not being able to travel back to before their creation, and
(2) no time machine has yet been created.

The absence of time travelers amongst us, therefore, provides no insight, one way or the other, on the eventually possibility of constructing a time machine.

1.7   For Further Discussion

> Observations of the background microwave radiation that permeates the universe is strong experimental evidence for the Big Bang, the singularity thought to be the origin of the universe. This singularity is not shielded from us by an event horizon, and so is not a naked singularity (note 110), which means it is potentially visible. In 1969 the English theoretician Roger Penrose, however, proposed a metaphysical 'law' called the *cosmic censorship principle*, which asserts that naked singularities are impossible. Discuss the obvious tension between Penrose's principle and the Big Bang singularity. (See, for example, P. Kosso, "Spacetime Horizons and Unobservability," *Studies in History and Philosophy of Science*, June 1988, pp. 161–173.)

---

[137]O. Saari, "The Time Bender," *Amazing Stories*, August 1937. In this story we read that the time traveler "could not travel into the past for the plate had to exist in all ages traveled, and it had not existed before he made it."

[138]E. L. Rementer, "The Time Deflector," *Amazing Stories*, December 1929.

[139]K. S. Thorne, "Do the Laws of Physics Permit Closed Timelike Curves?" *Annals of the New York Academy of Science*, August 10, 1991, pp. 182–193. Science fiction writer Damon Knight (1922–2001) anticipated Thorne's rebuttal in his story "Azimuth 1, 2, 3, ... ," *Isaac Asimov's Science Fiction Magazine*, June 1982.

In the text I mention the "transparent aluminum" bootstrap paradox that appears in the 1985 movie *Star Trek IV*. Even earlier, a movie bootstrap appeared in 1980 film *The Final Countdown*. There, the designer of a modern naval warship that temporarily time travels back to the Pearl Harbor of December 6, 1941, turns out to be a crew member who was (is) accidently left behind in the past when the ship returns to the present. In the past he *will be* able to design the ship because he already knows how it *was* designed—by himself! In the more recent 2014 film *Interstellar*, a wormhole near Saturn is discovered. By the end of the film we learn that it was put there by future humans, humans who exist because their ancestors (us!) were saved from a planet-wide ecological disaster when they used the wormhole to discover new worlds in far-flung regions of the universe. Decide whether or not the existence of the wormhole represents a bootstrap paradox, and defend your position.

One difficulty in using a black hole as a means of traveling from one region of the universe to another (with time travel as a special case) is simply getting to a black hole in the first place. The nearest one to Earth, as far as is known, is many light years distant. One reason for this may be an anthropic one (see note 13 in the "Introduction"). That is, a planet near a rotating black hole would either be eventually swallowed whole, or have its surface blasted by a firestorm of radiation produced by in-falling matter. In any case, no intelligent life able to recognize time travel would ever evolve on such a planet in the first place. That is, we are here to wonder about the absence of near-by black holes precisely *because* we aren't near a black hole. The lack of black holes near Earth is addressed in Joe Haldeman's 'Earth vs. Aliens' novel, *The Forever War*, by using the time dilation effect of special relativity (discussed in Chap. 3) that allows long travel distances to be covered in a reasonable time (as measured by clocks in rocket ships traveling near the speed of light). Still, while the travel time to reach a black hole distant from Earth by many light years may only be 6 months of ship time, back on Earth many *years* may pass. Once at the black hole the ship enters it and instantly 'jumps' to a vastly different region of the universe. In the novel, no time travel after the time dilation experienced in just getting to the black hole occurs, but Haldeman uses that to great effect as follows. Before entering into combat, Earth's soldiers are told that when they exit the hole into a new region of the universe, they may encounter alien warships equipped with *their* latest technology, technology that could be far in advance of the Earth warship's technology

(continued)

which dates from Earth's *past*. That is, humans will be fighting against technology that dates from the Earth warship's *future*. To quote the novel, "Relativity traps us in the enemy's past; relativity brings them from our future." Explain this.

In one of the quatrains of the *Rubaiyat*, the eleventh century Persian poet-philosopher Omar Khayyam wrote

The Moving Finger writes; and having writ,
Moves on; nor all your Piety nor Wit
Shall lure it back to cancel half a Line,
Nor all your Tears wash out a Word of it.

Nearly a 1000 years later the German theoretical physicist Hermann Weyl (1885–1955), a colleague of both Einstein and Gödel at the Institute for Advanced Study in Princeton, NJ, wrote the following in his book *Space-Time-Matter* (published in 1921, three decades *before* Gödel's 1949 time travel paper):

> *It is possible to experience events now that will in part be an effect of my possible future resolves and actions. Moreover, it is not impossible for a world-line (in particular, that of my body), although it has a time-like* [see the index] *direction at every point, to return to the neighborhood of a point which it has already once passed through. The result would be a spectral image of the world more fearful than anything the weird fantasy of E. T. A. Hoffmann* [an early nineteenth-century German writer of the eccentric] *has ever conjured up. In actual fact the very considerable fluctuations of the* [components of the metric tensor, to be discussed in Chap. 3] *that would be necessary to produce this effect do not occur in the region of the world in which we live ... Although paradoxes of this kind appear, nowhere do we find any real contradictions to the facts directly presented to us in experience.* Compare these two views and, in particular, discuss what each says about the idea of 'reliving the past.'

In the opening section of this book ("Some First Words") I mention how now and then science fiction has anticipated physics. One interesting example of this occurs in a story of a time traveler *almost* meeting himself, a story published 2 years *before* Gödel's 1949 time travel paper in which he suggests just such a possibility. The story opens with a man on a ship spotting the signal fire of a castaway on a Pacific island, as well as the tiny, distant figure

(continued)

of a man waving and jumping about. While sailing in to help, the ship hits a mine left over from the war, and the would-be-rescuer becomes a castaway, too. After swimming to the island, he can find no trace of who built the fire, although there are footprints all about in the sand. Exploring the island, he finds the remains of a crashed interstellar spaceship (!), powered by a drive unit based on 'temporal precession.' The man, curious, turns the drive on and thus sends himself backward in time by one day. He then spots a ship on the horizon, builds a fire, waves and jumps about, then *recognizes the ship as his own* .... And so the loop nearly but not quite closes. The man, apparently, rushes off into the jungle, terror-stricken at the thought of meeting himself. (You can find this tale by A. B. Chandler (1912–1984), "Castaway," in the November 1947 issue of *Weird Tales*, a publisher more of fantastic, supernatural, and horror stories than of science fiction. Perhaps easier to locate would be an anthology in which it has been reprinted: *Science-Fiction Adventures in Dimension* (G. Conklin, editor), Vanguard 1953.) Speculate on what happens to the man. In particular, does Wells' own criticism of *The Time Machine*, concerning "vain repetitions" of time travelers, apply here (see the "Introduction" again)?

A philosopher has argued against the force of Hawking's chronology protection conjecture as follows: "There is an old argument to the effect that while backward time travel may be *possible*, it will never actually occur—for if it were going to occur, we would *already* have encountered the time travellers involved, whereas in fact we have done no such thing. ... But consider an isolated society living in a remote part of the world. Some members of this society are engaged in a long-running debate concerning the possibility of human flight. Were a 747 to pass overhead, would the debaters necessarily recognize it as containing flying humans? The answer to their question might have been staring them in the face for years, without them realizing." (See Nicholas J. J. Smith, "Bananas Enough for Time Travel?" *British Journal for the Philosophy of Science*, September 1997, pp. 363–389, in particular note 3 on p. 364. The perhaps curious appearance of 'bananas' in the title of this paper will become clear when, in Chap. 4, we delve into the details of the famous grandfather paradox.) How would *you* answer Smith's question? Do you think it is plausible, as Smith implies, that we could right now be observing (without realizing it) effects in the present-day world that are the result of time travelers amongst us? What sort of effect (s) might raise this suspicion in *your* mind?

As discussed in Sect. 1.6, Tipler expressed some pessimism in his 1977 paper (note 134) about the possibility of actually constructing a time machine from a rotating cylinder. But that doesn't mean he didn't have some doubts, too, about theoretical 'proofs' of something being impossible. In his 1976 PhD dissertation, for example, he included an amusing reference to Simon Newcomb (a real-life mathematician that Wells' Time Traveller cites in *The Time Machine*—see note 102 in Chap. 2) who published mathematical 'proofs' that it would be impossible with known science to build a "practicable machine by which men shall fly long distances through the air." Why do you think Tipler did that? You can read more about Newcomb's 'proofs' in "Is the Airship Coming," *McClure's Magazine* (September 1901, pp. 432–435) and "The Outlook for the Flying Machine," *The Independent* (October 22, 1903, pp. 2508–2512).

In his autobiography, the Princeton physicist John Wheeler had this to say about time: "The smooth flow of time—or our smooth passage through it—is an illusion that is shattered when we … ask about time at the moment of the Big Bang, at a moment of gravitational collapse, at the moment of the Big Crunch. Students and others often ask what existed before the Big Bang. To say that we don't know is not to say enough. Even to say that we have no *way* of knowing is not enough. We really have to say that space and time came into existence, along with matter and energy and the laws of physics, at the moment of the Big Bang. If the universe expands to a maximum size, starts contracting, and eventually collapses to a fiery death—a fate that seems likely to me and to some other theorists … then time and space, too, will end in this Big Crunch. I can reach no conclusion other than this: there was no 'before' before the Big Bang, and there will be 'after' after the Big Crunch." (See J. A. Wheeler, *Geons, Black Holes, and Quantum Foam: a life in physics*, W. W. Norton 1998, pp. 349–350.) That is, when the Big Bang singularity occurred, time was *created*, and if the universe should collapse in the far future in a Big Crunch, time will be *annihilated*. This is a view of *nothingness* that transcends even that of the grave. Sharing Wheeler's dark view of the ultimate fate of reality, but instead giving the *victory to* time (rather than its annihilation), was the Irish writer Jonathan Swift (1667–1745) in his poem *Riddles* (circa 1724): "Ever eating, never cloying/All-devouring, all destroying/Never finding full repast/ Till I eat the world at last." How do you think theologians would respond to Wheeler and Swift?

The American philosopher Roy Sorenson was cited in note 13 in the discussion of the difficulties a time travel would have in convincing skeptics that he *had really* time traveled (short of bringing a fresh dinosaur egg back and hatching it!). This question was treated in early pulp science fiction ("The Sands of Time," see note 43 in "Some First Words") as follows: The time traveler takes a sealed box of pure radium (with his name written on the inside of the lid) into the distant past, and buries it in a secure location. Upon returning to the present he unearths the box; testing of the contents will show that some of the radium has radioactively decayed to lead. Indeed, the amount of decay would be a direct measure of how far back into the past the box had been transported. This issue was later elaborated on by the English philosopher Alasdair Richmond in his paper "Time Travel, Parahistory and the Past Artefact Dilemma," *Philosophy*, July 2010, pp. 369–373. There he imagined two possible ways a time travelling Shakespearean scholar might attempt to convince skeptical colleagues that he had discovered a draft of *Hamlet* dating from the year 1589 (10 years *before* the earliest accepted date of its composition by Shakespeare). The first attempt is to simply bring that draft directly back with him in the time machine, from 1589 to the present. Then, of course, many of the inherent clues as to the draft's authenticity, such as chemical composition of the ink, the weave of the paper, and orthography (the style of writing in 1589) would be consistent with the time traveler's claim, but other clues would *not*—the *age* of the paper and of the ink, for example, would be taken as evidence fatal to the claim, as they would not be nearly 430 years old. They would appear, in fact, to be practically new! The draft would, therefore, be dismissed as simply a clever forgery. The second attempt would try to get around this problem, as follows. After locating the draft in 1589, the time traveler *doesn't* bring it back to the present, but rather stashes it away in a secret hiding place. Then, once back in the present, he takes his colleagues to the secret hiding place and, with a flourish, reveals the draft which now *is* nearly 430 years old. Much to the time traveler's frustration, however, his colleagues still reject his time travel claim, this time saying he must have simply found the draft in the 'usual' way (under the floorboards in somebody's attic, for example), and is just pretending to have found it via time travel. Can you think of a way, using the *Hamlet* draft, the time traveler might be able to convince his skeptical colleagues?

The fan I quoted in "Some First Words," who wrote to *Astounding Stories* in 1931 to express his unhappiness with the appearance of women in that magazine's stories, was quite clear about his concerns—although, given the times, he carefully avoided any direct mention of sex. A modern, highly successful *female* writer of science fiction, Anne McCaffrey (1926–2011), didn't shy away from that, however, when she wrote the following in a hilariously funny essay: "Prior to the '60s, stories with any sort of love interest were very rare. True, it was implied in many stories of the '30s and '40s that the guy married the girl whom he had rescued/encountered/discovered during the course of his adventures. But no real pulse-pounding, tender, gut-reacting scenes. The girl was still a 'thing' to be used to perpetuate the hero's magnificent chromosomes. Or perhaps to prove that the guy wasn't . . . I mean, all those men locked away on a spaceship for months/years at a time. I mean . . . and you know what I mean even if I couldn't mention it in the sf of the '30s and '40s." (See Anne McCaffrey, "Hitch Your Dragon to a Star: Romance and Glamour in Science Fiction," in *Science Fiction, Today and Tomorrow*, R. Bretnor, editor, Harper & Row 1974, pp. 278–292.) Modern time travel science fiction has shown a huge change (for the better) on this score. Discuss, for example, the emotional power of a love story between a couple separated in time, as depicted in the 1975 novel *Bid Time Return* by Richard Matheson (made into the 1980 film *Somewhere in Time*). How do you think the 1931 fan would have reacted to Matheson's story? (Indeed, if that fan was a teenager—or even a few years older—in 1931, then 44 years later he would have been, at most, in his mid-60s and might well *have* read the novel.)

In the 2014 film *Interstellar*, a space probe dives into a black hole, gets a glimpse of the hole's singularity, measures some unspecified quantum effects, and then sends the measurements back to Earth (via the fifth dimension) as a signal in the form of spasmodic Morse code twitches of the second-hand on somebody's watch. This all leads (it is hinted) to a theory of quantum gravity. If you saw the second-hand on *your* watch suddenly begin to spasmodically twitch, would you then immediately think

(a) that a Morse code message was coming to you via the fifth dimension bearing the secrets of quantum gravity?

or

(b) that your watch needs a new battery?

(continued)

or

(c) something else?

Vigorously defend your answer.

William Grey, a philosopher at The University of Queensland, pointed out numerous conceptual difficulties with the idea of time travel in his paper "Troubles with Time Travel," *Philosophy*, January 1999, pp. 55–70. That paper quickly prompted a rebuttal from the philosopher Phil Dowe (at the University of Tasmania), who replied a year later with the paper "The Case for Time Travel," *Philosophy*, July 2000, pp. 441–451. We'll eventually take up all the issues discussed in those two papers but, for now, read both papers and summarize their respective arguments. Do you feel one of the writers won the day (for you, anyway)?

Chapter 2
Philosophical Space and Time

"I do not believe that there are any longer any philosophical problems about Time; there is only the physical problem of determining the exact physical geometry of the four-dimensional continuum that we inhabit." [1]

2.1 Time: What Is It, and Is It Real?

"Time is generally thought to be one of the more mysterious ingredients of the Universe." [2]

Before going any further with time *travel*, it will be well worth the effort to take a closer look at time itself, the 'stuff' or 'thing' or ... ? that we are interested in traveling 'through' or 'around' or 'across' or ... ? Oddly enough, I'll start with religion, as philosophical theologians had identified time as something unusual long before Newton's words on time in his *Principia* that I mentioned in the Introduction, and many thousands of years before science fiction writers and their time travel stories.

We can, in fact, trace the religious interest in time back at least sixteen centuries to the Christian theologian St. Augustine and his *Confessions* (in which he famously admitted "What, then, is time? I know well enough what it is, provided that nobody asks me: but if I am asked what it is and try to explain, I am baffled."). Certainly the seventeenth century Spanish Jesuit Juan Eusebius Nieremberg caught the spirit of wonder that time holds for the devout when he wrote, in his *Of Temperance and Patience*, that "*Time* is a sacred thing; it flows from Heaven ... It is an emanation from that place, where eternity springs ... It is a *clue* cast down from Heaven to guide us ... It has some assimilation to Divinity."

Going outside Christianity, we can easily find other equally strong reactions to the mystery of time. From Plutarch's *Platonic Questions* we learn that when the question of time's nature was put to Pythagoras, he simply uttered the mystical "time is the soul of the world." The *Laws of Manu* of Hinduism, the *Torah* of

---

[1] H. Putnam, "Time and Physical Geometry," *Journal of Philosophy*, April 1967, pp. 240–247.

[2] P. Horwich, *Asymmetries in Time*, MIT Press 1987.

© Springer International Publishing AG 2017
P.J. Nahin, *Time Machine Tales*, Science and Fiction,
DOI 10.1007/978-3-319-48864-6_2

Judaism, the *Koran* of Islam, and the revealed truths of Gautama Buddha are all full of references to time. It is, in fact, to the pagan gods of Greek mythology that we owe our 'modern' image of Chronos, or Father Time.

Not just the Greeks made time a god. In the *Bhagavad Gita* (*Song of the Lord*), the central religious-romantic epic of Hinduism that predates Christ by five centuries, one of the characters reveals his divine nature and declares his power thus: "Know that I am Time, that makes the worlds to perish, when ripe, and bring on them destruction." And in the even older Egyptian Book of the Dead, which dates back over three thousand years, the newly deceased was thought literally to become one with time itself. The merging of time and the resurrection of the body after death in the Book is shown in the line "I am Yesterday, Today and Tomorrow, and I have the power to be born a second time."

The Greek philosopher Plato (circa 400 B.C.) gave us a curious way to think of time: as a *closed loop*. While Plato did think of time as having a beginning, his conception did not have time extending off into the infinite future as does the modern, everyday view. Rather, Plato visualized time as curving back on itself—as *circular* in nature. This was, in fact, a reasonable reflection on what Plato could see everywhere in nature, with the seemingly endless repetition of the seasons, the regular ebb and surge of the tides (the old English word *tid* was a unit of time), the unvarying alternation of night and day, and the rotation of the visible planets in the sky. Whatever might be observed today, it seemed obvious to Plato, would happen again in nature. Circular time in science fiction was briefly mentioned in Chap. 1,[3] and it occurs outside that genre, too, as in James Joyce's novel *Finnegans Wake*, which opens in mid-sentence and ends with the first part of the same sentence. This view of time has a powerful, ancient visual symbol, the Worm Ouroborous, or World Snake, that eats its own tail endlessly.

Circular time, with its closed topology, was favorably presented in Stephen Hawking's famous book *A Brief History of Time*. In it he concludes that there is no need for God because in circular time there is no first event and hence no need for a First Cause. Vigorous philosophical rebuttals were quick to come, of course![4]

Turning to fiction, Ray Bradbury wrote a beautifully poetic passage about the mystery of time in "Night Meeting," one of the splendid sub-stories in his episodic 1950 masterpiece *The Martian Chronicles*. A man of A.D. 2002, who is one of the modern inhabitants of Mars, somehow meets the ghostly image of a long-dead Martian one cold August night. The conditions are just right for such a cross-time encounter. As the man thinks to himself, "There is the smell of Time in the air

---

[3] Another example from science fiction is the story by I. Hobana, "Night Broadcast," in which a television signal from the past is picked up by a gadget that is probing the *future*: "By going far enough into the future one comes upon what we call the past." You can find this tale in the *Penguin World Omnibus of Science Fiction*, Penguin Books 1986.

[4] See, for example, W. L. Craig, "What Place, Then, for a Creator?: Hawking on God and Creation," *British Journal for the Philosophy of Science*, December 1990, pp. 473–491, and R. Le Poidevin, "Creation in a Closed Universe Or, Have Physicists Disproved the Existence of God?," *Religious Studies*, March 1991, pp. 39–48.

tonight. . . . There was a thought. What did Time smell like? Like dust and people. And if you wondered what Time sounded like it sounded like water running in a dark cave and voices crying and dirt dropping down on hollow box lids, and rain. And, going further, what did Time *look* like? Time looked like snow dropping silently into a black room or it looked like a silent film in an ancient theater, one hundred billion faces falling like those New Year balloons, down and down into nothing. That was how Time smelled and looked and sounded. And tonight . . . tonight you could almost *touch* Time."

Well, lovely words, yes, but they don't really tell us what time *is*. Perhaps Einstein the physicist can tell us. In the *New York Times* of December 3, 1919, we find him quoted as follows: "Till now it was believed that time and space existed by themselves, even if there was nothing [Newton's view]—no Sun, no Earth, no stars—while now we know that time and space are not the vessel for the Universe, but could not exist at all if there were no contents, namely, no Sun, no Earth, and other celestial bodies." Less than 2 years later Einstein stated this view again (*New York Times*, April 4, 1921): "Up to this time the conceptions of time and space have been such that if everything in the Universe were taken away, if there were nothing left, there would still be left to man time and space." Einstein went on to deny this view of reality, saying that, according to his general theory of relativity, time and space would *cease to exist* if the universe were empty. This has the ring of one of Einstein's favorite philosophers, Spinoza, who declared in his *Principles of Cartesian Philosophy* that "there was no Time or Duration before Creation." In a correspondence with Samuel Clarke—Newton's friend who translated Newton's *Optiks* into Latin—the German philosopher Gottfried Leibniz (who began the correspondence in 1715) expressed similar ideas: "Instants, consider'd without the things, are nothing at all . . . they consist only in the successive order of things."

The pragmatic scientist would certainly agree with Leibniz. After all, what could it even mean to talk of time unless you can measure it? And what you use to measure time is a clock—some kind of changing configuration of matter involving spinning gears, ticking pendulums, and rotating dial pointers. Mere *unchanging* matter, alone, is not sufficient to measure time because a still clock measures nothing. *Changing* matter seems to be required. Yet, not surprisingly, not everybody agrees. The counterview, the view that time has nothing to do with change, was expressed in an interesting manner by a science fiction fan in a letter to the editor of *Wonder Stories* (January 1931): "Just one thing, you have these time-traveling yarns, good stuff to read all right, but bunk, you know; because if there's no such thing as time, which there isn't, *only change* [my emphasis], how can one travel in . . . something that doesn't exist. To our planet which goes around the Sun there is simply a turning and warming of one side and then the other, i.e., years, days, minutes, etc., is something purely artificial, invented by man to tell him when to do certain things, work and stop work . . ."[5]

---

[5]This fan's idea was not new. For Plato's most famous student, Aristotle, time was *motion* (in a world in which nothing moved, argued Aristotle, there would be no time), and he expressed this

Going even beyond the ideas of Einstein, Spinoza, Leibniz, Plato, Aristotle, and our science fiction fan, at least one metaphysician felt that time would have no meaning, even in a massive and changing universe, without the additional presence of conscious, rational beings.[6] That sounds very much like an echo of the French philosopher Henri Bergson who, in 1888, somewhat mysteriously declared that time is "nothing but the ghost of space haunting the reflective consciousness." A few years before Taylor, however, a fellow philosopher had argued for exactly the opposite view, that temporal passage is independent of the existence of conscious beings.[7]

All this divergence of opinion perhaps explains why even a lightweight Hollywood movie like Mel Brooks' 1987 *Spaceballs* can get a laugh from a time joke. Even kids know that the characters, when talking about time, haven't the slightest idea of *what* they are talking about. The movie, a spoof on such classic films as *Star Wars*, *The Wizard of Oz*, and *Raiders of the Lost Ark*, quickly reaches a point of crisis. To find out what to do next, the evil Lord Helmet and his chief henchman decide on a novel approach: they will look at an instant video of their own movie! (Instant videos are available *before* the movie is finished.) Perplexed at watching on a television screen everything that he is doing as he does it (the screen correctly shows an infinite regression of television screens, each being watched by a Lord Helmet), Lord Helmet initiates the following rapid-fire exchange. (It is, of course, a clever take-off on Abbott and Costello's "Who's on First?")

What the hell am I looking at? When does this happen in the movie?
Now! You're looking at now, sir. Everything that happens now, is
happening, now.
What happened to then?
We're past that.
When?
Just now, now.
Go back to then.
When?
Now.
Now?
Now.
I can't.
Why?
We missed it.

---

view in his famous metaphor "Time is the moving image of eternity." For Aristotle, then, time and change were inseparably intertwined. For Aristotle the world had existed for eternity, and the circularity of time was a central and powerful image; using his vivid illustration, it is equally true *in circular time* that we live both before *and* after the Trojan War.

[6]R. Taylor, "Time and Life's Meaning," *Review of Metaphysics*, June 1987, pp. 675–686.

[7]S. McCall, "Objective Time Flow," *Philosophy of Science*, September 1976, pp. 337–362.

When?
Just now. [The henchman then sets the video to rewind.]
When will then be now?
Soon.

We may laugh at this, even dismiss it as mere movie madness, but could any of us *really* do much better if, like Saint Augustine, we were backed into a corner and asked to explain time? Somehow, I think even the distinguished twentieth-century Harvard professor Hilary Putnam whose words open this chapter would find it difficult to know where to begin. He might even become as confused as the time traveler in the 1968 film *Je t'aime, Je t'aime*, whose oscillations in time, from present to past and back again, leave him so befuddled that he decides he'd rather be dead. What, then, *can* we say about time? Despite Putnam's bold words, I suspect that most people would come down on the side of Augustine.

The mystery of time was well captured by R. H. Hutton (1826–1897), the literary editor of the *Spectator*, when he wrote in his 1895 review (see note 1 in the Introduction) of Wells' *Time Machine* that "the story is based on that rather favorite speculation of modern metaphysicians which supposes *time* to be at once the most important of the conditions of organic evolution, and the most misleading of subjective illusions … and yet Time is so purely subjective a mode of thought, that a man of searching intellect is supposed to be able to devise the means of traveling in time as well as in space, and visiting, so as to be contemporary with, any age of the world, past or future, so as to become as it were a true 'pilgrim of eternity.'"

Novelist Israel Zangwill (1864–1926) wrote a similar but much more analytical review of Wells' novel for the *Pall Mall Magazine* (see note 1 in the Introduction). Zangwill was the only Victorian reviewer to attempt a scientific analysis of time travel. Although he thought Wells' effort was a "brilliant little romance," Zangwill also thought the time machine—"much like the magic carpet of *The Arabian Nights*"—was simply "an amusing fantasy." Zangwill continued in his review with what was even then a common idea about a way one might actually be able, at least in principle, to look backward in time; one could travel far out into space by going faster than light and then watch the light from the past as it catches up to you. (Note, carefully, that Zangwill was writing in 1895, 10 years before Einstein's special relativity put a limit on possible speeds.) In this way, Zangwill wrote, one could watch "the Whole Past of the Earth still playing itself out."

Indeed, even before Zangwill, the well-known French astronomer Camille Flammarion (1842–1925) had made this dramatic idea a centerpiece of his 1887 novel *Lumen*. That book, a best-seller in Europe even before its appearance in England, describes how a man just dead (in 1864) instantly finds his spirit on the star Capella, where he is able to watch the light then arriving from the Earth of 1793. In particular, he watches the French Revolution play itself out and sees himself as a child. Flammarion may have, in fact, been inspired to write his novel by an essay written several years earlier (in 1883) by the British physicist

J. H. Poynting (1852–1914). Poynting's essay,[8] which opens with the statement that it was, in turn, inspired by an anonymous pamphlet published "30 or 40 years ago" on the same topic, specifically mentions watching historical events from Capella.

By the beginning of the twentieth century the idea of watching the past by outrunning light had drifted down into juvenile literature, as in the 1904 novel *Around a Distant Star* by Jean Delaire (the pen name for Pauline Touchemoline (1868–1950)), in which a young man builds a spaceship that can travel at two thousand times the speed of light. With it, he and a friend travel to an Earth-like planet nineteen hundred light-years distant and use a super-telescope to watch the Crucifixion (and then the resurrection) of Jesus. Early magazine science fiction also found the idea of looking backward in time with delayed light to be an irresistible one, involving romance and murder.[9] In another tale incorporating human emotions, a scientist loses his wife to a rival who kidnaps her and then escapes in a faster-than-light rocket ship headed for parts unknown. After searching for them with his own brilliant invention of the 'ampliscope' (several quantum leaps beyond the telescope), the scientist locates the couple, skipping from planet to planet light-years distant. His only pleasure, then, is to use his own faster-than-light craft to outrun the images of his lost love and watch them over and over. Eventually, however, he comes to realize the ultimate futility of it all. As the final line of this sad tale says, "It would be senseless, I knew, chasing on and on after yesterdays."[10]

The reality of time received a new twist with the additional imagery of instants of time being likened to the points on a *straight* line. In the West it was the Christian theological doctrine of *unique* historical events that gave rise to *linear* time in the minds of the common folk. The creation of the world and Adam and Eve, the adventures of Noah and the cataclysmic Flood, the Resurrection—these were all events that occurred in sequence, *once*. None would happen again and so, for Christianity, circular time just would not do.[11] In addition, it has been argued that the major spiritual content of Christianity—a significant reason for its popular support even in the face of brutally harsh Roman suppression—is that it brought the *expectation of change* into the static world of ancient times. It was, in fact, in ancient religious teachings that our modern view of linear time had its origin, a view that most people today (including the most hardened agnostic physicist) find to be as natural as Plato and Aristotle found circular time.

---

[8] J. H. Poynting, "Overtaking the Rays of Light," in Poynting's *Collected Scientific Papers*, Cambridge University Press 1920.

[9] As in, for example, G. A. England, "The Time Reflector," *The Monthly Story Magazine*, September 1905.

[10] D. D. Sharp, "Faster Than Light," *Marvel Science Stories*, February 1939. The year before saw the appearance of a story with the same idea, a story that specifically cites Flammerion: M. Weisinger, "Time On My Hands," *Thrilling Wonder Stories*, June 1938.

[11] Still, just to show how one can find support for almost any view in the same religious dogma, Ecclesiastes 1:9 would seem to be a claim *not* for linear time but rather for circular time!: "The thing that hath been, it is that which shall be; and that which is done is that which shall be done; and there is no new thing under the sun."

Calvin and Hobbes   by Bill Watterson

Even though linear time was the norm after Christ, there were still enough questions about time to perplex the deepest of thinkers, and the next 2000 years resulted in plenty of thinking. Discourses on time by such philosophers as Descartes, Spinoza, Hobbes, Kant, Nietzsche, and Hegel can be found by the yard in any decent university library. Nearly all (if not indeed all) of these presentations have metaphysical, even theological, underpinnings. For example, Descartes is generally believed to have argued for a discontinuous, atomistic nature to time (recall the *chronon* from Chap. 1). This is the modern view of his thinking, because in his *Meditations* (1641), in particular in the third meditation on God's reality, Descartes appears to argue that God must continually recreate the world at each *separate* moment of its existence. That is, the world is recreated in a discontinuous succession of *individual* acts by God.[12]

Finally, with Newton's discussion of *absolute* time, which is the belief that time is the same everywhere in the universe, there was for the first time a *physicist* writing about time (although, as I mention in Chap. 1, Newton's views were also influenced *heavily* by theological considerations, in addition to mathematical physics). But, despite Newton's genius, the mystery of time remained a mystery.

In 1905 Einstein's name appeared among the contributors to the study of time, and so at last something besides metaphysical speculation on the subject was added to the body of human thought. Einstein's paper on special relativity introduced the revolutionary idea of *relative* time, which is the anti-Newton belief that the passage of time is *not* the same everywhere, but rather depends on local conditions. In retrospect, Einstein's 1905 work seems to be the perfect reply to the comment by Isaac Barrow (1630–1677)—Newton's teacher and the first Lucasian professor of mathematics at Cambridge (the chair once held by Stephen Hawking centuries

---

[12]For more on this, see R. T. W. Arthur, "Continuous Creation, Continuous Time: A Refutation of the Alleged Discontinuity of Cartesian Time," *Journal of the History of Philosophy*, July 1988, pp. 349–375.

later)—that "because *Mathematicians* frequently make use of Time, they ought to have a distinct idea of the meaning of the Word, otherwise they are Quacks."

Then, just 3 years after Einstein, along came a second astonishing paper by the Cambridge philosopher John Ellis McTaggart (1866–1925). This paper[13] claims to prove that whatever time might be *thought* to be (even by Einstein), it really isn't that because time isn't even real. (This would seem, I think you'd agree, to have potentially profound implications for time travel!) The method of the paper is to deny the reality of time via an infinite-regress argument that one philosopher[14] has called the *pons asinorum* ("bridge of asses") of the riddle of time. As McTaggart's own opening sentence freely admits, "It doubtless seems highly paradoxical to assert that Time is unreal, and that all statements which involve its reality are erroneous."

McTaggart began his analysis by observing that there are two separate and distinct ways of talking about events in time. Following his terminology, one can say that events are either future, present, or past (the so-called *A-series*), or one can say that events temporally ordered by each being later than some other events, earlier than others, and simultaneous with still others (the so-called *B-series*). He then continued by asserting that time requires change, and followed that with the observation that the A-series (but not the B-series) incorporates such change. That is, if event X is earlier than event Y, then X is *always* earlier than Y and thus there is no change in this (or in any other) example of a B-series. As a specific example, let Y be the birth of a child, and let X be the birth of its mother. In contrast, if X is first in the future, then is in the present, and finally is in the past, then we have an example of change (and hence of *time*) in the A-series; for example, let X be the next time you blink.

With this rather pedestrian start, McTaggart then pulled his rabbit out of the hat. It makes no sense, he argued, to talk of the 'future,' 'present,' and 'past' of an event because these terms are mutually exclusive. That is, no two of these predicates can apply at once, and yet, paradoxically, every event possesses all three and thus we have a contradiction. It therefore, concludes McTaggart, makes no sense to talk of future, present, or past. And because it makes no sense to talk of them, they do not exist, and so there can be no A-series and hence no change, and thus no reality to time. McTaggart apparently realized just how befuddling all that would appear to just about everybody who read it, and so he played devil's advocate (D.A.) in his paper by trying to anticipate the various objections people could raise. Of course, he always managed to refute the D.A. at every turn. It is worth the effort to go through the details of McTaggart's 'proof,' as that will make it clear what there is about

---

[13]J. E. McTaggart, "The Unreality of Time," *Mind*, October 1908.

[14]L. O. Mink, "Time, McTaggart and Pickwickian Language," *Philosophical Quarterly*, July 1960, pp. 252–263. The phrase *pons asinorum* has its origin in a plane geometry theorem: the angles opposite the equal sides of an isosceles triangle are themselves equal. Seeing the truth of this is said to separate the quick-witted from the dull. It isn't clear (to me, anyway), however, on which side of McTaggart's 'proof' the quick-witted were imagined to fall. You'll see what I mean in just a moment.

'traditional' philosophical reasoning that so irritates modern philosophers trained in mathematical physics (and what makes physicists roll their eyes when confronted with arguments like McTaggart's).

The predicates of future, present, and past are really not incompatible for any event, the D.A. says some will claim, because the real predicates we should use are '*was* future,' '*is* present,' and '*will be* past,' and these *can* be possessed all at once by any event. Nice try, counters McTaggart, but that will not solve the problem. By allowing such modified predicates, we must actually allow for all nine possibilities, some of which are still incompatible. That is, the 'was,' 'is,' and 'will be' could each be potentially attached to 'future,' 'present,' and 'past': for example, 'was past' is incompatible with 'will be future.'

Oh, counters the D.A., we can eliminate that concern by allowing even more complex predicates to arrive at a third level of structure, such as 'is going to have been past,' and 'was going to be future,' and those *are* compatible. But McTaggart swats that argument away, too, by displaying new incompatibles, as well as by showing that the process of ever-increasing predicate complexity is a vicious infinite regress that drags along the seeds of its own doom at every step.[15] There is simply no escape from incompatibility, he says, and so there is no time.

Well! What can one do when presented with such an argument, one that seems to claim philosophers can wrest free the secrets of nature by pondering the historical accidents of English syntax? As David Hume once said, "Nothing is more usual than for philosophers to encroach on the province of grammarians, and to engage in disputes of words, while they imagine they are handling controversies of the deepest importance and concern." One modern philosopher apparently agreed with Hume, at least in the case of McTaggart's 'proof,' and he was pretty blunt with his evaluation of it: "McTaggart's famous argument for the unreality of time is so completely outrageous that it should long ago have been interred in decent obscurity. And indeed it would have been, were it not for the fact that so many philosophers are not sure that it has ever really been given a proper burial, and so from time to time someone digs it up all over again in order to pronounce it *really*

---

[15]Here's a clever way to systematically generate McTaggart's infinite regress of complex predicates, as presented by M. Dummett, "A Defense of McTaggart's Proof of the Unreality of Time," *Philosophical Review*, October 1969, pp. 497–504): "Let us call 'past,' 'present,' and 'future' 'predicates of first level.' If, as McTaggart suggests, we render 'was future' as 'future in the past,' and so forth, then we have nine predicates of second level, where we join any of the three on the left with any of the three on the right:

past		past
present	in the	present
future		future

Similarly, there are twenty-seven predicates of third level ... "Dummett's construction clearly shows that, at the N-th level, there are 3^N predicates, most of which are incompatible.

dead. These periodic autopsies reveal that something more remains to be said."[16] That is certainly true, in as much as McTaggart's disarmingly innocent argument has caused disagreement and furrowed brows among philosophers for decades.

It is, in fact, easy to find examples of the continuing debate over McTaggart's analysis and, as silly as it strikes physicists, it still has a pulse in some quarters. While at least one philosopher has argued that McTaggart simply didn't really understand his own proof, this philosopher nevertheless agreed with McTaggart's conclusion about the unreality of time.[17] Another writer has illustrated how McTaggart's ideas have found their way into modern philosophical debates on the meaning of time in the cinema, particularly in the analysis of *anachrony*, the telling of a story out of normal time sequence, such as occurs in time travel movies.[18]

Other sorts of metaphysical proofs for the unreality of time have been offered besides McTaggart's. For example, it has be argued that time is unreal, at least in a world empty of consciousness, because the concepts of past, present, and future could not possibly have any meaning unless events could be remembered, experienced, and anticipated. Or, for a second example, some have held time to be unreal, at least in a deterministic world (as some argue four-dimensional spacetime to be), because any event whose occurrence follows from present conditions, and from physical laws, would exist (they say) *now*. This view, which seems to assert that everything should happen at once, I personally find to be sufficiently obtuse as not to be bothered by it.[19] Debates between those who believe in the common-sense idea that present, past, and future are attributes of events (the 'tensers') and those who deny it (the four-dimensional spacetime, block universe 'detensers') continues to now and then still flair up on the pages of philosophy journals. At least one philosopher likes both views![20] Most modern physicists, I think, simply don't care about this line of inquiry.

On the other hand, less than a month before his death Einstein revealed his feelings about the meaning of present, past, and future, and his words appear to be ones that show some sympathy to the philosophers. In a letter written on March 21, 1955, to the children of his dearest friend who had just died, Einstein wrote—with full knowledge that his own illness would be his last—"And now he has

---

[16]F. Christensen, "McTaggart's Paradox and the Nature of Time," *Philosophical Quarterly*, October 1974, pp. 289–299.

[17]Q. Smith, "The Infinite Regress of Temporal Attributions," *Southern Journal of Philosophy*, Fall 1986, pp. 383–396. To this came a rebuttal a year later by L. N. Oaklander, in the same journal (Fall 1987, pp. 425–431).

[18]G. Currie, "McTaggart at the Movies," *Philosophy*, July 1992, pp. 343–355.

[19]But if, upon reflection, it starts to bother *you*, see R. Gale, "Some Metaphysical Statements About Time," *Journal of Philosophy*, April 1963, pp. 225–237. We'll soon get to some of the more common philosophical questions on the nature of four-dimensional spacetime, such as 'is it deterministic or is it fatalistic?,' and 'does free-will have any meaning in four-dimensional spacetime?' Even physicists are interested such questions!

[20]R. Weingard, "Space-Time and the Direction of Time," *Nous*, may 1977, pp. 119–131.

preceded me briefly in bidding farewell to this strange world. This signifies nothing. For us believing physicists, the distinction between past, present, and future is only an illusion, even if a stubborn one."[21] Later in this chapter I'll return to these curious words and speculate on what Einstein may have meant by them.

I started this opening section on a religious note, and I'll end it on one. If you think the philosophical speculations on the nature of time that I've so far cited are 'really far out,' here's yet another one that leaves all the rest in the dust. In a paper that took real nerve to write (or, perhaps, simply a wicked sense of humor—and I write that in pure admiration) we read of how a spacetime that supports time travel can give the start for a *physics* explanation to the theological concept of Hell! After introducing just a bit of elementary spacetime physics (which I'll skip describing here because we'll do it later in the book), the author[22] shows how to 'construct' a compact region in spacetime (Hell) with the following properties:

1. While "so small even the Hubble Telescope couldn't image it" it can hold an infinity of physical beings;
2. Each of the beings in it are doomed, because of its time travel property, to an infinitely long personal future of damnation;
3. Each of the beings in it, because of its time travel property, can view all the future stages of their own personal damnation and so be "continually presented with a reminder of the impossibility of escape—a refinement no causally normal Hell can seemingly offer." In other words, and not to be too ironic about it, 'Theological Progress Through Physics!';
4. Each of the beings in it are continually being compressed together ("brought into dismaying proximity" with themselves) and so will spend eternity "listening to a cacophony" of their own cries of despair from *their personal future.*

There's more, but that's probably enough for you to get the idea. Richmond does admit that, as it stands, his time travel creation of Hell is not compatible with either quantum theory or even general relativity. Still, it *is* something to ponder, don't you think, when the subject of time travel comes up!

2.2 Linear Time and the Infinity of Past and Future

"A thousand years is a huge succession of yesterdays beyond our clear apprehension."[23]

—H. G. Wells

---

[21]Quoted from B. Hoffmann, *Albert Einstein: Creator & Rebel*, New American Library 1972, pp. 257–258.

[22]Alasdair M. Richmond, "Hilbert's Inferno: Time Travel and the Damned," *Ratio*, September 2013, pp. 233–249.

[23]This line appears in Wells' 1944 doctoral thesis, written for the University of London. You can find an abridgement of the thesis in *Nature*, April 1, 1944, pp. 395–397.

The modern concept of linear time as a straight line extending from the dim past through the present and disappearing into the misty future gives rise immediately to twin questions: "Did time have a beginning?" and "Will time ever end?" As one philosopher put it (long before physicists became seriously interested in singularities like the Big Bang) "Endings and beginnings are rooted in the very conception of time itself."[24] Starting at the beginning, we'll ask if the past has been forever? Early Biblical scholars, of course, believed the answers to both questions to be *no*.

They believed that the world came into being because of a First Cause, God's creation of everything. Those scholars expended vast quantities of energy (and, need I say it, time itself) in calculating the date of creation. Martin Luther, for example, argued for 4000 B.C. as roughly when everything, including time, began. Johannes Kepler adjusted this by a notch, to 4004 B.C., and later the Calvinist James Ussher, Archbishop of Armagh and Primate of All Ireland, tweaked it again. His date is the most impressive of all, at least in detail: the first day of the world was 4003, 70 days, and 6 h before the midnight that started the first day of the Christian era. Six days after that first day of the world, Adam was made, and as a final dash of specificity, this last date was declared to be Friday, October 28! Ironically, then, though Christian theology may be given credit for introducing linear time, it certainly did not provide very much of it. The beginning of time was just 6000 years or so ago, and of course The End—in the form of the Battle of Armageddon—has been awaited (with varying degrees of eagerness) for the last 1000 years.

The discovery in the seventeenth century of geological time cast a certain amount of skepticism on those early calculations concerning the duration of the past. With the discovery that the very Earth itself could be decoded for its history, the lure of trying to decode a mere book of admittedly finite age declined for most people although it cannot be denied that modern Creationists still find such a task to have its rewards). Geological time was discovered to a *chasm* of time extending backward for billions of years, a duration that is really incomprehensible for the human brain. It has become fashionable for geologists to refer to such enormous durations with the apt term *deep time*, a subtle play on the metaphor of the "ocean of time."

It is nothing less than humbling to historians who pause to think on how little of the past is known, that is, recorded. As the ever anonymous wit once put it, "History is a damn dim candle over a damn dark abyss." Still, even as enormous as is the age of the Earth, it is not infinite. But of course our planet is very old, and the universe is many billions of years older. Is the age of the universe also the duration of the past? Or is the past itself actually *infinite*?

An implicit assumption of the infinity of the past (and of the future, too) can be found in Book Three of Lucretius' science poem *De Rerum Natura* (*On the Nature of Things*) where, just before the birth of Christ, Lucretius argues for the irrationality of fearing death: "The bygone antiquity of everlasting time before our birth

---

[24]I. Stearns, "Time and the Timeless," *Review of Metaphysics*, December 1950, pp. 187–200.

was nothing to us. Nature holds this up to us as a mirror of the time yet to come after our death. Is there anything in this that looks appalling, anything that means an aspect of gloom? Is it not more untroubled than any sleep?"

One philosopher[25] has traced the origins of rational support for the finite duration of the past to as far back as the sixth century A.D. The argument presented then by the Christian philosopher Joannes Philoponus of Alexandria (who is otherwise known as John the Grammarian) is simply that the world could *not* have been forever because that implies an infinity of successive acts could have taken place which (according to Philoponus) is impossible. A variation on this is the claim that if the past were infinite in extent, then everything would have happened by now! Infinity was just too big for the ancient mind (Zeno's hoary pre-Christian paradoxes, as is well-known today, are based on subtle errors in the use of infinity).

This view on the impossibility of an infinite past seems to have been the prevalent view; even as late as the twelfth century the debate among Christian theologians was not about the possibility of an infinite past, but instead about whether the Biblical 'six days of Creation' actually had taken place simultaneously. For many, the past was 'obviously' finite in duration.[26] Not all Christians accepted that conclusion, however, and the following century saw St. Thomas Aquinas (a follower of Aristotle) arguing for the opposite view of an infinite past.

Thomas' contemporary, St. Bonaventure, however, argued again for a *finite* past, and it is with Bonaventure that we start to see some mathematical sophistication.[27] Bonaventure argued that in a world infinitely old, the Sun would have made an infinite number of its annual trips around the ecliptic. But for each such trip the Moon would have made twelve monthly trips around the Earth, and so this second infinity would be twelve times as great as the first one, and how could that be? Infinity is infinity, and how can something be twelve times bigger than infinity? This argument doesn't have any strength today because of the nineteenth century German mathematician Georg Cantor's work on the concept of infinity,[28] but it *is* clever. Agonized, convoluted theological analyses of God, infinity, and eternity continued long after Aquinas and Bonaventure. Two examples should capture the spirit of those times.

---

[25]G. J. Whitrow, "On the Impossibility of an Infinite Past," *British Journal for the Philosophy of Science*, March 1978, pp. 39–45. Whitrow adds modern scientific support to the idea of a finite past by citing the prediction from general relativity of a singularity in spacetime at some finite past time; that is, the theory's prediction that time—and everything else—had its beginning in the now famous Big Bang.

[26]C. Gross, "Twelfth-Century Concepts of Time: Three Reinterpretations of Augustine's Doctrine of Creation *Simul*," *Journal of the History of Philosophy*, July 1985, pp. 325–338.

[27]See, for example, L. Sweeney, "Bonaventure and Aquinas on the Divine Being as Infinite," *Southwestern Journal of Philosophy*, Summer 1974, pp. 71–91, and S. Baldner, "St. Bonaventure on the Temporal Beginning of the World," *New Scholasticism*, Spring 1989, pp. 206–228.

[28]For simple high school-level presentations on Cantor's astonishing infinity results, see my book *The Logician and the Engineer*, Princeton 2013, pp. 169–171.

Consider first this one, on the supposed immortality of the soul. If $A = B$, then $2A = 2B$. Next, let $A =$ 'half alive' and $B =$ 'half dead,' where $A = B$ in the same sense that a glass half-full is also half-empty. Thus, to be completely dead is to be completely alive, and so the soul is immortal. Outrageous? *Yes*, in my opinion, but I do also have to admit the 'reasoning' does have a certain charm!

For my second example, let me begin by setting the historical stage. After publication of the English political philosopher Thomas Hobbes' *Leviathan* in 1651, with its arguments against the power of the Church and for civil power (with some criticism tossed in, as well, for universities), Seth Ward counterattacked. Ward, who was both a minister (later a bishop) in the Anglican Church and Savilian Professor of Astronomy at Oxford, was greatly offended by the secular nature of *Leviathan*. Even before Leviathan, in fact, Ward certainly would not have liked Hobbes' earlier denial of the existence of immaterial substances (such as souls). Ward's 1652 book *A Philosophical Essay Towards An Eviction of the Being and Attributes of God, the Immortality of the Souls of Men, the Truth and Authority of Scripture*, was the first of a two-punch reply to Hobbes. The second came in 1654 with the appearance of Ward's *Vindiciae academiarum*. In both of these works Ward attempted to undermine Hobbes' credibility by attacking his mathematical ability. (Hobbes had long been fascinated by, and was considered an expert on, the ancient problem of 'squaring the circle,' a task that has been known to be impossible only since 1882.[29]) In his *Essay*, Ward also attempted to defend the view that the world has a finite age—that is, it had a specific moment of creation, presumably by God. In an opening note, in fact, Ward cites Hobbes' rejection of immaterial substances as the motivation for his writing *Essay*.

To support his view of a finite age for the world, Ward invoked infinity in an interesting way. He argued that nothing is permanent, certainly not humans. Each is created; one can imagine tracing a chain of creation events backward in time through successive generations. Now, there are only two separate and distinct possibilities to where this chain could lead to in the past. First, it could terminate, after a finite number of generations, at a *first* generation, that is, with the 'creation' of the first human. If that is the case, then, said Ward (in effect), 'case closed.' If that is not the case, however, then the chain of successive generations never terminates, that is, the chain is infinitely long. But that, argued Ward, is nonsense—how could anything *infinitely* long have an end (our present *now*)?

Why Ward thought this an unanswerable paradox is hard to understand; after all, one can imagine a line in some coordinate system *beginning* at the origin and yet still being infinitely long (an example is the positive x-axis). This counter-example was not put forth by Hobbes in his own self-defense, but rather was offered by one of Ward's own colleagues at Oxford, John Wallis, the Savilian Professor of Geometry. As for Hobbes, he was little bothered by Ward's argument. As he pointed out (surely with a smile on his face), Ward was in danger of impaling

---

[29]The problem of 'squaring the circle' is, given a circle of area A, to construct (using only compass and straightedge) a square of area A.

himself as a theologian on his own sword: Ward's argument 'proved' the finite age not only of the world but of *everything*, including God (thus raising the awkward question of who, or what, made God?).

Similar problems with infinity lay behind Kant's rejection of an infinite past. It is interesting to note that Kant, somewhat paradoxically, thought an infinite *future* a possibility. Why did Kant think time could be infinite in one direction but not in the other? One philosopher tells us[30] that Kant "failed to make himself clear," and I think that *understates* the case. I say that because Kant's argument was that the duration of the future is less problematic than is that of the past because it is only the past that influences the present. The best I can do in 'explaining' this is to speculate that if the present depends on an *infinite* past, then perhaps Kant thought that the possibility of so much influence was simply too much for the present to handle! In any case, Kant's view falls apart if we consider the possibility of backward time travel and the resulting implication that the future could also influence the present.

There is, as will come as no surprise, a philosopher for every conceivable point of the compass, and so a paper by one on the logical possibility of an infinite past soon prompts a rebuttal by another.[31] In illustration of this, you'll recall the quote from Augustus De Morgan in the opening section of this book, concerning the philosophers of his times; De Morgan went on in his critique to amusingly summarize the metaphysics of those times as follows: "Here we go up, up, up,/And there we go down, down, down,/Here we go backwards and forwards/And there we go round, round, round."

So, with De Morgan's words in mind, here are a few more examples of how people have struggled with the issue of the past. One quite interesting, *scientific* twist on the duration of the past was pointed out before the exchange between Smith and Ells. In a paper[32] observing that although general relativity and its predicted spacetime singularity in the distant past may indeed allow for a finite past, that does not completely close the door to the possibility that the Big Bang was a continuation from a previous contraction phase of the universe, and so on, *ad infinitum*. (You'll recall the discussion in Chap. 1 of this idea in science fiction: see note 53 in that chapter.) To quote T. S. Eliot (from his "Little Gidding"):

---

[30]J. Bennett, "The Age and the Size of the World," *Synthese*, August 1971, pp. 127–146. See also Q. Smith, "Kant and the Beginning of Time," *New Scholasticism*, Summer 1985, pp. 339–346.

[31]See, for example, Q. Smith, "Infinity and the Past," *Philosophy of Science*, March 1987, pp. 63–75, and then read E. Ells, "Quentin Smith on the Infinity of the Past," *Philosophy of Science*, March 1988, pp. 453–455. Smith's paper "The Uncaused Beginning of the Universe" appeared in this same issue (pp. 39–57), stating that he believed, *really*, only in the *logical* possibility of an infinite past and that the universe had in fact originated in an uncaused (no God required) Big Bang singularity. And, indeed, he *had* so argued for a finite past, in "On the Beginning of Time," *Nous*, December 1985, pp. 579–584.

[32]R. Weingard, "General Relativity and the Length of the Past," *British Journal for the Philosophy of Science*, June 1979, pp. 170–172.

"What we call the beginning is often the end
And to make an end is to make a beginning.
The end is where we start from."

Even without entertaining such an oscillating, accordion-like universe that endlessly expands and shrinks, it is possible to have a universe that originated in a *single* Big Bang a finite time ago in the past but yet *has no first instant*! This astonishing statement shocks most at first encounter, but it is simply the cosmological version of a well-known mathematical result. The instant $t = 0$ is not actually part of spacetime, because the Big Bang was quite literally a singular event for which the laws of spacetime physics fail. Thus, all instants in time are greater than zero—and there is no smallest number greater than zero. If you name a positive number, no matter how small, I can name a positive number still smaller, such as one-half of yours. (Of course, if there really is merit to the idea of a quantum of time, the chronon, this argument goes out the window.)

In an ingenious observation that seems to have been missed by most philosophers, E. A. Milne, a professor of mathematics at Oxford, suggested in his 1948 book *Kinematic Relativity*, that with general relativity it is conceivable to have both a single Big Bang a finite time ago *and* an infinite past. Pointing out that to talk meaningfully of time implies that we have a clock to measure it by, Milne looked for a Universal Clock that would be far more durable than our heartbeats, or anything else that exists only transiently. He suggested the expansion rate of the universe itself as the ideal clock. As we go back in time to the Big Bang, the expansion rate rises towards infinity and, as another analyst put it, "We see the Universe ticking away quite actively. *The Universe is meaningfully infinitely old because infinitely many things have happened since the beginning.*"[33]

The debate over the length of the past in modern times can be just as contentious as it was in medieval times. For example, in his editorial ("Down with the Big Bang") of August 10, 1989, the then editor of *Nature* (John Maddox) declared the standard explosive model of the universe to be "philosophically unacceptable," because "the implication is that there was one instant at which time literally began and so, by extension, an instant before which there was no time." For Maddox, this meant that the Big Bang "is an *effect* [my emphasis] whose *cause* [my emphasis] cannot be identified or even discussed." The usual (non-time travel) use of the words *cause* and *effect* is that the cause happens first and then the effect occurs—but if the Big Bang (the effect) is the origin of time, then how (asked Maddox) could there be a cause of the Big Bang *before* that beginning?[34]

---

[33]C. W. Misner, "Absolute Zero of Time," *Physical Review*, October 1969, pp. 1328–1333. In this view cosmic time is taken as proportional to the negative of the logarithm of the normalized volume of the universe ($V = 1$ represents maximum volume, and so time 'stops' at the end of the universe's expansion). Thus, because V goes to zero as we go backward in time, time runs ever faster as we travel ever further into the past. This puts the Big Bang (with $V = 0$) infinitely long ago.

[34]This was not a new insight, of course, as Aristotle had long ago (in his *Physics*) declared an instant in time with no predecessor to be an absurdity.

The answer is obvious *for creationists*, of course—God did it. Creationists avoid the question of God's cause, however, saying only that 'He needs no cause,' or even that 'He made Himself'! It is these standard (ridiculous) responses from creationists that Maddox said had prompted his editorial against the Big Bang, because creationists *embrace* the Big Bang as it seems to endorse their position of 'science by imagination.' Whatever the truth of that, I think juxtapositioning the *scientific* Big Bang model of the universe with theological metaphysics and the pseudo-science nonsense of creationism to be terribly unfair.

When will the philosophical debates on the age of the past end? Not until the end of the (infinite?) future, is my wager!

2.3   Cause and Effect

"There are few paradoxes which have been resolved so often as the time-asymmetry paradox."[35]

The philosophical literature is full of discussions about potential causal relationships between events. One of the most famous of these discussions, illustrating that cause and effect can be pretty slippery concepts, asks what at first appears to be an almost trivial question: Did the death of Socrates cause the widowhood of Xanthippe? The quick and easy answer is "Of *course*—she was his wife and it was his death that causes us to say she was then a widow. What could be more obvious?" One philosopher has provided some interesting commentary, however, that might make you reconsider, or to at least become aware of how different are the questions concerning time that are of interest to physicists and philosophers.[36]

Suppose we agree that there are two events to be considered; Socrates ceasing to live, and Xanthippe becoming a widow. Those events occurred at different places (in prison, and wherever Xanthippe happened to be). Then, as Kim asserted, "the two events occur with absolute simultaneity ... [and so] we would have to accept this case as one in which causal action is propagated instantaneously through spatial space." (As we'll discuss in Chap. 3, the *relativity* of distant simultaneity weakens this assertion, but we'll take that up later.) For now, it is the conclusion that Kim draws from the assertion that interests us here: just *what* is propagating instantly? If it isn't mass-energy (as 'widowhood' would appear not to be!) then special relativity isn't bothered and physicists are happy. But those same physicists might also scratch their heads over *why* philosophers even wonder about such a question, because isn't becoming a widow just another way of saying that Socrates died and so we really don't have *two* events, but just one? In other words, for physicists this really isn't a question about cause and effect at all!

---

[35]J. Hurley, "The Time-Asymmetry Paradox," *American Journal of Physics*, January 1986, pp. 25–28.

[36]J. Kim, "Noncausal Connections," *Nous*, March 1974, pp. 41–52.

The central puzzle of time travel to the past is its apparent denial of causality—that is, its denial of the belief that we live in a world where every effect has a cause and that the cause happens first. *First* we flip the switch and *then* the kitchen light comes on. It is *never* the other way around. So deeply embedded is the temporal ordering of cause and effect in our feelings about how the world—and all the rest of the cosmos—works, that the Australian philosopher John Mackie (1917–1981) called causation the "cement of the universe" (and used that wonderful phrase as the title of a 1980 book). Without causality, said Mackie, everything would come unglued and fall apart. For example, when electrical engineers design an electronic system that they intend to actually construct (as opposed to doing a mere theoretical 'paper design') they insist that the design be a *causal* one. By that they mean the system must have no output before an input is applied. That is, the system must not be able to anticipate (foresee) the application of an input. To put it bluntly, our engineers are insisting that they are *not* building a time machine!

Now all that might seem to be self-evident, but there *are* some subtle problems. For example, it has become almost a cliché to say that nothing can go faster than light; that's what physicists mean by *relativistic causality*. In other words, no cause can produce an effect at a distant location sooner than the time lapse required for a light pulse to make the trip. Classical mechanics, however, the science of Newton's laws that engineers use all the time, is *not* relativistically causal. Push the left end of a rigid rod, for example, and the right end moves *instantly*. Most of the time the lack of this form of causality causes no problems, but the fact remains that the mechanics all engineers (and physicists, too!) learn first in school is flawed on a fundamental level. A rigid rod is an impossibility in Einstein's mechanics.

Indeed, it is interesting to speculate about how, after a discussion of causality, a traditional engineering professor would respond if challenged on this issue by a bright student. Causality might not look so obvious, after all, if such a student stuck up her hand in class and said "Professor, you've told us that everything that happens in nature is due to a cause. That what we see happening all around us, as the world unfolds, is the domino-process of cause-effect-cause-effect, and so on, into the future. But suppose, Professor, that at some instant, somehow, every particle in the world suddenly reversed its velocity vector. Wouldn't that mean, given the time-reversible nature of the classical equations of motion, the world would then run backward in time along the same path it had followed up until the instant of reversal? Wouldn't that mean what was effect is now cause, and that what was cause is now effect? And if cause and effect can change roles like that … well, Professor, just what do our words *mean*?"

An amusing, and instructive, cartoon illustration of the student's idea of reversing all the velocity vectors in a system appeared on the cover of the November 1953 issue of *Physics Today*. That issue contains an article on the 1949 nuclear magnetic resonance experiments performed by the American physicist Erwin Hahn, which in a certain sense dealt with just such reversed systems. In that illustration a group of runners on a multi-lane circular race track begin at the starting line in a coherent state, that is, all lined up together. Then, as they run around the track at various speeds, they gradually spread out into what appears to be an incoherent state.

But that incoherence is an illusion because if, at some instant (signaled in the cartoon by a pistol shot), they all turn around and run in reverse, they will all arrive back at the starting line *together, at the same instant*. The initial coherence of the runners was actually never lost, despite the superficial appearance of disorder, and the coherent state can be recovered at any time by a reversal of velocity vectors.

This isn't mere theoretical speculation, as an almost magical application of velocity vector reversal is actually used in what is called *optical phase conjugation*, a process to 'time-reverse' the severe distortion suffered by light beams during atmospheric propagation. For example, by effectively reversing the velocity vectors of photons, one can remove the turbulence blurring in satellite pictures of the Earth's surface as seen from space.[37]

Let me immediately short-circuit one possible answer our beleaguered professor might give in desperation, a response based on the fact that equations of physics are *not* all time reversible. Indeed, it was discovered decades ago that, in certain very rare, fundamental particle decay processes involving neutral K-mesons, there is the hint that perhaps nature *can* indeed distinguish between the past and the future. In particular, K-mesons should violate what is called *CP-symmetry*, and the so-called TCP theorem[38] says that then *T-symmetry* must also fail. In 1968/69 direct, experimental observation of the failure of T-symmetry in K-meson decays was reported. In an astonishing example of science fiction prescience, the use of K-mesons in a machine for affecting the past had appeared years earlier in a 1955 (!) story.[39]

So, could K-mesons account for the physical processes that we see evolve in time in one direction (past to future) but not in the other? As Hurley (note 35) put it so nicely, "The decay of the neutral K-meson is not time-reversal invariant; perhaps it is this ubiquitous meson which is responsible for the cream diffusing uniformly throughout our coffee in the morning. Possibly, but again this conjecture cannot account for the computer models [of diffusion processes that, like cream in coffee, also display a bias for one temporal direction over the other—in Chap. 3 I'll show you such a computer model] which have no neutral K-mesons." Still, the tiny chink that K-mesons appear to have made in the once-solid rock of time direction indistinguishability is an active area of research and speculation.

Even with that chink the fact that the classical laws appear to be insensitive to a direction of time, whereas the real world—which seems in no way dependent on the arcane properties of K-mesons—seems distinctly asymmetric, is a puzzle of the first rank. As one philosopher wrote, "The Universe seems asymmetric with respect to

---

[37]C. R. Giuliano, "Applications of Optical Phase Conjugation," *Physics Today*, April 1981, pp. 27–35.

[38]The TCP-theorem says that the 'mirror-image' of a physical process is a legitimate process, too, *if* the 'mirror' reverses time (T), electric charge (C)—so that particle and anti-particle are interchanged, and parity (P)—which is the measure of left and right. There is strong reason to believe in the validity of the TCP theorem because quantum field theory is compatible with special relativity only if the TCP theorem holds.

[39]F. Pohl, "Target One," *Galaxy Science Fiction*, April 1955.

the past and future in a very deep and non-accidental way, and yet all the laws of nature are purely time symmetric. So where can the asymmetry come from?"[40]

There have of course been attempts to answer that question. For example, one philosopher[41] discusses some curious mathematical examples he interprets as meaning, in the context of classical mechanics, that there are physical systems that are temporally irreversible *in principle*. A reply[42] from a fellow philosopher, however, argues that Hutchinson has, at most, shown only that classical mechanics is perhaps not deterministic. And that, Savitt argues, is not equivalent to showing a failure of time reversibility. There is, in fact, powerful experimental evidence that, with the rare exceptions of K-mesons, the classical laws of physics (including general relativity and quantum mechanics) *are* time-reversible.

Perhaps the most compelling of such evidence comes from the *reciprocity theorem* that electrical engineers routinely use when designing radio antennas. The theorem is easy to illustrate. Suppose two electrical engineers, Bob in Boston and Lois in Los Angeles, send radio signals to each other. Bob sends his messages by exciting his antenna with a time-varying current, which thus launches electromagnetic radiation into space. Lois' distant antenna intercepts some of that radiation, which then creates a (very tiny) signal current in her antenna.

The reciprocity theorem states the following: Suppose Bob makes a tape recording of his excitation signal and mails it to Lois, who then plays Bob's tape back into her transmitter as the excitation to *her* antenna. Then the signal current induced in Bob's antenna, as it intercepts Lois' launched radiation, will be the very same (very tiny) signal that Lois measured in her antenna as a result of Bob's transmission. This result is completely independent of the details of the two antennas, which can be utterly different in design, as well as independent of the details of the propagation path between Boston and Los Angeles (as long as those details don't change with time). The reciprocity theorem *is* true—it can be *measured* to be true as accurately as one wishes to perform this experiment—because of the reversibility of physics right down to the electronic level. In fact, the answer to the professor's problem of explaining why we don't see velocity vectors suddenly reverse, and then everything 'run backwards,' has not yet been found in any law of physics.

Now, to make things even more interesting, consider the problem of *mutual* or *simultaneous* causation, which can quickly lead to several interesting questions. When two leaning dominoes, A and B, hold each other up, is A nearly upright because of B, or is it B that is nearly upright because of A? When two children bob up and down on a see-saw, whose motion is the cause and whose is the effect? There are other puzzles, too, that involve mutual causation.

---

[40]J. Earman, "The Anisotropy of Time," *Australasian Journal of Philosophy*, December 1969, pp. 273–295.

[41]K. Hutchinson, "Is Classical Mechanics Really Time-Reversible and Deterministic?" *British Journal for the Philosophy of Science*, June 1993, pp. 307–323.

[42]S. F. Savitt, "Is Classical Mechanics Time-Reversal Invariant?" *British Journal for the Philosophy of Science*, September 1994, pp. 907–913.

For example, causation is usually thought to be transitive: if A causes B, and if B causes C, then A causes C. But if A and B are mutually causative, then 'A causes B' coupled with 'B causes A' leads to 'A causes A' (and to 'B causes B'). That is, mutual causation, together with transitivity, seems to imply *self*-causation! Except for those theologians who like this sort of result (it lets them answer the question 'Who made God?' with 'He made Himself'), hardly anyone likes self-causation. But how do we avoid the conclusion that perhaps the mutual causation of two leaning dominoes, coupled with transitivity, represents experimental proof that God could have made himself? Well, of course this is certainly outrageous stuff, but don't you wonder how our poor professor would respond if asked?

This last example is actually a far more esoteric one than we need to illustrate how our ordinary, everyday concept of cause and effect can be turned inside out by going only a little bit beyond the routine. Consider, for example, the problem of the data processing of recorded time signals, such as the information written onto magnetic tapes, hard drives, or disks. Typical applications that produce such recordings include the strata-probing seismic echoes from dynamite explosions set by oil exploration geologists; arms control compliance monitoring stations that listen for the acoustic rumbles generated by both earthquakes and underground nuclear tests—and then try to tell one from the other; and the gathering by various military intelligence agencies of turbine shaft/propeller noise signatures emitted by different types of submarines. In each of those situations, the raw information is recorded and then later processed with a certain degree of unhurried calm and leisure. That pool of oil, after all, has been underground for several hundred million years, and waiting a few more days or weeks for a computer analysis of the explosion echo isn't going to make much difference.

Such after-the-fact processing of recorded data is said to be done 'off-line, in non-real time.' When we play a disk back in the lab, however, we can do all sorts of neat things, like speed up the playback (make time 'run fast'), or slow it down (make time 'run slow'), or even play it *backwards* (make time 'run in reverse'). For various technical reasons, generically called *spectrum shifting*, such tricks are often quite useful. Now, the way we retrieve magnetically recorded information from (for example) a magnetic tape, is to run it through a playback machine with a 'read-head' that senses the magnetic flux variations. The electrical signal produced by the read-head is just like the original signal and, in fact, we can pretend we don't know it is really coming off a tape, but rather that it *is* the original signal. For high-quality digitally recorded tapes and disks, in fact, it is virtually impossible to distinguish the original from a playback.

Now, suppose we construct our playback machine with *two* read-heads, with the new head sensing the recording slightly *before* the old head does. The two heads produce the same electric signal, of course, but the signal from the new head is *ahead* in time compared to the signal from the old head. The new head is, in a certain sense, 'seeing the future' of the old head! We can use these two signals, the old head representing 'now' time and the new head representing 'future' time, to build real systems that are *not* causal. The causality violation occurs in non-real time, of course, not *our* time, but no matter; some absolutely astonishing signal processing can be achieved this way. The universe is about fifteen billion years old, and pretending that time has shifted a few milliseconds or so doesn't seem to be too much violence to reality.

Two heads are often used on radio call-in talk shows to catch inappropriate remarks from intemperate callers and prevent them from being broadcast. A short time delay is introduced by first recording remarks 'live' on tape with a write head and, then a few seconds 'up-stream,' a read head regenerates the remarks for broadcast. A 5 s delay is generally sufficient, so what is heard on a radio receiver *now* actually occurred 5 s ago in the *past*. A caller can get terribly confused if she doesn't turn her own receiver off, because one ear hears the present on the telephone while the other ear listens to the past over the radio.[43] The 1956 British film *Timeslip* incorporates a similar situation, with an atomic scientist's perception advanced 7 s into the future as the result of an accidental radiation exposure. His resulting confusion and disorientation is the center of the film.[44]

2.4 Backward Causation

"Causation as a topic of philosophical discussion refuses to die. Each year, books and articles on causation continue to pour forth. Of course, all this activity may simply be a symptom of the necrophilia that infests so much of philosophy."[45]

All of the previous discussion has fueled countless arguments about what is called *backward*, *reverse*, or even *retro* causation. What is generally meant by forward causation is, of course, that any event that occurs at time *t* is caused by events that all occurred at some earlier time(s). Backward causation says that at least one of the causing events occurs after time *t*—this should make it clear that backward causation is a close relative of time travel. Indeed, one philosopher uses the terms *time traveler* and *retro-causal engineer* interchangeably.[46] The topic, understandably, is at the root of many hot philosophical debates, though not everybody (as this section's opening quote makes clear) thinks those debates are illuminating.

Just *why* does Professor Earman take his harsh position? He offers, as one reason, his disdain for the common philosophical 'proof' of the impossibility of

---

[43] A science fiction use of this idea is in B. W. Aldiss, "Man In His Time," *Science Fantasy*, April 1965, the story of an astronaut who returns from a trip to Mars and finds himself 3.3077 min ahead of everybody else.

[44] Science fiction had used a twist on this idea long before the film; see E. Binder, "The Man Who Saw Too Late," *Fantastic Adventures*, September 1939, a tale of what it might be like to have a 3 min *delay* in your vision.

[45] J. Earman, "Causation: A Matter of Life and Death," *Journal of Philosophy*, January 1976, pp. 5–25.

[46] B. Brown, "Defending Backwards Causation," *Canadian Journal of Philosophy*, December 1992, pp. 429–443.

backward causation: By definition, a cause is always before its effect. Yes, that's the entire 'proof.' One can, of course, win *any* argument by *defining* the answer to be what it is you wish to believe. More interesting, and certainly more pertinent to time travel, is the argument that if backward causation were possible then one could change the past—but that cannot be done because the past is dead and gone and thus unchangeable. That does seem to be a pretty solid argument against backward causation,[47] but Earman rebuts it by pointing out that the very same logic could be applied to the future, and so the usual, uncontested forward causation would also be denied. That is, one could argue that whatever the future will be, *will be* (literally 'by definition'), so one cannot change the future. A similar argument was presented even earlier,[48] in which we find "suppose that someone says 'I can change the future. I can do *this* or I can do *that*.' Well, then, suppose that he does *that*. Has he changed the future? No, because doing *that* was the future."

The reversal of the 'usual' causal order of events by backward time travel has been a mainstay of science fiction almost from the start of the genre. Consider, for example, this tale.[49] A man on vacation by himself, without his wife along, meets a young lady—and they fall in love. The man loves his wife, too, though, and he realizes (as the young lady leaves him for the last time), never to return, that it is all for the best. But she really hasn't gone that far away from him, as the reader soon discovers. She is a time traveler from the future, and after leaving him she goes even further back in time, back an additional 20 years. She does this because she has learned that he met his wife 20 years ago, and so she goes back to be *that* woman! Thus, the usual causal order of the two events 'a long marriage' and the 'pre-marriage courtship' has been reversed (if we accept the fact that the man doesn't remember what his wife looked like when they married).

Actually, even our everyday uses of cause and effect are not nearly so straight-forward as one might think, even when they are under far less stress than backward causation and time travel inflict. Consider, for example, the endless problems that are easy to imagine in the legal world. If a man falls off the roof of a ten-story building and is electrocuted as he plunges through power lines while still twenty feet above ground, was gravity or electricity the cause of death? Or was it both? As this example and others demonstrate,[50] one clearly does not have to discuss time travel to get into a serious argument about cause and effect. But with time travel, and the resultant backward causation, things can become even more perplexing. For example, we normally think it foolish to prepare, now, for an event that has already

---

[47]See, for example, D. H. Mellor, "Fixed Past, Unfixed Future," in *Michael Dummett: Contributions to Philosophy* (B. M. Taylor, editor), Martinus Nijhoff 1987.

[48]J. J. C. Smart, "A Review of *The Direction of Time*," *Philosophical Quarterly*, January 1958, pp. 72–77.

[49]R. F. Young, "The Dandelion Girl," *The Saturday Evening Post*, April 1, 1961.

[50]See also P. Mackie, "Causing, Delaying, and Hastening: Do Rains Cause Fires?" *Mind*, July 1992, pp. 483–500.

happened, but the prudent time traveler about to visit an ice age in the distant past would be wise to pack a fur coat before getting into his time machine!

One philosopher provides, I think, a good start at explaining why so many other philosophers (and not just a few physicists) have adopted the 'common sense' position of rejecting backward causation. As he writes, "Part of the answer, no doubt, is a confusion between affecting and altering [the past—a distinction we'll discuss at length later in this book]. We cannot alter the past. But then we cannot alter the future either, although we can affect it. However, I take the common-sense rejection of backward causation to be, for the most part, quasi-empirical. It is based on a thought experiment. Think how you would set about affecting the past. By building a time-machine, perhaps? But how would you build one? We have no idea how to start. Yet, by contrast, we can work out how to affect the future ... we just move our bodies."[51] But, as he goes on to argue, if we accept that we can't *change* the past (which means there is no way we could actually observe backward causation), then there still exists the possibility that past events were as they were because of events in the future.

Are there actual phenomena that justify a belief in the possibility of effect before cause in real time (not just in tape recorder time)? The only example I know of, and a controversial one at that, is a theoretical result from a reformulation of electrodynamics by the great English physicist Paul Dirac (1902–1984). Classical theory models electric charges as point objects of zero size, which causes problems when one tries to calculate certain details, such as the total field energy of a single electron. The answer comes out as infinity. In an attempt to find more reasonable (that is, finite) answers to such questions, Dirac modified the zero size of a charge to one taking them to be extended objects (while retaining the validity of Maxwell's equations for electrodynamics right down to a point). To calculate how such extended objects will behave mechanically, however, one has to include what are called the *self-interaction* forces, such as the force one side of an electron exerts on the other side.

When it was all worked through, Dirac arrived at a third-order differential equation of motion, an equation that involves a force term proportional not to the usual first time derivative of the velocity (that is, to the acceleration), but rather to the second derivative.[52] This force is proportional to the first derivative of the acceleration, and is a quantity of direct interest mostly to the designers of automobile suspensions, who call it the *jerk*. There is no force in physics, at least not in Newtonian physics, that shows that sort of dependence, and there are some curious consequences. For example, in Dirac's theory an electron experiencing no external force can still continually accelerate, exhibiting what is called a 'runaway solution.'

Dirac showed how the runaway solution can be eliminated by picking a particular value for what up to then was an arbitrary constant of integration in the

---

[51]P. Forrest, "Backward Causation in Defense of Free Will," *Mind*, April 1985, pp. 210–217.

[52]P. A. M. Dirac, "Classical Theory of Radiating Electrons," *Proceedings of the Royal Society A*, August 1938, pp. 148–168.

analysis, but that trick causes, in turn, a new problem called 'pre-acceleration.' That is, if an electron experiences an external disturbance (Dirac considered a passing pulse of electromagnetic radiation), then the electron will start to move *before* the pulse reaches it! Now that does seem to be a pretty clear example of backward causation. The time interval during which the pre-acceleration occurs is very short, on the order of the time it takes light to travel across the spatially extended electron (about 10^{-24} s), but no matter. The apparent crack in the door of causality may be slight, but it was enough to satisfy some philosophers seeking scientific support for backward causation.

Not everybody liked this, however. One physicist was clearly uneasy about it, calling pre-acceleration "unpleasant" acausal behavior.[53] On the other hand, one can find believers, too.[54] Others have argued that the whole business is simply a non-problem. One philosopher, in fact, raised a very interesting technical point, arguing that Dirac's equation is non-Newtonian (remember the *jerk* force) and so we have no reason for coupling force and acceleration together as a cause-and-effect pair.[55] In Newtonian mechanics we do use that particular coupling, yet we do not think of force and velocity as a cause-and-effect pair because there is an integration operation involved in getting from to the other. Similarly, in Dirac's theory we have an integration operation separating force and acceleration.

One curious aspect to the debate on pre-acceleration is that many commentators seem not to have paid much attention to what Dirac himself had to say about it. As a Nobel laureate, it hardly seems likely that he would let such a result pass unnoticed and, indeed, his paper contains the following physical explanation: "It would appear that we have a contradiction with elementary ideas of causality. The electron seems to know about the pulse before it arrives, and to get up an acceleration ... The behavior of our electron can be interpreted in a natural way, however, if we suppose the electron to have a finite size. There is then no need for the pulse to reach the center of the electron before it starts to accelerate. It starts to accelerate ... as soon as the pulse meets its outside. Mathematically, the electron has no sharp boundary."

Two physicists suggested a fascinating connection between travel backward in time and Dirac's relativistically correct, quantum mechanical description of an electron.[56] They showed that in flat, two-dimensional spacetime the assumption of time travel to the past leads in a natural way to Dirac's equation. If, on the other hand, time travel only into the future is assumed, then additional assumptions are required to derive Dirac's equation. This connection between Dirac's equation and

[53]P. C. W. Davies, *The Physics of Time Asymmetry*, University of California Press 1977.

[54]J. Earman, "An Attempt to Add a Little Direction to 'The Problem of the Direction of Time'," *Philosophy of Science*, March 1974, pp. 15–47.

[55]A. Grunbaum, "Is Preacceleration of Particles in Dirac's Electrodynamics a Case of Backward Causation? The Myth of Retrocausation in Classical Electrodynamics," *Philosophy of Science*, June 1976, pp. 165–201.

[56]D. G. McKeon and G. N. Ord, "Time Reversal in Stochastic Processes and Dirac's Equation," *Physical Review Letters*, July 6, 1992, pp. 3–4.

time travel to the past makes some philosophers and physicists nervous, but it didn't seem to bother Dirac. In fact, he went on in his paper to show how the pre-acceleration implies the possibility of building a device for sending a faster-than-light signal backward in time. Science fiction writers were, of course, quick to grasp that idea and such gadgets were dubbed "Dirac radios."[57]

One of the more perplexing aspects of backward causation is that it seems to allow for the possibility of *causal loops*, and for the breaking of such loops, a central feature in many of the very best time travel stories. For example, suppose there is a gadget such that if I push its control button *now*, then today's lecture notes will have appeared in the gadget's output tray *yesterday*. Indeed, yesterday I found today's notes there and, in fact, I am about to go to class to deliver that lecture. A mighty good one it is, too, so I think I think I'll send it back to yesterday in just a few minutes with the help of the gadget. But I haven't yet pushed the button. What if I now decide *not* to push the button? Why did the notes appear so I could use them today? Philosophers call this potential breaking of a causal loop a *bilking paradox*. Later in the book I'll discuss how such paradoxes have regularly appeared in the physics and philosophy literature since the 1940s.

By contrast, such paradoxes had been discussed in the science fiction magazines long before World War II. For example, in a letter to the editor at *Astounding Stories* (June 1932) a fan clearly stated his objection to time travel with the aid of a bilking paradox. He suggested the following experiment: Immediately publish an open offer to the inventor of time travel (who will be born, presumably, at some future date) to travel back to one week before the offer is published. But of course (argued the fan) we'd have a pretty problem if we then decided not to publish the offer after the inventor showed up! As that fan wrote, "Paradoxical? I'll say so, if time travel is possible." That fan didn't know about what seems to be a generic limitation on time machines, however: that one can't travel back to a date before the date of the time machine's creation. Thus, that fan's particular bilking paradox actually has no force.[58]

For another fictional example of a bilking paradox, consider the story[59] of time travelers who, just before they begin a trip into the future, see Earth invaded by Martians. At first the invaders are unbeatable, but then the defending military forces of Earth suddenly and mysteriously acquire a fantastically powerful new weapon. It isn't long before the time travelers realize where it came from—they themselves will go into the far future, obtain the weapon, and then return with it to what is now their own past (when the weapon first appeared). But then they wonder what might happen if they don't go, if instead they 'cheat time.' After all, they reason, why

---

[57]See, for example, J. Blish, "Beep," *Galaxy Science Fiction*, February 1954.

[58]A similar bilking paradox had actually appeared the year before in the 1931 novel *Many Dimensions* by the English writer Charles Williams (1886–1945), which reads like a suitable script for an Indiana Jones movie.

[59]E. Binder, "The Time Cheaters," *Thrilling Wonder Stories*, March 1940. There is an amusing reference in this tale to Orson Welles' famous radio-drama-hoax, from just 2 years earlier, of just such an alien invasion based on H. G. Wells' *War of the Worlds*.

bother now to hunt for the weapon when the invasion has already been defeated? We are told that this potential bilking paradox is a "sinister conception, crawling evilly within their brains, like an unanswerable enigma."

Some philosophers, and practically all physicists, agree with that last assessment about bilking paradoxes, and so they believe there is simply nothing more to say. That is, bilking puzzles like the one in "The Time Cheaters" show that causal loops (and backward causation) must be impossible. Many feel this way about time loops, and backward causation, because (as is well known) time travel to the past can create all sorts of paradoxes. But such paradoxes are offensive only to human, culturally-biased intuitions on 'how things ought to work,' and not to the laws of physics which are indifferent to a reversal in the direction of time—which of course underlies what time travel is all about.

As the great American chemist G. N. Lewis expressed it, "Our common idea of time is notably unidirectional, *but this is largely due to the phenomena of consciousness and memory* [my emphasis]."[60] Lewis' words caught the eye of the editor at one science fiction magazine, who summed it up for his readers in a half-page essay that contained dramatic words hinting at backward causation: "A new theory of time ... reveals the possibility that events now occurring are among the factors that decided Caesar nearly 2,000 years ago to cross the Rubicon."[61]

Lewis' willingness to accept causality violations is not a universally popular view today. For example, one physicist has written[62] that "It is fair to say that most conservative physicists have very serious reservations about the admissibility and reality of causality-violating processes. Causality violation (i.e., the existence of a 'time machine') is such an extreme violation of our understanding of the cosmos that it behooves us to be as conservative as possible about introducing such unpleasant effects into our models." He then goes on to declare closed timelike loops to be verboten because "the existence of closed timelike loops leads us to such unpleasant situations as meeting oneself 5 min ago." He sums up his philosophical position nicely with "any theory that is 'just a little bit causality violating' is 'just a little bit inconsistent.'"

Agreeing with this physicist is at least one philosopher who believes that the "association of causality with a particular temporal direction is not merely a matter of the way we speak of causes, but has a genuine basis in the way things happen" and that there is indeed an asymmetry with respect to past and future that is bound up with our concept of intentional action.[63] He then goes even further when he continues with the claim that being an agent of cause is not a necessary condition for seeing the asymmetry; being an observer is enough, as even an immobile yet

[60]G. N. Lewis, "The Symmetry of Time in Physics," *Science*, June 6, 1930, pp. 569–577.

[61]Editorial essay, "Two-Way Time," *Astounding Stories*, September 1931.

[62]M. Visser, "Wormholes, Baby Universes, and Causality," *Physical Review D*, February 15, 1990, pp. 1116–1124.

[63]M. Dummett, "Bringing About the Past," *Philosophical Review*, July 1964, pp. 338–359.

intelligent tree (!) could detect the difference between past and future. (How he knows this about certain trees is left unexplained.)

The everyday views of causality that we have formed through our limited experiences when living in a world in which time travel is 'uncommon' may actually be incomplete. As the British philosopher Bertrand Russell (1872–1970) said with some humor long ago, in his 1912 Presidential Address ("On the Notion of Cause") to the Aristotelian Society, "The law of causality, I believe, like much that passes muster among philosophers, is a relic of a by-gone age, surviving, like the monarchy, only because it is erroneously supposed to do no harm." And I do agree with his fellow philosopher who, decades later, declared "The concept of cause is powerless to solve the problems posed by the concept of time. The fundamental laws of physics present our most careful, best established and most sophisticated understanding of time. Notoriously, nothing in these laws endorses the idea of a *flow of time nor of the direction* [my emphasis: we'll return to both of these issues later in this chapter] which is basic to our conception of it. Nor are these laws causal (in the sense of singling out causes) even when they are deterministic. The concept of cause is not a fundamental one and cannot illuminate the darker corners in our understanding of the fundamental concept of time."[64]

2.5   The Fourth Dimension

"We are facing an invasion of fourth dimensional creatures ... We are being attacked by life which is one dimension above us in evolution. We are fighting, I tell you, a tribe of hellhounds out of the cosmos. They are unthinkably above us in the matter of intelligence. There is a chasm of knowledge between us so wide and deep that it staggers the imagination."[65]

"Fourth dimension. Time factor. *You* know ..."[66]

The idea of a fourth dimension to *space* has long been a staple of science fiction, but it has also long been viewed with suspicion. Indeed, many quite sophisticated scientists have thought it to be quite mysterious. For example, in his 1897 Presidential Address to the American Mathematical Society, the Canadian/American

---

[64]G. Nerlich, "How to Make Things Have Happened," *Canadian Journal of Philosophy*, March 1979, pp. 1–22.

[65]From "Hellhounds of the Cosmos," *Astounding Stories*, June 1932, by Clifford Simak (1904–1988). Simak went on to write a number of much better tales, but this passage lends credence to the editorial introduction to the 1957 anthology *Famous Science-Fiction Stories* (Random House) that declared so much in the early pulp science fiction was "science that was claptrap and fiction that was graceless."

[66]Uninformative 'explanation' given to a befuddled, inadvertent time traveler who emerges miles away and one hour backward in time after a wild ride through the fourth dimension in a gadget (constructed from a bicycle tire!) in the shape of a three-dimensional Möbius strip (see note 99 in Chap. 1). From the story by H. Nearing, Jr., "The Maladjusted Classroom," *The Magazine of Fantasy and Science Fiction*, June 1953.

astronomer-mathematician Simon Newcomb (1835–1909) declared "The introduction of what is now very generally called hyperspace, especially space of more than three dimensions, into mathematics has proved a stumbling block to more than one able philosopher." Einstein stated Newcomb's view in blunter terms when he wrote "The non-mathematician is seized by a mysterious shuddering when he hears of 'four-dimensional' things, by a feeling not unlike that awakened by thoughts of the occult."[67]

To see just how right Einstein was with this observation, consider the reaction one Egyptian philosopher had (in 1929) to Einstein's own writings: "We have no doubt in our mind that nobody can understand it (the fourth dimension), including Einstein himself. The incomprehensibility of these assumptions [of general relativity] is due to their nature. They deal with the fourth dimension . . . and the reality of time and space. They can only be described by a mathematician's hypothesis or by religious faith."[68] This reaction is easy to understand—after all, anybody can 'see' that there are exactly three spatial dimensions, and that is that!

The 1901 novel *The Inheritors*, by the English writer Ford Madox Ford (1873–1939), like Simak's, is the tale of an insidious hyperspace invasion of our world. It illustrates Einstein's assertion about how many people react to the fourth dimension with an example from the time before the science fiction magazines. When the novel's narrator is bluntly told by an invader that she (the invader) is from the fourth dimension—an idea inspired by Ford's appreciation of how much success his acquaintance H. G. Wells had enjoyed with it—he recoils from that claim with the words "If you expect me to believe you inhabit a mathematical monstrosity, you are mistaken." And who can really blame that skeptical narrator? How can there be *four* spatial dimensions? No less an authority than Aristotle, writing in 350 B.C., had declared in his essay "On the Heavens" that "the three dimensions are all that there are."

Others were not so sure. In 1873, for example, we find an essay in *Nature* that refers to well-known mathematicians who even earlier had shown that they had an inner assurance of the reality of transcendental space (hyperspace).[69] The American philosopher Charles Sanders Peirce (1839–1914) was also an early advocate for the four-dimensionality of space. Just what he thought the nature of the fourth dimension to be is somewhat unclear, but the context of what he said suggests he took it to be spatial. He thought three-dimensional space to be "perverse" because of the existence of incongruous counterparts (such as left- and right-handed gloves), and this was apparently strong evidence for him that space could not be three-dimensional. Now, incongruous counterparts exist in all n-dimensional spaces, but Peirce preserved the special purity of the fourth dimension by suggesting that all physical objects, although capable of motion in the fourth direction, could

---

[67] A. Einstein, *Relativity: the Special and General Theory*, Crown 1961, p. 33.

[68] From A. A. Ziadat, "Early Reception to Einstein's Relativity in the Arab Periodical Press," *Annals of Science*, January 1994, pp. 17–35.

[69] G. F. Rodwell, "On Space of Four Dimensions," *Nature*, May 1, 1873, pp. 8–9.

themselves have no extent in that direction (remember, Peirce was a philosopher, not a physicist, and he offered no experimental support for any of this).[70]

But is it *really* possible that there could be *four* spatial dimensions? We experience three independent directions, each lying at a right angle to the other two—but why just *three*, and not ten or fifteen? Indeed, in an 1888 talk to the Philosophical Society of Washington, Simon Newcomb dismissed the view that space must necessarily be three-dimensional as an "old metaphysical superstition." Yet, despite Newcomb's open-mindedness, it has been shown that in the framework of classical physics there are, in fact, several powerful reasons for why there must be *exactly three* spatial dimensions.

The beginning of a scientific explanation for the dimensionality of space appears in Kant, who believed the *three* dimensions of space and Newton's inverse-square law for gravity are intertwined (but he offered nothing beyond philosophical speculation). The origin of Kant's view is actually quite old, dating back to the ancient Greeks, who had already begun to suspect that there was something special about *three* dimensions, at least as far as geometry was concerned. They knew of the infinity of regular two-dimensional polygons, but that there were just five regular polyhedrons in three dimensions (the so-called *Platonic solids*). This early observation was trapped in mystical speculations, however, and it wasn't until the development of physics as a science that non-mystical discussions on the dimensionality of space began to appear.

Beginning with the work of Einstein's friend, the Austrian/Dutch physicist Paul Ehrenfest (1880–1933) in 1917, we can find the idea that the Poisson-Laplace equation, a second-order partial differential equation that describes the potential functions for both Newtonian gravity and electrostatics, does not allow for stable planetary or electronic orbits in *any* space with dimensionality greater than three. Further, the distortionless, reverberation-free propagation of both electromagnetic and sound waves is possible only in spaces of dimensions one and three. These conclusions have been shown to hold even when we go beyond nineteenth century physics into general relativity and quantum mechanics.[71]

Using a slightly different approach, a biological-topological argument for why space cannot have fewer than three dimensions exists. In all of our common experience, complex intelligent life is always found to occur as an aggregate of a vast number of elementary cells, interconnected via electrical nerve fibers. Each cell is connected to several others, *not all immediate neighbors*, by these fibers. If space had only one or two dimensions, then such highly interconnected nets of cells would be impossible because the overlapping nerve fibers would have to intersect, which would result in their mutually short-circuiting one another.

---

[70]R. R. Dipert, "Peirce's Theory of the Dimensionality of Physical Space," *Journal of the History of Philosophy*, January 1978, pp. 61–70.

[71]See, for example, I. M. Freeman, "Why Is Space Three-Dimensional?" *American Journal of Physics*, December 1969, pp. 1222–1224, and L. Gurevich and V. Mostepanenko, "On the Existence of Atoms in n-Dimensional Space," *Physics Letters A*, May 31, 1971, pp. 201–202.

It wasn't long before these views on the dimensionality of space found their way into science fiction. An early use of space as four-dimensional occurs in an awkward rewrite of Jules Verne's *Around the World in Eighty Days*, in which a professor and his crew fly into hyperspace and around the world and to the moon and back, in less than a day.[72] They do this with a plane equipped with a four-dimensional rudder! More interesting is the tragic story (originally published in 1926) of a math professor who learns how to move into hyperspace and back.[73] A colleague catches him at it and, once over his astonishment, asks what is behind it all. The professor replies, "My assumption is that the fourth dimension is just another dimension—no more different in kind from length, say, than length is from breadth and thickness, but perpendicular to all three. Now suppose that a being in two dimensions—a flat creature, like [a moving shadow on a surface]—were suddenly to grasp the concept of a third dimension and so step out of the [surface]. He might move only an inch, but he would vanish completely from the sight of the world."

The professor has similarly learned how to step out of 3-space and into 4-space but, when asked to explain *how*, all he can say is "How can I explain? It's just the *other* direction. It's *there!*" His colleague can't see it, but nonetheless is quick to grasp the practical implications: "This is power! Think of it! A step, and you are invisible! No prison cells can hold you, for there is a side to you on which they are as open as a wedding ring! No ring is secure from you: you can put your hand *round the corner* and draw out what you like. And, of course, if you looked back on the Universe you had left, you would see us in sections, open to you! You could place a stone or a tablet of poison right in the bowels of your enemies!"

What the professor's colleague is getting at involves a comparison with a prison in planar 2-space, which would merely be a circle around the captive. Knowledge of the third dimension would make it possible to escape, however, by simply moving along that new direction, over the circle, and then back into the plane. To a 2-space guard it would seem that the prisoner had suddenly vanished from view *inside* the circle and then just as suddenly materialized again *outside* the circle. Similarly, to escape from a 3-space prison, one would merely move along the fourth dimension, and in the same way one could remove the yolk from an egg without damaging the shell; indeed, one could remove the yolk directly from the chicken without damaging the chicken![74]

---

[72]B. Olsen, "Four Dimensional Transit," *Amazing Stories Quarterly*, Fall 1928.

[73]R. Hughes, "The Vanishing Man," reprinted in *The Mathematical Magpie* (C. Fadiman, editor), Simon and Schuster 1962.

[74]This astounding insight appeared in early pulp science fiction in, for example, M. J. Breuer, "The Appendix and the Spectacles," *Amazing Stories*, December 1928. The concept appeared even earlier in Bob Olsen, "The Four-Dimensional Roller-Press," *Amazing Stories*, June 1927, and then later in Olsen's "The Great Four Dimensional Robberies," *Amazing Stories*, May 1928 to rob locked safe deposit boxes, and "The Four Dimensional Escape," *Amazing Stories*, December 1933, in which a man sentenced to die by hanging at San Quentin Prison is rescued, while standing on the gallows' trap, by an inventor who pulls him through the fourth dimension.

In a later tale[75] we meet another professor who dramatically uses this very feature of the fourth dimension. His right hand has been modified through an accident to exist in four-dimensional hyperspace and so, to finance his research, he uses his 'talent' to become the perfect pickpocket, able to reach into any wallet no matter how well secured. He also can, indeed, reach right into the very bowels of his fellow man. And he *does*. When he demonstrates his hand to the policeman who has arrested him for being a thief, the astonished officer chokes on a lemon drop. Dr. Fuddles then, of course, does the right thing and removes the drop from the poor fellow with ease. There is one additional aspect to Dr. Fuddles' hand, however, that the story missed. If he had turned his right hand over in the fourth dimension, then he would have had *two* left hands!

It was discovered in 1827 by Möbius (of the strip) that any three dimensional object can be converted into its mirror image by flipping it over through the fourth dimension. Thus, a left-handed glove can be made by pure geometry (no scissors, thread, or needle required) into a precise copy of its right-handed mate. If a living organism is so flipped, however, there may be a problem, as everything in the body would be reversed, including the optically active organic molecules discovered by Pasteur in 1848, which are involved in vital biological processes. These molecules, called *stereoisomers*, exist in two versions in nature (the left-handed and the right-handed versions, if you will), but our bodies have developed the ability to use only one version. To be flipped through the fourth dimension would make some reversed stereoisomers unable to participate in the digestion of food and we would starve to death.

For modern science fiction writers the fourth dimension (and hyperspace, in general), is still a major concept. One physicist, writing in *Analog* (today's premier 'hard science' fiction magazine), summed up nicely what was so fascinating in early pulp, and still is today, about the idea of an extra dimension or two, or perhaps even more, at least from a fictional point of view: "Are there hidden dimensions not accessible to us, dimensions in which we could go adventuring, dimensions within which malevolent hyper-dimensional aliens may be lurking, ready to pierce our flimsy paper-thin three-space bodies with their terrible hyper-sharp claws?"[76] The early pulp science fiction magazines encouraged this lurid imagery. Witness the editorial blurb that opened one many-dimensional monster story as follows: "It was a strange world in which Lester and Florence found themselves. A world of sudden

---

[75]N. Bond, "Dr. Fuddle's Fingers," in *Mr. Mergenthwirker's Lobblies and Other Fantastic Tales*, Coward-McCann 1946.

[76]J. Cramer, "The Other Forty Dimensions," *Analog*, April 1985. 'Monsters in hyperspace' stories were numerous in pulp science fiction. Three examples (in no particular order of literary merit!) are: M. J. Breuer, "The Einstein See-Saw," *Astounding Stories*, April 1932; P. Ernst, "The 32nd of May," *Astounding Stories*, April 1935; "The Monster from Nowhere," *Fantastic Adventures*, July 1939.

Fig. 2.1 An experiment in hyperspace goes astray. The young man is pulling on "hyper-forceps" in an attempt to retrieve a surgeon who has fallen out of 3-space (along with his patient, a professor of non-Euclidean geometry, who suffers from gallstones). The hyper-forceps allow the removal of the gallstones without cutting into the body. Illustration by Frank R. Paul, ©1928 by Experimenter Publishing Co. for "Four Dimensional Surgery" (*Amazing Stories*, February 1928) by Bob Olsen, reprinted by permission of the Ackerman Science Fiction Agency, 2495 Glendower Ave., Hollywood, CA 90027 for the Estate

death and strange science, ruled by inhuman beasts."[77] But as outrageous as that might sound, the real physics of hyperspace is even more amazing.

Hyperspace is, in general, simply any space with more dimensions than the one we obviously seem to live in. In particular, our universe appears to be a four-dimensional (three spatial and one temporal) hyperspace called *spacetime*. This four dimensional world can, at least mathematically, be thought of as the boundary surface of a five dimensional hyperspace. This is analogous to the way the two-dimensional space of the surface of a sphere bounds the three-dimensional space of the sphere itself. This interesting imagery appeared quite early in pulp science fiction. For example, in one remarkably sophisticated story, an eccentric scientist at one point exclaims "A mathematical physicist lives in vast spaces ... where space unrolls along a fourth dimension on a surface distended from a fifth."[78]

There are some interesting geometrical implications to hyperspace which play big roles in time travel considerations. For example, for beings in the two-dimensional world of a sphere's surface there are *two* ways to travel from

---

[77]M. Duclos, "Into Another Dimension," *Fantastic Adventures*, November 1939. See the illustration for this story in "Some First Words."

[78]M. J. Breuer, "The Gostak and the Doshes," *Amazing Stories*, March 1930.

pole to pole. There is the usual way, *on* the surface of the sphere, and the hyperspace way which takes them *through* the sphere along the polar diameter. In imagery motivated by thinking of the sphere as an apple, and of the hyperspace path as a tunnel bored by a worm through the apple, it has become popular to call all such shortcuts, through any hyperspace of any dimension, *wormholes* (a word coined in the 1950s by the Princeton physicist-wordsmith John Wheeler). Wheeler used wormholes to show how electric charge could be thought of as lines of force trapped in the changing topology of a multiply connected space (indeed, Wheeler claimed that the observation of what we call electricity is experimental evidence that space is *not* simply connected).[79]

The general theory of relativity predicts the existence of wormholes in spacetime and, in fact, they were first 'discovered' theoretically in the mathematics of relativity as early as 1916 by the Viennese physicist Ludwig Flamm (1885–1964). Later analyses were done by Einstein, himself.[80] Wormholes have been discussed as a possible model for pulsars (as opposed to the more usual model as rotating neutron stars).[81] It has also been suggested that the interior of a charged black hole may be the entrance to a wormhole.[82] All of these various solutions to the gravitational field equations are generically called "Einstein-Rosen bridges" in the physics literature (see note 81, for example), and the term soon appeared in fiction, too.[83]

The use of hyperspace wormhole portals for explaining some observed physical phenomenon appeared in the scientific literature long before Wheeler's electricity example. In his 1928 book *Astronomy and Cosmogony*, for example, the British theoretician Sir James Jeans devoted a chapter to what were then called nebulae, the island-universes we now call galaxies. At the end of his discussion on the arms of spiral galaxies, Jeans offered the following speculation: "Each failure to explain the spiral arms makes it more and more difficult to resist a suspicion that the spiral nebulae are the seats of types of forces entirely unknown to us, forces which may possibly express novel and unsuspected *metric properties of space* [my emphasis]. The type of conjecture which presents itself, somewhat insistently, is that the centers of the nebulae are of the nature of 'singular points,' at which matter is poured into our universe from some other, and entirely extraneous, special

---

[79]A space is simply connected if *all* the points on the straight line that joins *any* two points in the space are also in the space. The interior of a sphere is simply connected. The interior of a sphere with a hole in it is *not* simply connected.

[80]A. Einstein, "The Particle Problem in the General Theory of Relativity," *Physical Review*, July 1, 1935, pp. 73–77.

[81]J. M. Cohen, "The Rotating Einstein-Rosen Bridge," in *Relativity and Gravitation* (C. G. Kuper and A. Peres, editors), Gordon and Breach Science Publishers 1971.

[82]A. Ori, "Inner Structure of a Charged Black Hole: An Exact Mass-Inflation Solution," *Physical Review Letters*, August 12, 1991, pp. 789–792.

[83]See, for example, J. G. Cramer, *Einstein's Bridge*, Avon 1997 (this is the same Cramer cited in note 76). The *Rosen* comes from the American-Israeli physicist Nathan Rosen (1909–1995), who was a collaborator of Einstein's.

dimension, so that, to a denizen of our universe, they appear as points at which matter is being continually created." This, in everything but name, is a wormhole.

What would hyperspace be like? It is intuitively obvious that in the case of the 2-D surface of a 3-space sphere, the 'hyperspace' wormhole path is shorter than the surface path. Even if this 'shorter path' view holds for wormholes in our 4-D spacetime, however, getting around in science fiction hyperspace may not be a simple task. One tale, for example, tells the story[84] of how one of the first space-ships to explore hyperspace gets lost. The trouble with hyperspace travel is that "You go in at one point, you rocket around until you think it's time to come out, and there you are. Where is 'there'? Why, that's the surprise that's in store for you, because you never know until you get there. And sometimes not even then." The same idea plays a central role in Robert Heinlein's 1957 novel *Tunnel in the Sky*, in which a 'hyperspace gate' is discovered by accident during failed time travel experiments.

Another story[85] asks the same question about hyperspace, and arrives at the same answer: "When you took the Jump ... how sure were you *where* you would emerge? The timing and quantity of the energy input might be as tightly controlled as you liked ... but the uncertainty principle reigned supreme and there was always the chance, even the inevitability of a random miss ... a paper-thin miss might be a thousand light-years."

A common way to visualize hyperspace wormhole shortcuts is to imagine the beginning and the end of a journey as points A and B on the 2-D surface of a piece of paper. Then imagine that the paper is folded so as to position A over B, perhaps with A almost touching B. The distance from A to B *through hyperspace* (the 3-D space in which the folding took place) can clearly be much less than is the distance through 'normal' space (the distance covered by a trip that always remains in the 2-D surface). This is the specific example used in one tale to explain the instantaneous "space-warp" (wormhole) device invented by the story's hero.[86] Such imagery actually appeared quite early in science fiction, as in one story in which a gadget is used to "bend space" so that Earth and Venus touch![87]

The idea of hyperspace folding has broken free from science fiction and can now be found in modern stories in other genres. For example, in one Stephen King story ("Mrs. Todd's Shortcut") a woman keeps finding ever shorter ways to drive from Castle Rock, Maine to Bangor. As the crow flies it is 79 miles, but she gets the journey down to 67 miles, and later to 31.6 miles. When doubted, she replies: "Fold the map and see how many miles it is then ... it can be a little less than a straight line if you fold it a little, or it can be a lot less if you fold it a lot." The doubter remains unconvinced: "You can fold a map on paper, but you can't fold *land*."

---

[84]F. Pohl, "The Mapmakers," *Galaxy Science Fiction*, July 1965.

[85]I. Asimov, "Take a Match," in *New Dimensions II: Eleven Original Science Fiction Stories* (R. Silverberg, editor), Doubleday 1972.

[86]G. O. Smith, "The Möbius Trail," *Thrilling Wonder Stories*, December 1948.

[87]E. L. Rementer, "The Space Bender," *Amazing Stories*, December 1928.

For the purpose of wormhole creation in spacetime, we actually have to imagine much more: the folding of four-dimensional spacetime through a five dimensional hyperspace. The folding imagery has even appeared in the movies: spacetime folding is demonstrated with a piece of paper in both *Event Horizon* (perhaps the worst movie of 1997) and the 2014 *Interstellar*.

Another feature of hyperspace that science fiction has taken a liking to is its vastness. An interesting fictional treatment of this idea was given by a writer who, in real life, was an academic psychologist at the University of Michigan. He put himself in a story[88] of a starship captain who is explaining to the crew psychologist how he feels about hyperspace (or *subspace*, as it is called in the story): "God forsaken. That's just what it is. Completely black, completely empty. It frightens me every time we make the jump through it . . . it frightens me because—well, because a man seems to get lost out there. In normal space there are always stars around, no matter how distant they may be, and you feel that you've got direction and location. In subspace, all you've got is nothing—and one hell of a lot of that. It's incredible when you stop to think about it. An area—an opening as big as the whole of our Universe, big enough to pack every galaxy we've ever seen in it—and not a single atom of matter in it . . . until we came barging in to use it as a shortcut across our own Universe."

The vastness of hyperspace got a more humorous treatment from the early pulp science fiction writer Bob Olsen (1884–1956), who wrote the following verses[89] in the introduction to one of his many stories of the fourth dimension:

I read a yarn the other day—
A crazy concept, I must say.
It states that objects have extension
In what is called the "Fourth Dimension."
In hyperspace one could, no doubt,
Make tennis balls turn inside out;
And from a nut remove the kernel
And not disturb the shell external.
A crook could pilfer bonds and stocks,
Then laugh at prison bars and locks,
One step in this direction queer,
And presto! He would disappear!
Let's hope, in planning new inventions,
They'll give us cars with four dimensions.
When searching for a parking place
We sure could use some hyperspace!

It is not just science fiction that takes hyperspace seriously. We find a mathematician, for example, writing that "most science fiction addicts are familiar with

---

[88] J. V. McConnell, "Avoidance Situation," *If*, February 1956.

[89] B. Olsen, "The Four-Dimensional Auto-Parker," *Amazing Stories*, July 1934. "Bob Olsen" was the pen-name for Alfred Johannes Olsen.

the notion of 'hyperspace,' a higher dimensional space-time bounded by Space-Time through which, in the far distant future, interstellar voyages shortcut the (otherwise unsurmountable) distances between the stars. The purpose of this article[90] is to demonstrate that any ... relativistic space-time model is the boundary of some ... five-dimensional hyperspace." That is just what Breuer's magazine character (see note 78) said—in 1930!

The concept of *time* as a fourth dimension has long been a popular concept, and science fiction in particular has embraced it with enthusiasm. We find a little joke on the idea in a story where a young couple, visited by time travelers from 500 years in the future, are said to live in Apartment 4-D.[91] One physicist[92] traced the idea back to the late eighteenth century, finding references to the idea in pre-1800 works of the great French mathematical physicists Jean le Rond d'Alembert (1717–1783) and Joseph-Louis Lagrange (1736–1813). In fact, a philosopher[93] has found a 1751 passage written by d'Alembert that appears to indicate that it is some unknown, earlier person to whom the credit should really go: "I have said [that it is] not possible to imagine more than three dimensions. A clever acquaintance of mine believes, however, that duration could be regarded as a fourth dimension and that the product of time and solidity would be in some way a product of four dimensions; that idea can be contested, but it seems to me that it has some merit, if only that of novelty."

Still, it wasn't until a curious letter appeared in *Nature* in 1885 that the concept of time as the fourth dimension was mentioned seriously in an English-language scientific journal. The author, mysteriously signing himself only as "S.," began by asking "What is the fourth dimension? ... I [propose] to consider Time as a fourth dimension ... Since this fourth dimension cannot be introduced into space, as commonly understood, we require a new kind of space for its existence, which we may call time-space."[94] Who was this prophetic writer that, if he had just made a simple swap, would have been the first to use space-time as a word? Nobody knows. Bork speculates that it was an acquaintance of H. G. Wells, but Wells himself is on record that it certainly wasn't him.

In his 1934 *Experiment in Autobiography*, Wells wrote "In the universe in which my brain was living in 1879 there was no nonsense about time being space or anything of that sort. There were three dimensions, up and down, fore and aft and right and left, and I never heard of a fourth dimension until 1884 [when Wells was

---

[90]G. S. Whiston, "'Hyperspace' (The Cobordism Theory of Space-Time)," *International Journal of Theoretical Physics*, December 1974, pp. 285–288.

[91]L. Padgett, "When the Bough Breaks," *Astounding Science Fiction*, November 1944.

[92]A. M. Bork, "The Fourth Dimension in Nineteenth-Century Physics," *Isis*, October 1964, pp. 326–338.

[93]E. Meyerson, *The Relativistic Deduction*, volume 83 of *Boston Studies in the Philosophy of Science*, D. Reidel 1985, p. 78.

[94]S., "Four-Dimensional Space," *Nature*, March 26, 1885, p. 481. The editorial staff at *Nature* has informed me that, more than a century-and-a-quarter later, there is no longer any record of the identity of S. in the journal's archives.

eighteen] or thereabout. Then I thought it was a witticism." He had, in fact, said this before. In a 1931 edition of *The Time Machine* (Random House), for example, he wrote in the Preface that the idea for the novel "was begotten in the writer's mind by students' discussions in the laboratories and debating society of the Royal College of Science in the eighties and already it had been tried over in various forms by him before he made this particular application of it."

The idea of time as the fourth dimension entered the popular mind around 1894–95, with the publication of the first of Wells' so-called "scientific romances," *The Time Machine*. Then, after that pioneering use of time as the fourth dimension, science fiction quickly adopted the idea as the basis for one of its most popular subgenres. One of the great "golden age of science fiction" writers, 'Murray Leinster' (1896–1975)—the pen-name for William Jenkins—used it as the basis for his first published story.[95] It is the incredible tale of a Manhattan skyscraper (and its 2000 occupants) sent backward in time several 1000 years because its foundation slips (in an unexplained way) along the fourth dimension. The scientific sophistication of the story is primitive, with just one of the many logical flaws being a vivid description of the time travelers living forward-in-time even as their wrist watches run backward. Indeed, when pulp pioneering editor Hugo Gernsback reprinted the tale in one of the early issues of *Amazing Stories*, a reader complained about that very point. Gernsback felt compelled to defend the story, but could muster only a weak rebuttal based on an author's right to "poetic license."[96]

More technical is the discussion in the story of a clerk who transforms the main entrance to a department store into a time machine by building a tesseract (a four-dimensional cube).[97] The claim is made there that the fourth dimension of the cube/doorway is time. That tale appeared just 5 months after a classic of science fiction by Robert Heinlein (1907–1988) had appeared, also using a tesseract, in which the fourth dimension is taken as *spatial*.[98]

Some writers wanted to have the fourth dimension both ways, as space *and* time in the same story. One wonderful example of this is a classic,[99] written by one of the giants of science fiction. In that tale an electrical engineer named Nelson is caught in the middle of an enormous electromagnetic field surge produced by a short circuit in a power plant. As a physicist explains to the shocked board of directors of the utility, "It now appears that the unheard-of-current, amounting to millions of amperes . . . must have produced a certain extension into four dimensions . . . I have been making some calculations and have been able to satisfy myself that a

---

[95]M. Leinster, "The Runaway Skyscraper," *Argosy*, February 1919.

[96]H. Gernsback, "Plausability in Scientifiction," *Amazing Stories*, November 1926.

[97]W. P. McGivern, "Doorway of Vanishing Men," *Fantastic Adventures*, July 1941.

[98]R. Heinlein, "—And He Built a Crooked House," *Astounding Science Fiction*, February 1941. Here we read of a Los Angeles architect who builds a house in the shape of a tesseract as it would appear if collapsed into normal three-dimensional space. It isn't stable in 3-space (we are told), however, and so a California earthquake is sufficient to topple the house into a stable 4-D configuration, along with its occupants.

[99]A. C. Clarke, "Technical Error," *Fantasy No. 1*, December 1946.

'hyperspace' about ten feet on a side was, in fact, generated: a matter of some ten thousand quartic—not cubic!—feet. Nelson was occupying that space. The sudden collapse of the field [when the overload breakers finally broke the circuit] caused the rotation of that space."

Being rotated through 4-space has inverted the unlucky Nelson [see *For Further Discussion* at the end of this chapter for more on this point], and to bring him back to normal he must be flipped again. The physicist brushes aside a question about the fourth dimension as time, asserting that the only issue is one of space. Poor Nelson is, therefore, again subjected to a stupendous power overload—only now he disappears! Too late, the physicist realizes that the fourth dimension is both space *and* time and that Nelson has been spatially flipped *and* temporally displaced into the future. To understand the particularly monstrous fate of Nelson, just ask yourself what the result would be if he should materialize *inside* matter sometime in the future!

The interpretation of the fourth dimension as time is, of course, the one of interest to prospective time travelers, to physicists studying time travel, and to philosophers of time, and so for us, too. The sort of science fiction that is of greatest interest to us is like the one in which one of the characters, displaced in time, asks for an explanation from a higher-dimensional being who appears on the scene: "'Just where is Tuesday?' he asked. 'Over there [and when the being extends its hand, the hand disappears].' 'Do that again.' 'What? Oh—Point toward Tuesday? Certainly.'" The being explains the physics of the situation to the astonished time traveler thus: "It is a direction like any other direction. You know yourself there are four directions—forward, sideward, upward, and—*that* way! ... It is the fourth dimension—it is duration."[100]

And how about stories like the one in which a mad inventor discovers how to make a substance whose atoms resist being pushed by "pushing back at right angles to all the other [spatial] directions." That is, to push on this exotic stuff is to risk experiencing a back reaction, of being pushed "off into the fourth dimension [which we are told is time] ... into the middle of the week after next."[101] Now wouldn't *that* really be something?!

But of course it was H. G. Wells who, in fiction, pioneered time travel and its connection to the fourth dimension as it is popularly thought of today (with the caveats about Wellsian time machines kept firmly in mind). We are therefore quite interested, as *The Time Machine* opens, to listening-in as the Time Traveller expounds to a group of friends at a dinner party in his London home. He starts with the assertion "There is no difference between Time and any of the three dimensions of Space except that our consciousness moves along it." When asked to say more about the fourth dimension, he replies, "It is simply this. That Space, as our mathematicians have it, is spoken of as having three dimensions, which one may call Length, Breadth, and Thickness, and it is always definable by reference to

---

[100]T. Sturgeon, "Yesterday Was Monday," *Unknown Fantasy Fiction*, June 1941.

[101]M. Leinster, "The Middle of the Week After Next," *Thrilling Wonder Stories*, August 1952.

three planes, each at right angles to the others. But some philosophical people have been asking why *three* dimensions particularly—why not another direction at right angles to the other three?—and have even tried to construct a Four-Dimensional geometry. Professor Simon Newcomb was expounding this to the New York Mathematical Society only a month or so ago."[102]

2.6 Spacetime and the Block Universe

"And now he has preceded me briefly in bidding farewell to this strange world. This signifies nothing. For us believing physicists, the distinction between past, present, and future is only an illusion, even if a stubborn one."

—Albert Einstein[103]

The poet Henry Van Dyke wrote, in his 1904 "The Sun-Dial at Wells College," words that echo the spirit of Omar Khayyam's *Rubaiyat* from nine centuries before:

The shadow by my finger cast
Divides the future from the past:
Before it, sleeps the unborn hour,
In darkness, and beyond thy power:
Behind its unreturning line,
The vanished hour, no longer thine:
One hour alone is in thy hands,—
The NOW on which the shadow stands.

The very next year Einstein's theory of special relativity appeared and, 3 years later, came Minkowski's spacetime interpretation of special relativity. Van Dyke's beautiful poetry was dealt a mighty blow by those developments in mathematical physics, and in the rest of this chapter we'll see how that came to pass.

The modern view of reality, that the past, present, and future are joined together into a four-dimensional entity called *spacetime*, is due to Hermann Minkowski (1864–1909), Einstein's mathematics teacher when he was a student in Zurich. Minkowski gave spacetime (the visual imagery of Einstein's mathematics) to the world during a famous address at the 80th Assembly of German Natural Scientists and Physicians in Cologne, on September 21, 1908. Entitled "Space and Time," his

---

[102] And so Newcomb actually was. Wells, it is certain, routinely read *Nature* (one of his college friends, Richard Gregory, eventually became the journal's editor), and Wells must have read Newcomb's address of December 28, 1893 to the New York Mathematical Society when reprinted in the February 1, 1893 issue (on pp. 325–329), where he called hyperspace "the fairyland of geometry." From the Time Traveller's own words, then, that wonderful Victorian dinner party must have taken place in January or February of 1894.

[103] From a letter written by Einstein on March 21, 1955, to the children of Michele Besso, his dearest friend, who had just died. Einstein's use of the word *briefly* was due to his knowledge that he was nearly out of time, too (he died just a month later).

remarks were electrifying then and still are today.[104] He began dramatically: "Gentlemen! The views of space and time which I wish to lay before you have sprung from the soil of experimental physics, and therein lies their strength. They are radical." Then came the famous line, quoted in so many freshman physics texts and philosophy papers, concerning the nature of spacetime: "Henceforth space by itself, and time by itself, are doomed to fade away into mere shadows, and only a kind of union of the two will preserve independence." Minkowski explained what spacetime is in these words to his audience:

"A point of space at a point of time ... I will call a *world point*. The multiplicity of all thinkable x, y, z, t systems of values we will christen the *world*. With this most valiant piece of chalk I might project upon the blackboard four world axes ... Not to leave a yawning void anywhere, we will imagine that everywhere and everywhen there is something perceptible. To avoid saying 'matter' or 'electricity' I will use for this something the word 'substance.' We fix our attention on the substantial point which is at the world point x, y, z, t, and imagine that we are able to recognize this substantial point at any other time. Let the variations dx, dy, dz, of the space coordinates of this substantial point correspond to the time element dt. Then we obtain, as an image, so to speak, of the everlasting career of the substantial point, a curve in the world, a *world-line*. ... The whole Universe is seen to resolve itself into similar world-lines, and I would fain anticipate myself by saying that in my opinion physical laws might find their most perfect expressions as relations between these world-lines ... *Thus also three-dimensional geometry becomes a chapter in four-dimensional physics* [my emphasis]."

With those words Minkowski gave mathematical expression to the philosophical exposition of Wells' Time Traveller to his dinner party friends. Taking the Minkowskian view of the primacy of spacetime as the ultimate building block stuff of reality was Princeton professor of physics John Wheeler, who wrote[105] "There is nothing in the world except empty curved space. Matter, charge , electromagnetism ... are only manifestations of the bending of space. *Physics is Geometry*." This idea was echoed in fiction, in the 1987 novel Moscow 2042 by Vladimir Voinovich, where we find a time traveler who declares "Anyone with even a nodding acquaintance with the theory of relativity knows that nothing is a variety of something and so you can always make a little something out of nothing."

But not everybody understood Minkowski. In a little-known yet quite erudite essay, published just after a stunning experimental verification of general relativity (the bending of starlight passing through the Sun's gravitational field[106]), an anonymous author presented an optical analogy to help those who thought relativity

---

[104]For a study that includes the original German text, careful English translations, and photographs of Minkowski's agonized corrections to his pre-address manuscript, see P. L. Galison, "Minkowski's Space-Time: From Visual Thinking to the Absolute World," *Historical Studies in the Physical Sciences* (volume 10), 1979, pp. 85–121.

[105]C. W. Misner and J. Wheeler, "Gravitation, Electromagnetism, Unquantized Charge, and Mass as Properties of Curved Empty Space," *Annals of Physics*, December 1957, pp. 525–603.

[106]General relativity had already explained the long-puzzling excess precession of the perihelion (point of closest approach to the Sun) of Mercury's orbit. The excess was an observational (and so experimental) fact which Newton's gravity *cannot* completely explain.

simply "a mathematical joke." Signing himself only as "W.G.," he included the following passage[107]:

> "Some thirty or more years ago [it was forty] a *jeu d'esprit* was written by Dr. Edwin Abbott entitled *Flatland* ... Dr. Abbott pictures intelligent beings whose whole experience is confined to a plane, or other space of two dimensions, who have no faculties by which they can become conscious of anything outside that space and no means of moving off the surface on which they live. He then asks the reader, who has consciousness of the third dimension, to imagine a sphere descending upon the plane of Flatland and passing through it. How will the inhabitants regard this phenomenon? They will not see the approaching sphere and will have no conception of its solidity. They will only be conscious of the circle in which it cuts their plane. This circle, at first a point, will gradually increase in diameter, driving the inhabitants of Flatland outward from its circumference, and this will go on until half the sphere has passed through the plane, when the circle will gradually contract to a point and then vanish, leaving the Flatlanders in undisturbed possession of their country ... Their experience will be that of a circular obstacle gradually expanding or growing, and then contracting, and they will attribute to *growth in time* what the external observer in three dimensions assigns to a movement in the third dimension. Transfer this analogy to a movement of the fourth dimension through three-dimensional space. Assume the past and future of the Universe to be all depicted in four-dimensional space, and visible to any being who has consciousness of the fourth dimension. If there is motion of our three-dimensional space relative to the fourth dimension, all the changes we experience and assign to the flow of time will be due simply to this movement, *the whole of the future as well as the past existing in the fourth dimension* [my emphasis]."

W.G.'s words are a clear and unequivocal statement of the so-called *block universe* concept of four-dimensional spacetime. One can find the block universe concept in the writings of the ancients, too. Consider, for example, the fifth-century B.C. Greek philosopher Parmenides' view of reality: "It is uncreated and indestructible; for it is complete, immovable, and without end. Nor was it ever, nor will it be; for now it *is*, all at once, a continuous *one*." And in Thomas Aquinas' *Compendium Theologiae*, written in the thirteenth century, we find "We may fancy that God knows the flight of time in His eternity, in the way that a person standing on top of a watchtower embraces in a single glance a whole caravan of passing travelers." This is the block universe idea, too, but whereas for Parmenides it was metaphysics and for Aquinas it was theology, for Einstein and Minkowski it was physics.[108]

---

[107]W. G., "Euclid, Newton, and Einstein," *Nature*, February 12, 1920, pp. 627–630. As with the mysterious S. (note 94), the editorial staff at *Nature* has informed me that, nearly a century later, there is no longer any record of the identity of W. G. in the journal's archives.

[108]And for some it was all nonsense. The British philosopher Peter Geach (1916–2013), for example, declared the Minkowskian view to be "very popular with philosophers who try to understand physics and physicists who try to do philosophy." See P. T. Geach, "Some Problems About Time," in *Studies in the Philosophy of Thought and Action* (P. F. Strawson, editor), Oxford University Press, 1968. In his introduction to Geach's essay, editor Strawson put in his two cents by stating the four-dimensional view of reality to be nothing but "fanciful philosophical theorizing."

The block universe concept may explain the enigmatic statement made by Einstein at the death of Michele Besso (note 103). As interpreted[109] decades later:

> "It seems that Einstein's view of the life of an individual was as follows. If the difference between past, present, and future is an illusion, i.e., the four-dimensional spacetime is a 'block Universe' without motion or change, then each individual is a collection of myriad of selves, distributed along his history, each occurrence *persisting on the world line, experiencing indefinitely the particular event of that moment* [my emphasis]. Each of these momentary persons, according to our experience would possess memory of the previous ones, and would therefore believe himself identical with them; yet they would all exist separately, as single pictures in a film. Placing the past, present and future on the same footing this way, destroys the notion of the unity of the self, rendering it a mere illusion as well."

It appears by his words that Einstein was indeed in agreement with the block universe concept, and that he was attempting to give his friend's family some reason to believe that their father still lives 'somewhen.' The makers of the 2002 film *Minority Report* made use of the block universe concept, even if not intentionally; there we see police stopping crime *before* it happens because they can 'see the future.'

Not everybody believed that this view of spacetime was Einstein's, however. Karl Popper (1902–1994), an Austrian philosopher of science, wrote 28 years after the scientist's death that "Einstein was a strict determinist when I first visited him in 1950: he believed in a 4-dimensional Block-Universe. But he gave this up."[110] Shortly before he wrote those words, however, Popper must have learned something new to convince himself of his final comment, because just 2 years earlier he had declared[111] Einstein to (still) be a determinist. Popper presents no evidence to support his claim of Einstein's philosophical conversion, however, and it would seem that the Besso letter still offers the best insight into his actual view of spacetime shortly before his death. I say this because I think Popper's labeling of Einstein as a determinist is wrong. Determinism says 'If you do A, then B will happen, and if you do not do A then (perhaps) something other than B will happen.' A deterministic universe has plenty of room for free will, because you can *choose* to do A or not to do A, and what you decide makes a difference. A fatalistic universe, however, as is the block universe, simply says 'You will do A and B will happen.' To accept the block universe, as did Einstein, is to be a fatalist, not a determinist.

---

[109]L. P. Horwitz, R. I. Arshansky, and A. C. Elitzur, "On the Two Aspects of Time: The Distinction and Its Implications," *Foundations of Physics*, December 1988, pp. 1159–1193. See also Einstein's own book (note 67) where he wrote "From a 'happening' in three-dimensional space, physics becomes, as it were, an 'existence' in the four-dimensional 'world'."

[110]See the *Seventh International Congress of Logic, Methodology and Philosophy of Science*, volume 4 (Salzburg, Austria, 1983), p. 176. Popper describes his early discussions with Einstein on the reality of time and the four-dimensional Parmenidean block universe in some detail in his autobiography: see volume 1 of *The Philosophy of Karl Popper* (P. A. Schilpp, editor), The Library of Living Philosophers, Open Court 1974, pp. 102–103.

[111]In the Foreword to the book by B. Gal-Or, *Cosmology, Physics and Philosophy*, Springer-Verlag 1981.

Einstein's final position on this, then, *might* have been like that of the fictional time traveler who takes a little girl 25,000 years back into the past, where she sees an ancient ancestor of humanity.[112] She then asks if the ancestor is really alive. The time traveler replies, "Every man who ever lived is still alive, child. In time there is no real death. When a man dies he's still alive 10 min ago, 10 years ago. He's always alive to those who travel back through time to meet him face to face."

Did Einstein *really* believe this? Not everybody thinks so. At the 1922 meeting of the French Philosophical Society, for example, the philosopher of science Emile Meyerson asked Einstein whether the spatialization of time (the idea that time is a dimension on the same footing as the spatial ones) is a legitimate interpretation of Minkowski's spacetime. Einstein's terse answer was that "it is certain that in the four-dimensional continuum all dimensions are *not* [my emphasis] equivalent."[113]

Use of the term *block universe* is generally thought to have originated with the Oxford philosopher Francis Herbert Bradley (1846–1924) who, in his 1883 book *Principles of Logic*, wrote "We seem to think that we sit in a boat, and are carried down the stream of time, and that on the bank there is a row of houses with numbers on the doors. And we get out of the boat, and knock at the door with number 19, and, re-entering the boat, then suddenly find ourselves opposite 20, and having then done the same, we go on to 21. And, all this while, the firm fixed row of the past and future stretches in a *block* [my emphasis] behind us, and before us." The house numbers would seem to be Bradley's way of referring to the centuries. Note that he wrote these words 12 years before *The Time Machine*, and that they preceded Minkowski's famous address by a quarter-century.

But this origin of *block universe* may not be as clear-cut as I have made it appear. Bradley, who was frequently criticized by the Harvard psychologist William James (1842–1910)—a man who argued for free will[114] and indeterminism, concepts disallowed in a block universe—may have been mocked on the idea by James during an address to the students of the Harvard Divinity School in March 1884 ("The Dilemma of Determinism"), the year after Bradley's book had been published. In his address James spoke of a deterministic world as being a "solid" or "iron block" (this are *not* characteristics of determinism, but rather of fatalism, and so James makes the same mistake as did Popper). However, writing the year before Bradley's book, in the April 1882 issue of *Mind*, James wrote (with obvious disdain) of "the universe of Hegel [the German philosopher Georg Hegel (1770-1831)]—the *absolute block* [my emphasis] whose parts have no loose play," as having "the oxygen of possibility all suffocated out of its lungs" and as being a universe in which "there can be neither good nor bad, but [only] one dead level of

---

[112]F. B. Long, "Throwback in Time," *Science Fiction Plus*, April 1953.

[113]A. Einstein, "La Théorie de la Relativité," *Bullentin de la Société Francaise de Philosophia* (volume 17), 1922, pp. 91–113.

[114]A famous line from James, one that perhaps illustrates his sort of reasoning about free will, is "My first act of free will shall be to believe in free will." If only proving theorems in math and physics were that easy.

mere fate." So, perhaps, the chain of evolution of the term block universe is actually from Hegel to James and *then*, finally, to Bradley.

We can actually find the block universe in fiction *before* Minkowski (and so certainly before pulp science fiction) came on the scene. In an 1875 (!) story[115] we read of a man who sees, years in advance, his own death in the American Civil War. In the following extract, this man speaks to an unnamed friend (who is the narrator):

> "Do you know," said Bernard, presently, "I sometimes think prophecy isn't so strange a thing ... I really see no reason why any earnest man may not be able to foresee the future, now and then ..."
>
> "There is reason enough to my mind," I replied, "in the fact that future events do not exist, as yet, and we cannot know that which is not, though we may shrewdly guess it sometimes ..."
>
> "Your argument is good, but your premises are bad, I think," replied my friend, ... his great, sad eyes looking solemnly into mine.
>
> "How so?" I asked.
>
> "Why, I doubt the truth of your assumption, that future events do not exist as yet ... Past and future are only divisions of time, and do not belong to eternity ... To us it must be past or future with reference to other occurrences. But is there, in reality, any such thing as a past or a future? If there is an eternity, it is and always has been and always must be. But time is a mere delusion ... To a being thus in eternity, all things are, and must be present. *All things that have been, or shall be, are* [my emphasis]."

When the block universe concept did eventually appear in science fiction, it did so early. In a 1927 story, for example, a time traveler from the future and a man in the present (who is the narrator) have the following exchange:

> "I have just been five years into your future."
> "My future!" I exclaimed. "How can that be when I have not lived it yet?"
> "But of course you have lived it."
> I stared, bewildered.
> "Could I visit my past if you had not lived your future?"[116]

So, while the block universe has a bit of a history to it, the history of the concept of *mathematical* spacetime in physics has a much clearer origin: it derives from Minkowski, not from Hegel, Bradley, James, or even Einstein (who often gets credit for it even though he didn't use the concept in special relativity in 1905, 3 years before Minkowski's address.). Eventually, of course, Einstein did come to appreciate the power and conceptual beauty of four-dimensional spacetime, and it came to play a central role in his ideas about gravity. Indeed, in Einstein's general theory of relativity gravity *is* (curved) spacetime. The starting point for general relativity (and so a *scientifically plausible* theory of time travel) was Minkowski's creation of spacetime, and he is truly deserving of the title 'father of the fourth dimension.'

---

[115]G. C. Eggleston, "The True Story of Bernard Poland's Prophecy," *American Homes*, June 1875. George Cary Eggleston (1839–1911) had served as a soldier in the Confederate Army.

[116]F. Flagg, "The Machine Man of Ardathia," *Amazing Stories*, November 1927.

Of course, it is true that Newton's physics also talks about an analytical (as opposed to merely philosophical) space and time long before either Minkowski or Einstein, but 'Newtonian spacetime' is something very different from Minkowski's self-described "radical" view.[117] In the Newtonian view there is a universal time, a *cosmic* time, which is the same time for everyone, everywhere, in the universe. At every instant, a cosmic simultaneity exists for Newton. Newton's space is Euclidean; that is, through any point exterior to a line exactly one parallel line can be constructed and those two lines will never meet, all triangles (no matter their size) have an interior angle sum of $180°$, and so on. For Newton, space and time were absolutely and uniquely separable. They were, as philosophers like to say, "distinct individuals."

Minkowski changed all that. For him space and time are only relatively separable, and the separation is different for observers in relative motion. For Newton, space and time are the *background* in which physical processes in the world evolve. For Minkowski, spacetime *is* the world.

In a famous philosophical paper[118] by an advocate of the block universe view of reality, we find the words "I ... defend the view of the world ... which treats the totality of being, of facts, or of events as spread out eternally in the dimension of time as well as the dimensions of space. Future events and past events are by no means present events, but in a clear and important sense they do exist, now and forever, as rounded and definite articles in the world's furniture." The title of Williams' paper comes from an ancient dilemma stated by Aristotle in his *De Interpretatione*, where he asked a question now classic in philosophy: "Will there be a sea fight tomorrow?"

Aristotle began his famous answer by first posing the following premise: If a statement about some future event is, eventually, shown to be true (or false), then that statement was true (or false) from the moment it was made. Consider, then, the following two assertions: (A) "It is true that there will be a sea fight tomorrow" and (B) "It is true that there will *not* be a sea fight tomorrow." Surely, argued Aristotle, (A) and (B) cannot both be true, but equally surely, one of them must be true. Suppose it is (A) that is true. Then there is nothing that can be done to prevent the sea fight, and so the future is fated. Suppose, however, it is (B) that is true. Then there is nothing that can be done to cause the sea fight, and so the future is fated. The conclusion is the same no matter which assertion is the true one; thus, the future is fated.

---

[117]See, for example, H. Stein, "Newtonian Space-Time," *Texas Quarterly*, Autumn 1967, pp. 174–200; G. Berger, "Elementary Causal Structures in Newtonian and Minkowskian Space-Time," *Theoria* (volume 40), 1974, pp. 191–201; J. Earman and M. Friedman, "The Meaning and Status of Newton's Laws of Inertia and the Nature of Gravitational Forces," *Philosophy of Science*, September 1973, pp. 329–359.

[118]D. C. Williams, "The Sea Fight Tomorrow," in *Structure, Method and Meaning*, The Liberal Arts Press 1951. Donald Williams (1899–1983) was a professor of philosophy at Harvard.

As might be expected, those who like the fatalistic block universe like this conclusion, but, ironically, Aristotle wasn't one of them—he disliked it so much that he struggled to find a way around it. On the other hand, there are philosophers, like Professor Williams (who believed in a fatalistic universe), who reject Aristotle's rejection of his own logic! Professor Williams went so far, in fact, to calling Aristotle's reasoning "a tissue of error" and "swaggeringly invalid." Possibly so, but the philosophical debates over the sea fight question, and the fatalistic (or not) nature of the world, have not ceased to this day.

In an even more famous paper, Professor Williams makes clear his belief that the passage of time is a myth; he poetically declared "the total of world history is a spatio-temporal volume, of somewhat uncertain magnitude, chockablock with things and events."[119] Professor Williams did, indeed, embrace four-dimensional spacetime, and this is demonstrated by the following incredible passage, perhaps his best-remembered words: "It is then conceivable, though doubtless physically impossible, that one four-dimensional area of the time part of the manifold be slewed around at right angles to the rest, so that the time order of that area, as composed by its interior lines of strain and structure, run parallel with a spatial order in its environment. It is conceivable, indeed, that a single whole human life should lie thwartwise of the manifold, with its belly plump in time, its birth at the east and its death in the west, and its conscious stream running alongside somebody's garden path."

Good Lord!

Now, I am willing to admit that Professor Williams probably wrote that wonderful passage mostly for effect,[120] but I ask you—what, if anything, does it *mean*? It is marvelous to read and yet it remains (for me) mysterious.[121] It should come as no surprise that Professor Williams originally presented his papers to the Metaphysical Society of America, rather than to the American Physical Society. But this passage was perhaps not without impact in areas far removed from metaphysics; some years later there appeared a science fiction story[122] that reads as though it had been inspired by Williams. In it, a scientist discovers how to bend his perception of the four dimensions so as to view verticality as duration and duration as verticality. Thus, he is in October while sitting, but when he stands up he is in November! As bizarre as this may seem, such coordinate interchanges actually do occur in the

---

[119]D. C. Williams, "The Myth of Passage," *Journal of Philosophy*, July 1951, pp. 457–472.

[120]In a footnote, Williams sort of admits this when he writes "I should expect the impact of the environment on such a being to be so wildly queer and out of step with the way he is put together, that his mental life must be a dragged-out monstrous delirium." I think this a great understatement.

[121]As it was for some of Williams' fellow philosophers, one of whom bluntly called the 'myth-of-passage' paper "an interesting piece of science fiction": see M. Capek, "The Myth of Frozen Passage: The Status of Becoming in the Physical World," in *Boston Studies in the Philosophy of Science* (volume 2), Humanities Press 1965. Capek's title reflects his view of the block universe as simply a giant refrigerator and so, turning the tables on Williams, we have 'passage' changed to 'frozen passage.' See also note 136.

[122]G. Wolfe, "The Rubber Bend," *Universe 5* (T. Carr, editor), Random House 1974.

mathematical theory of time machines; we'll see this later, for example, when we discuss Tipler's rotating cylinder time machine.

By the 1930s the block universe had found a home in pulp science fiction. The block universe view that past and present coexist with the present got dramatic treatment in one story of a high school teacher who invents a "spacetime warp" theory, and who is then tricked by an evil industrialist into implementing it in the form of a gun. The weapon produces incredible effects when it is tested; for example, an allosaurus appears, which we are told is "a carnivorous dinosaur of the Jurassic Age, the most frightful engine of destruction that ever walked the Earth!"[123] At the story's end, the teacher explains what has happened to a crowd of breathless newspaper reporters:

> "Spacetime was warped slightly ... The Einsteinian spacetime continuum buckled ... Because it was superficial, only a little of the past, a little of the future broke through. The folds of the warp distorted spacetime evanescently, erratically skirting the vast gulf where the past lies buried and lightly tapping the vast stores of the future. It is a truism of modern speculative physics that the past and the future exist simultaneously and coextensively in higher dimensions of space. De Sitter has speculated as to the possibility of seeing an event before it happens. It is quite possible, gentlemen. Events of the far future already exist in spacetime."

That 'explains' the dinosaur. In the teacher's words, "You tell me that two men saw an incredible beast. ... They swear it looked like a dinosaur. I think it was a dinosaur, gentlemen. It broke through when the warp tapped the past."

And just 2 years later, Robert Heinlein made world lines the central concept in the first of his many classic tales.[124] The story draws an analogy between a world line and a telephone cable: the beginning and end points in spacetime for the world line of a person (birth and death) are associated with breaks (faults) in a telephone cable. By sending a signal up and down the cable, and measuring the time delay until the arrival of the echo produced by such discontinuities, a technician can both detect and locate the faults. In the same manner, Heinlein's story-gadget sends a signal of unspecified nature up and down a world line and thus locates the birth and death 'discontinuities.' Knowledge of the death date, in particular, causes financial stress among life insurance companies, and an examination of *that* tension (not strange physics) is the fictional point of the story.

And then, 2 years after Heinlein's tale with its serious tone, a far less serious story[125] (featuring an Attila the Hun character who roams up and down the corridors of time kidnapping beautiful women for his harem!), we find an 'editorial' footnote telling its young readers that "scientists—especially the new order of meta-physical scientists—are agreed on the principles of Space-Time. The future is not a thing which *will exist*. Rather it is a thing which *does exist*—all events from

---

[123]F. B. Long, "Temporary Warp," *Astounding Stories*, August 1937.

[124]R. Heinlein, "Life-Line," *Astounding Science Fiction*, August 1939.

[125]R. Cummings, "Bandits of Time," *Amazing Stories*, December 1941.

the Beginning to the End, exist in a record upon the scroll of Time." This story, itself, was silly, but the block universe metaphysics was up-to-date.

Somewhat surprisingly, I think, is that even before pulp science fiction embraced the block universe, the concept had already made a deep impression on a broader audience. For example, in a 1928 New York stage play[126] the action alternately takes place in the years 1784 and 1928 and, to explain how that can be, one character (a time traveler) tells another:

> "Suppose you are in a boat, sailing down a winding stream. You watch the banks as they pass you. You went by a grove of maple trees, upstream. But you can't see them now, so you saw them in the *past*, didn't you? You're watching a field of clover now; it's before your eyes at this moment, in the *present*. But you don't know yet what's around the bend in the stream ahead of you; there may be wonderful things, but you can't see them until you get around the bend, in the *future*, can you?"

Then, after this prologue about the stream of time, comes the block universe idea:

> "Now remember, *you're* in the boat. But *I'm* up in the sky above you; in a plane. I'm looking down on it all. I can see *all at once* the trees you saw upstream, the field of clover that you see now, and what's waiting for you around the bend ahead! *All at once!* So the past, present, and future of the man in the boat are all *one* to the man in the plane."

Then, finally, the obvious theological conclusion: "Doesn't that show how all Time must really be one? Real Time—real Time is nothing but an idea in the mind of God!"

To end this section, the block universe conception was cleverly used by one science fiction fan who argued in support of time travel, in reply to another fan how had claimed that a failure of mass/energy conservation was fatal to the plausibility of time travel. Their exchange began with a letter to the editor at *Astounding Stories* in November 1937, written in response to a recent story[127]:

> "Let us say that there is, at a certain time, 'x' amount of matter in the Universe, and 'e' amount of energy. Then if a man of 'a' mass travels backward in time to this particular instant aforementioned, the total amount of matter is thus 'x' plus 'a', while if no other such mass changing occurrences take place, the amount of matter in the future is 'x' minus 'a'. Only a corresponding loss and gain respectively in the amount of energy could explain this conservation of energy, advocates [of time travel] say what they may. But you can't rob or add energy to a Universe nilly-willy! Or perhaps time doesn't enter in on the matter. Perhaps you can add matter in a Universe provided you take it away on some future date."

This fan's concern clearly made an impression on science fiction writers, and the case for conservation of energy is stated in many of the time travel stories that appeared after the publication of this letter.[128]

---

[126]"Berkeley Square" by J. L. Balderson. This play was made into a 1933 movie of the same name, and again in 1951 as the film *I'll Never Forget You*.

[127]O. Saari, "The Time Bender," *Astounding Stories*, August 1937 (see also note 137 in Chap. 1).

[128]Examples include the novels *Lest Darkness Fall* (Henry Holt 1941) by L. Sprague de Camp, and *The Time Hoppers* (Doubleday 1967) by Robert Silverberg.

A reply was soon received by the magazine in a letter (January 1938) from another fan:

"[A recent letter] implies that the idea of time travel is incompatible with the law of conservation of mass and energy. I believe [the] reasoning is wrong [and that the] difficulty lies primarily in the assumption that a body moved in time is transported into a different Universe. According to Einstein, time and the three normal dimensions are so related as to form a continuous, inseparable medium we call the spacetime continuum. Time is in no way independent of the other components of our Universe. Hence a fixed mass [a time traveler and his machine] moved in time is by no means lost from the Universe, the action being analogous to a shift along any other dimension."

The block, or frozen, universe of Minkowski is clearly reflected in those words.[129]

2.7 Philosophical Implications of the Block Universe

"Is the future all settled beforehand, and only waiting to be 'pushed through' into our three-dimensional ken? Is there no element of contingency? No free will? I am talking geometry, not theology."[130]

I should tell you now that, despite the enthusiastic embrace of the block universe by Williams and others (including Einstein), there are those who have been harsh in their criticism of Minkowski's spacetime. The major philosophical problem with the block universe interpretation of four-dimensional spacetime is that it looks like fatalism disguised as physics. It seems to be little more than a mathematician's proof of a denial of free will dressed up in geometry. One philosopher illuminated this concern with the following story, one that vividly illustrates the compelling need many humans have to deny a fatalistic world:

"In a moving picture version of *Romeo and Juliet*, the dramatic scene was shown in which Juliet, seemingly dead, is lying in the tomb, and Romeo, believing she is dead, raises a cup containing poison. At this moment an outcry from the audience was heard: 'Don't do it!' We laugh at the person who . . . forgets that the time flow of a movie is unreal, is merely the unwinding of a pattern imprinted on a strip of film. Are we more intelligent than this man when we believe that the time flow of our actual life is different? Is the present more than our cognizance of a predetermined pattern of events unfolding itself like an unwinding film?"[131]

---

[129]In the context of mathematical physics (*not* science fiction) it has been shown that time travel does *not* imply any fatal violation of conservation of energy. See, for example, J. L. Friedman *et al.*, "Cauchy Problem in Spacetimes with Closed Timelike Curves," *Physical Review D*, September 15, 1990, pp. 1915–1930, and D. Deutsch, "Quantum Mechanics Near Closed Timelike Lines," *Physical Review D*, November 15, 1991, pp. 3197–3217.

[130]The lament of Victorian physicist Oliver Lodge (1850–1940) in his essay "The New World of Space and Time," *Living Age*, January 1920.

[131]H. Reichenbach, *The Direction of Time*, University of California Press 1956, p. 11.

Most people in the Western world would answer *yes* to Reichenbach's question. Most people do find Omar Khayyam's *Rubaiyat* to be a beautiful poem, yes, but still they reject its fatalistic message: "And the first Morning of Creation wrote/What the Last Dawn of Reckoning shall read." Indeed, William James quoted these very words in his 1884 address to the students of the Harvard Divinity School when he argued against fatalism and the block universe.

Besides fatalism, another reason for the stinging words by critics of Minkowski's spacetime is that, in it, events don't *happen*—they just *are*. That is, there seems to be no temporal process of *becoming* in Minkowski's spacetime. Everything is already there and, as what we perceive to be the passing of time occurs, we simply become conscious of ever more of Minkowski's "world points," or events, that lie on our individual world lines. Hermann Weyl (1885–1955), a German mathematical physicist who in his last years was a colleague of Einstein and Gödel at the Institute for Advanced Study in Princeton, expressed this very interpretation in words that have become famous, words that sound very much like those of Wells' Time Traveller: "The objective world simply *is*, it does not *happen*. Only to the gaze of my consciousness, crawling upward along the life line of my body [Minkowski's world line], does a section of the world [spacetime] come to life as a fleeting image in space which continuously changes in time [creating what we call the *now* or the *present*]."[132]

Weyl was skillful in finding poetic ways to express the world line view of reality, but not everybody is convinced by the poetry because it seems to deny the common sense idea of time 'flowing,' of temporal passage; it effectively says time is mind-dependent, a mere *illusion*, as the time traveler in "Berkeley Square" declared (note 126). One philosopher who was particularly opposed to Weyl's view was the British-American academic Max Black (1909–1989), and he expressed his opinion in no uncertain terms: "The picture of a 'block Universe,' composed of a timeless web of 'world-lines' in four-dimensional space, however strongly suggested by the theory of relativity, is a piece of gratuitous metaphysics."[133] Another philosopher who was unhappy with Weyl's view of the block universe was just as blunt: "While philosophers may be forgiven intellectual extravagances of this kind, I think it is a pity when they receive encouragement from theoretical physicists."[134]

Weyl's views had supporters, too, however. Consider, for example, the Time Traveller's speech to his friends at the fateful dinner party that opens *The Time*

---

[132]H. Weyl, *Philosophy of Mathematics and Natural Science*, Princeton University Press 1949, p. 116. Sir James Jeans had already said the same, somewhat less elegantly, in his 1935 Sir Halley Stewart Lecture: "The tapestry of spacetime is already woven throughout its full extent, both in space and time, so that the whole picture exists, although we only become conscious of it bit by bit—like separate flies crawling over a tapestry ... A human life is reduced to a mere thread in the tapestry." Jeans then immediately *rejected* this fatalistic view: see his *Scientific Progress*, Macmillan 1936, p. 20.

[133]From a book review in *Scientific American*, April 1962, pp. 179–185.

[134]H. A. C. Dobbs, "The 'Present' in Physics," *British Journal for the Philosophy of Science*, February 1969, pp. 317–324.

Machine: "There is no difference between Time and any of the three dimensions of Space except that our consciousness moves along it ... here is a portrait of a man at 8 years old, another at fifteen, another at seventeen, another at twenty-three, and so on. All these are evidently sections, as it were, Three-Dimensional representations of his Four-Dimensional being, *which is a fixed and unalterable thing* [my emphasis]." Remember, these words were written in 1895, 13 years before Minkowski and his world lines, and of course decades before Weyl's famous words.

Wells' passage made a considerable impression on at least one well-known physicist of the time, who references it in his early book on relativity.[135] And in another book on relativity, published the same year, we find the same interpretation of Minkowski's spacetime as a block universe: "With Minkowski, space and time become particular aspects of a single four-dimensional continuum ... All motional phenomena ... become timeless phenomena in four-dimensional space. The whole history of a physical system is laid out as a changeless whole."[136]

The claim that time is an illusion has some thought-provoking implications concerning the concepts of omniscience and free will, concepts that occur in any discussion of time travel. Some old theology on God's omniscience, as discussed in Aquinas' *Summa Theologiae*, is seemingly lent at least some support by Minkowski's spacetime: "Now although contingent events come into actual existence successively, God does not, as we do, know them in their actual existence successively, but all at once; because his knowledge is measured by eternity, as is also his existence; and eternity which exists as a simultaneous whole, takes in the whole of time ... Hence all that takes place in time is eternally present to God." Somewhat paradoxically, however, Aquinas did make a distinction between past and future. In that same work he declares that "God can cause an angel not to exist in the future, even if he cannot cause it not to exist while it exists, or not to have existed when it already has." For Aquinas, then, whereas the past is rigid and unchangeable, the future is plastic, which is *not* the block universe view of spacetime.

As one theologian has observed,[137] this does not mean that Aquinas thought God had to view all events simultaneous with all others.[138] Rather, our theologian says that Aquinas could have thought of the relationship between God and events as being similar to that between the center of a circle and all the points on the circumference. That is, each point on the circumference has its own identity, coming before and/or after any other point, but the center is related to each and

---

[135]L. Silberstein, *The Theory of Relativity*, Macmillan 1914, p. 134.

[136]E. Cunningham, *The Principle of Relativity*, Cambridge University Press 1914, p. 191. The use of the words *timeless* and *changeless* explain the characterization of the block universe as being *frozen* (in note 121).

[137]W. L. Craig, "Was Thomas Aquinas a B-Theorist of Time?" *New Scholasticism*, Autumn 1985, pp. 475–483. For the B-theory of time, look back at the discussion in the first section of this chapter.

[138]A science fiction story by Norman Spinrad, "The Weed of Time" (*Alchemy and Academe*, Doubleday 1970) graphically describes what a nightmare that could be!

every point on the circumference in precisely the same way. The center, then, is 'eternity' and the circumference is the temporal series ('one thing after another') of reality. Saying that God is *eternal* is thus very different from saying he is *everlasting*. The first means outside of time, whereas the second means he is a temporal entity but has neither beginning nor end.

Our theologian supports the first interpretation, invoking Aquinas' own words from *Summa Contra Gentiles*: "The divine intellect, therefore, sees in the whole of its eternity, as being present to it, whatever takes place through the whole course of time. And yet what takes place in a certain part of time was not always existent. It remains, therefore, that God has a knowledge of these things that according to the march of time do not yet exist."

The issue of God's eternity and his place in spacetime has long been a hot topic among theologians with a scientific inclination. Practically every issue of the learned journal *Religious Studies*, for example, carries an article on the subject, often invoking relativity theory to support some argument. The Bible, itself, can be a confusing guide on this matter. For example, consider the Old Testament story of King Ahab (First Kings 21). Ahab, King of Sumeria, coveted Naboth's vineyard, but Naboth would not sell. The King retreated, but his wife Jezebel arranged for Naboth's downfall and judicial murder and thus caused the arrival of all his property into her husband's hands. This angered God, who commanded Elijah to prophesy disaster on Ahab's house. Ahab responded with sackcloth, and at that God shifted the disaster to the house of Ahab's son. The point, here, is that God, declared to be omniscient, seems to have been *surprised* at Ahab's penitence. God is aware of everything in this tale, but only as it happens. That is, God's knowledge is subject to growth. This Hebrew concept of God as a participant in history is at odds with the contemporary Christian conception of divine knowledge of all that has been, all that is, and all that will be, a view which has its own Biblical support (for divine eternality). For example, Malachi 3:61 ("For I am the Lord, I change not"), and James 1:17 ("the Father ... with whom is no variableness").

When *The Time Machine* was serialized in the *New Review*, it included a passage that does not appear in the now classic version of the story in which the Time Traveller explains his view of the connection between omniscience and the block universe to his dinner guests:

"I'm sorry to drag in predestination and free-will, but I'm afraid those ideas will have to help ... Suppose you knew fully the position and properties of every particle of matter, of everything existing in the Universe at any particular moment of time: suppose, that is, that you were omniscient. Well, that knowledge would involve the knowledge of the condition of things at the previous moment, and at the moment before that, and so on. If you knew and perceived the present perfectly, you would perceive therein the whole of the past. If you understood all the natural laws the present would be a complete and vivid record of the past. Similarly, if you grasped the whole of the present, knew all its tendencies and laws, you would see clearly all the future. To an omniscient observer there would be no forgotten past—no piece of time as it were that had dropped out of existence—and no blank future of things yet to be revealed ... Present and past and future would be without meaning to such an observer ... He would see, as it were, a Rigid Universe filling space and time ... If 'past' meant anything, it would mean looking in a certain direction, while 'future' meant looking the opposite way."

Wells' "Rigid Universe" certainly sounds like the block universe, and he (or least, the Time Traveller) seems to have believed that it held important implications for the concept of free will.

The 'Rigid Universe' got an interesting science fiction treatment in a story[139] that imagined an event in the present that occurs 'before it should' (a heart patient learns that her obituary notice will be in next week's *New York Times* when that paper arrives 'early'). As one character explains to the sister of the lady who is soon to die, "The future mustn't be changed . . . For us the events of . . . the future are as permanent as any event in the past. We don't dare play around with changing the future, not when it's already signed, sealed and delivered in that newspaper. For all we know the future's like a house of cards. If we pull one card out, say your sister's life, we might bring the whole house tumbling down. You've got to accept the decree of fate . . . You've got to."

With Einstein's discovery of the *relativity of simultaneity*,[140] we run into the question of 'How can there be any sense to the concept of divine, universe-wide knowledge in a four-dimensional spacetime?' That's because in some frames of reference it is possible for event A to be observed before event B, whereas in other frames the temporal order could be reversed, and so some theological questions prompted by spacetime physics are: 'What is God's frame of reference if he is to be actively involved in human affairs? Could God have a special frame of reference in which he is exempt from the relativity of simultaneity, a frame in which he imposes an absolute order on the sequence of becoming of events? Does it make any sense, that is, to say God enjoys what might be called 'divine immediacy'? And if so, what should we think of a God who follows rules of nature different from those that govern all he is supposed to have made?'

Theologians have debated questions like these for decades, and surely will continue to do so for many more decades to come. Alas, I suspect that physicists who study time travel have either been unaware, unimpressed, or just plain uninterested. That's too bad, because one doesn't have to be religious to appreciate the pure intellectual challenges presented by such questions. For example, consider the following debate between two philosophers, one who believes free will and divine foreknowledge are not compatible, and another who thinks the first has made a fundamental error in blurring the distinction between changing and affecting the past. (This distinction is of *great* importance

---

[139]R. Silverberg, "What We Learned From This Morning's Newspaper," *Infinity Four*, November 1972.

[140]This refers to the discovery that two events, which occur simultaneously for one observer in a spacetime, may not be simultaneous for another observer in the same spacetime. This will be discussed in more detail in the next chapter.

in any discussion of time travel.) This second philosopher presented some of his arguments in terms of a time traveler to the past[141]:

> "Consider the following. Parsons (P) has invented a special machine which allows him to go back in time. He enters the machine in 1986 and finds himself in the presence of or, perhaps better, observing, Quigly (Q) in 1876. P is an authority on Q, and knows immediately the situation Q is in. Not only that, but he remembers reading about the particular decision or act which Q made in that situation. Thus one might argue that from P's perspective what Q decides is as if already done. It is not already done, since P is standing there waiting for Q to do it. He has gone back in time. Yet from P's perspective, which is of one come back from the future, it is as if already done, since he knows what Q does decide. Since P strongly believes in the unalterability of the past, it is not within Q's power to do something other than what Q in fact does in that situation. From Q's perspective his decision is not already made nor is the action taken, so that it is in his power at that time to do either x or y. From his perspective, that he will do x rather than y is indeterminate; it is not yet done, though at the same time he can grant that P knows what he will do because for him it is as if he has already done it."

The first philosopher doesn't buy any of this, and dismisses it with "It should be abundantly clear ... that the fact that such stories are in some way imaginable and intuitively graspable says nothing about their logical coherence." Given the interest among modern physicists in time travel, however, I think the first philosopher wouldn't write that today.

One possible reply to all of these theological issues that spacetime physics prompts can perhaps be found in a paper[142] (written by a philosopher and two mathematicians) that describes a five-dimensional spacetime in which the fifth dimension is initially given the provocative label of the 'eternity' axis. But then the authors lost their nerve and elected to rename it 'anti-time.' It is interesting to note that pulp science fiction anticipated that terminology by decades, as in one story[143] we read "Beyond the fourth there is a fifth dimension ... Eternity, I think you would call it. It is the line, the direction perpendicular to time." For some, the eternity axis would appear to be perfect to serve as the temporal dimension for God, an axis distinct from the time axis of mere mortals.

The idea of supernatural beings existing outside of mortal time is an old one in theology, and it can also be found in secular literature long before science fiction got hold of it. For example, in the first act of Lord Byron's 1821 poem *Cain*, the fallen angel Lucifer tells Cain and his wife that

---

[141]For the complete exchange between these two philosophers, see W. Hasker, "Foreknowledge and Necessity," April 1985, pp. 121–157, B. Reichenbach, "Hasker and Omniscience," January 1987, pp. 86–92, and W. Hasker, "The Hardness of the Past: A Reply to Reichenbach," July 1987, pp. 337–342, all in the journal *Faith and Philosophy*. Hasker is the 'first' philosopher, and Reichenbach is the 'second' one. See also D. P. Lackey, "A New Disproof of the Compatibility of Foreknowledge and Free Choice," *Religious Studies*, September 1974, pp. 313–318.

[142]J. G. Bennett *et al.*, "Unified Field Theory in a Curvature-Free Five-Dimensional manifold," *Proceedings of the Royal Society of London A*, July 1949, pp. 39–61. A theological interpretation is given in G. Stromberg, "Space, Time, and Eternity," *Journal of the Franklin Institute*, August 1961, pp. 134–144.

[143]L. A. Eshbach, "The Time Conqueror," *Wonder Stories*, July 1932.

With us acts are exempt from time, and we
Can crowd eternity into an hour,
Or stretch an hour into eternity.
We breathe not by a mortal measurement,
But that's a myst'ry.

Before Minkowski, the debates over fatalism (as in Silverberg's story in note 139) and free will had been the exclusive province of philosophers, theologians, and lawyers (if a person has no control over his or her actions, then can we morally and ethically punish that person if those actions happen to be criminal?[144]). After Minkowski, the physicists (at least a few of them) joined the debates. According to one philosopher (note 118) the major motivation driving these debates is "the age-old dread that God's foreknowledge of our destiny can in itself impose the destiny upon us." The implication is, of course, that God is 'outside of time' and so can take in the entire Minkowskian block universe at a glance (hence his foreknowledge).

The relativistic view of the universe as a timeless four-dimensional spacetime seems to provide scientific, mathematical support for the conclusion that not only is the past fixed, but so is the future. Does that mean the future is what it will be—and if so, then why bother agonizing over the many apparent decisions each of us faces every day? If the future will be what it will be, then Christian theologians are left with the puzzling task of explaining what could possibly be meant by the Biblical exhortation (Deuteronomy 30:19) "I call Heaven and Earth to record this day against you, that I have set before you life and death, blessing and cursing; therefore *choose* [my emphasis] life, that both thou and thy seed may live."

This issue has bothered philosophers for a very long time. The so-called Master Argument (the name reflects its supposed invulnerability to rebuttal), for example, comes down to us from its origins in ancient times, in the *Discourses* of the first century A.D. Roman Stoic philosopher Epictetus. That argument can be summarized[145] as follows:

1. The future follows from the past;
2. The past is unchangeable;
3. What follows from the unchangeable is unchangeable;

   Therefore,

4. The future is unchangeable.

This certainly does seem to be fatalistic, in effect arguing that all events in a block universe spacetime are recorded in a 'Book of Destiny.' Since ancient times many great works of literature have adopted that view, recounting tales of the foretold

---

[144]For more on this, in the context of time travel, see the penultimate question in the *For Future Discussion* questions at the end of this chapter.

[145]See, for example, the two papers by R. L. Purtill, "The Master Argument," *Apeiron*, May 1973, pp. 31–36, and "Foreknowledge and Fatalism," *Religious Studies*, September 1974, pp. 319–324.

fates of men, such as Sophocles' *Oedipus*. It is, in a block universe, as though our conscious experience of the world is no different from that of the man watching the projected film images of *Romeo and Juliet*.

That view is the central issue in the early sixth century A.D. Roman philosopher Boethius' influential *De Consolatione Philosophiae* (circa A.D. 500) which was written during a year of imprisonment before his execution for treason; perhaps he wondered during that year if his fate could have been anything different. Certainly he must have taken some consolation in fatalism, but in fact he tried to argue that God's vision of *all* temporal reality does not limit the freedom to act. According to Boethius, "The expression 'God is ever' denotes a single Present, summing up His continual presence in all the past, in all the present ... and in all the future." That is, God sees in one timeless and eternal moment all that has been and will be freely chosen.[146]

When the fourteenth century English poet Geoffrey Chaucer prepared a translation of *Consolatione* he was obviously inspired by it when he wrote his very long, famous poem (*Troilus and Criseyde*) on the nature of love (Book IV.140):

> *Some say "If God sees everything before*
> *It happens—and deceived He cannot be—*
> *Then everything must happen, though you swore*
> *The contrary, for He has seen it, He."*
> *And so I say, if from eternity*
> *God has foreknowledge of our thoughts and deed,*
> *We've no free choice, whatever books we read.*

Two modern, purely philosophical rebuttals[147] to Chaucer, however, argue that his poetry misstates Boethius' philosophy when Troilus declares that divine foreknowledge is incompatible with free will. That is, in their view God's omniscience (a fundamental teaching in the theistic religions of Christianity, Judaism, and Islam) *is* compatible with free will (also a fundamental belief in those same religions). Both of these scholarly papers, though, depend much more on the nuances of grammar than most physicists will like.

The connection between spacetime physics and free will was made explicitly by the philosopher who wrote "For philosophers in either field, philosophy of science and philosophy of religion are too often viewed as mutually irrelevant ... This is unfortunate, because sometimes the problems can be quite parallel and a consistent resolution is required. One especially intriguing case in point concerns, in

---

[146]In his *The Sirens of Titan*, a 1959 novel meant to be a parody of God's omniscience, Kurt Vonnegut gave the curious name of *chrono-synclastic infundibulated vision* to God's power to see the past and future.

[147]G. I. Mavrodes, "Is the Past Unpreventable?" April 1984, pp. 131–146, and A. Plantinga, "On Ockham's Way Out," July 1986, pp. 235–269, both in *Faith and Philosophy*.

Fig. 2.2 The common view
of time

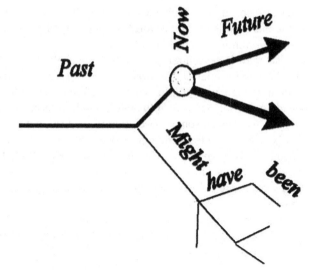

philosophy of science, the possibility of ... time travel and, in philosophy of
religion, the relationship between divine foreknowledge and human freedom."[148]

That philosopher could well have included science fiction writers in his group of
people interested in both spacetime physics and free will. In one story,[149] for
example, a man in the twenty-fifth century is about to travel back into the past to
escape criminal prosecution. He is aked where he'd like to go, and he replies "I do
not understand the paradoxes—what if I choose to build gravity-deflectors in
Ancient Rome?" When he is told (correctly) that he couldn't do that because it
didn't happen, he persists: "But if I can choose any period, it means I can alter
history at will—which presumes that the present can also be changed." Then, at last,
he gets the explicit answer that bothers nearly everyone: "The real answer is that in
the final analysis your decision to choose a certain time period is already made, and
the things you will do [in the time traveler's personal time] are already determined.
Free will is an illusion; it is synonymous with incomplete perception." The same
idea appears in another tale (note 57); when one character says, "What you are
saying is that the future is fixed, and that you can read it, in every essential detail,"
the response is "Quite right ... both those things are true."

However, no matter how hard we try—and by *we* I include even those physicists
and philosophers who embrace the block universe with its support of time travel to
the past—it is very difficult to break free of the view of time as shown in Fig. 2.2.
That is, as the passage of time up to the present or *now* (with all to the left of that

---

[148]W. L. Craig, "Tachyons, Time Travel, and Divine Omniscience," *Journal of Philosophy*, March
1988, pp. 135–150. Tachyons are hypothetical faster-than-light particles that theoretically travel
backwards through time. They will be discussed in Chap. 5.

[149]W. Kubilius, "Turn Backward, O Time," *Science Fiction Quarterly*, May 1951.

instant as the past), while to the right of the *now* we have multiple possible futures (depending on our free will choices). Lying to the side of all that (in our thoughts and imaginations) are all that 'might have been' if we had made different choices than we did at earlier times in the past.[150]

With all that said, even if events are really laid out in the spatial and temporal web that constitutes the four-dimensional block universe, there still remains the great mystery of why we see them unfold in the particular sequence that we do. Why not in reverse order? Why, indeed, do we see what we call *time* run from what we call the past to what we call the future and, indeed, what do we really mean by *past* and *future*? As you'll see in the next chapter, these are not easy questions, and nearly everybody who has thought about them believes we are not yet even close to knowing the answers.

On that perhaps gloomy note, it seems appropriate to end here with a few more words from St. Augustine's *Confessions*, with words that follow those that helped open this chapter: "I confess to you, Lord, that I still do not know what time is. Yet I confess too that I do know that I am saying this in time, that I have been talking about time for a long time, and that this long time would not be a long time if it were not for the fact that time has been passing all the while. How can I know this, when I do not know what time is? Is it that I do know what time is, but do not know how to put what I know into words? I am in a sorry state, for I do not even know what I do not know!"[151]

2.8 For Further Discussion

In the comics one of Superman's more interesting adversaries is Mr. Mxyzptlk (pronounced *mix-yez-pitle-ick*), a being with seemingly magical powers from the Land of Zrfff in the fifth dimension. Mr. Mxyzptlk's powers are not really because of magic, however, but are 'merely' the result of his hyperspace world with its extra dimension. Mr. Mxyzptlk, for example, in one of his misadventures with Superman in 1954, begins selling a

(continued)

---

[150]Figure 2.2 is based on a similar one in C. K. Raju, "Time Travel and the Reality of Spontaneity," *Foundations of Physics*, July 2006, pp. 1099–1113.

[151]There is another view of time even darker than St. Augustine's, which denies the existence of both future and past, and doesn't offer us much either for that special moment we call the present (or *now*). This view, called *presentism*, was hauntingly expressed in some lyrics I heard in the final episode of the second season (2015) of the HBO series *True Detective*: "There is no future/There is no past/In the present nothing lasts." Now *that* is depressing! Still, there are philosophers who believe even this view can support time travel: see S. Keller and M. Nelson, "Presentists Should Believe in Time-Travel," *Australasian Journal of Philosophy*, September 2001, pp. 333–345.

newspaper called the *Daily Mpftrz* in competition with the *Daily Planet.* Unlike a traditional newspaper that reports what *has* happened, the *Daily Mpftrz* (your guess is as good as mine!) prints what *will* happen. As Mr. Mxyzptlk says, "You see, as a resident of the fifth dimension, I can get all the news I want from the *fourth* dimension!" The science editor at the *Daily Planet* explains the meaning of that to his boss, Perry White: "That's right, Mr. White … many physicists consider *time* the fourth dimension … so if Mr. Mxyzptlk can travel from the fifth dimension to our three-dimensional world, he most likely *is* able to see the future!" (This leaves unanswered the question of *why* he continues to challenge Superman when he knows he will always be defeated—*as he always is*!) Presumably a five dimensional world would have our three spatial and one temporal dimension (for a total of four), and so the question now is: what is the nature of the additional (fifth) dimension? Is it spatial or is it temporal? (There is a brief appearance of the fifth dimension in the 2014 movie *Interstellar*, but we aren't told much of anything about its possible structure.) Discuss and compare the world of four space dimensions and one time dimension, with the world of three space and *two* time dimensions. (In Chap. 5 we'll discuss a possible connection between two-dimensional time and time travel.)

In the text it is stated that "If A and B are mutually causative, then 'A causes B' coupled with 'B causes A' seems to lead to 'A causes A.'" Suppose, however, that we imagine two adjacent sunken pools of water, **a** and **b**, on the same horizontal surface, with each pool filled to the brim. An overflow from one pool will flow into the other pool. Now, define the events A and B as 'A is the overflow of pool **a**' and 'B is the overflow of pool **b**.' Thus, A causes B and B causes A. Does the conclusion 'A causes A' make physical sense in this specific case? Discuss at length.

When reading A. C. Clarke's story "Technical Error" (see note 99), we learn that a rotation through 4-space *inverts* "the unlucky Nelson." The 'solution' to this awkward situation is to flip Nelson through 4-space a second time and so back to 'normal.' (When *Thrilling Wonder Stories* reprinted this tale in June 1950, after its original publication in 1946, the title was changed to the more appropriate "The Reversed Man.") Clarke may have missed an

(continued)

important technical 'detail,' however, in that when first flipped through 4-space *everything* inverts, and so matter becomes anti-matter and Nelson would have instantly been annihilated in a 100 % conversion of matter to energy (that is, the flipped Nelson would have initiated a very large explosion). Compare this to Alice's concern in her flipped world (Lewis Carroll's *Through the Looking Glass*) when she wonders "Perhaps Looking-glass milk isn't good to drink." Explain why Lewis Carroll certainly was *not* thinking of matter/anti-matter explosions when he wrote his novel. What *do* you think he might have had in mind?

A time travel story, even earlier than Clarke's, that uses spacetime 'rotations,' was authored by Edmond Hamilton (1904–1977), one of the pioneering pulp fiction writers. In his "The Man Who Saw the Future" (*Amazing Stories*, October 1930), a man is hauled before the Inquisitor Extraordinary of the King of France to explain his mysterious disappearance, and subsequent reappearance, in an open field, amid thunderclaps and in plain sight of many onlookers. As the story unfolds, we learn that the man was transported five centuries into the future, from A.D. 1444 to 1944, by scientists working in twentieth-century Paris. The thunderclaps were produced by spacetime 'rotations,' as the atmospheres of 1944 and 1444 were reversed. A skeptical Inquisition naturally finds this tale preposterous and the first time traveler is burned at the stake as a sorcerer. Can you think of *why* such 'atmospheric swaps' might produce thunderclaps?

A trip around a Möbius strip reverses the 'handedness' of a plane figure (left and right are swapped). You can see this for yourself by making a Möbius strip, and then sliding an arrow (pointing *across the width* of the strip) around the strip. (Cut a notch in the side of the strip to mark the starting point, with the arrow pointing at the notch.) When you get back to the notch, the arrow will point *away* from the notch. Notice that the arrow never left the surface of the strip, or crossed any 'weird' boundary. Then, read H. G. Wells' short story "The Plattner Story" and comment on its use of 'handedness.'

The *autoinfanticide paradox*, which results when a time traveler tries to kill his younger self, continues to fascinate both physicists and philosophers, and papers regularly appear in the scholarly literature on the topic: see, for example, Kadri Vihvelin, "What Time Travelers Cannot Do," March 1996, pp. 315–330 (which introduces Suzy the time traveler); Ira Kiourti, "Killing Baby Suzy," June 2008, pp. 343–352; Peter B. M. Vranas, "What Time Travelers May Be Able to Do," August 2010, pp. 115–121; and Joshua Spencer, "What Time Travelers Cannot *Not* Do (but are responsible for anyway)," October 2013, pp. 149–162, all in *Philosophical Studies*. All deal with an issue that is psychologically fascinating: *moral responsibility*. Spencer, in particular, opens with this definition: Someone is morally responsible for an action *only* if she could have done otherwise. As he goes on to write, "If I have been attacked and both of my legs have been broken, then it seems illegitimate to criticize me for failing to run away; I could not have done otherwise." And yet all of these papers are on a point that (I think) physicists would soon lose interest in: is the question 'If Suzy is a time traveler, can Suzy kill baby Suzy, given that Suzy doesn't kill baby Suzy?' the same question as 'If Suzy is a time traveler, can Suzy kill baby Suzy, given that Suzy is now alive?' The answer to the first question is, from pure logic, NO, while the answer to the second question is just bit squishier: it all depends on what the word *can* means. For the second question, Suzy *can* kill baby Suzy if she has a weapon (knife, gun, poison, etc.) and she is in the past next to baby Suzy, but it is just that she doesn't because otherwise Suzy wouldn't be alive now (which is a given). Such debates seem unlikely to produce any insights into the physics of time travel. Compare this situation to the old schoolboy conundrum "What happens when an irresistible force meets an unmovable object?', which is a *self-inflicted* 'paradox.' That is, the words *irresistible* and *unmovable* are mutually exclusive and so, used this way, it should be no surprise that we have a conflict. Are the two time travel questions above, concerning Suzy, confusing through a similar mushy use of grammar? Or are they deeper than that? Vigorously defend your position.

In addition to H. G. Wells, another nineteenth-century writer who was highly influential in bringing the fourth-dimension out of academia and into public consciousness was the mathematician Charles Howard Hinton (1853–1907). Hinton was no angle-trisecting crank, having earned an M.A. at Oxford, an appointment in the mathematics department at Princeton, and then another at the University of Minnesota. Later, with the help of the eminent astronomer

(continued)

Simon Newcomb, he obtained a position at the Naval Observatory in Washington, D.C., and was on the staff of the United States Patent Office at the time of his sudden death. Hinton was a man to be taken seriously. His first published essay "What Is the Fourth Dimension?" appeared in 1880, and then in book form in 1884 as part of his *Scientific Romances* (a phrase used by Hinton before it became associated with Wells' science fiction many years later). That book received a generally favorable review in *Nature* (March 12, 1885, p. 431). At one point he wrote "We might then suppose that the matter we know extending in three dimensions has also a small thickness in the fourth dimension," an idea that was used a few years later by the well-known British mathematician W. W. Rouse Ball (1850–1925) in an attempt to explain gravity. Hinton was extremely inventive, and he also proposed four-dimensional-space models for static electricity. Find out more about Hinton's life and work: a good source to start with is *Speculations on the Fourth Dimension: Selected Writings of Charles H. Hinton* (R. Rucker, editor), Dover 1980. Take a look, too, at J. E. Beichler, "Ether/Or: Hyperspace Models of the Ether in America," in *The Michelson Era in American Science 1870—1930* (S. Goldberg and R. H. Stuewer, editors), American Institute of Physics 1988.

Chapter 3
The Physics of Time Travel: Part I

"... within forty-eight hours we had invented, designed, and assembled a chronomobile. I won't weary you with the details, save to remark that it operated by transposing the seventh and eleventh dimensions in a hole in space, thus creating an inverse ether-vortex and standing the space-time continuum on its head."

—almost certainly *not* the way to build a time machine[1]

3.1 The Direction of Time

"Of all the problems which lie on the borderline of philosophy and science, perhaps none has caused more spilled ink, more controversy, and more emotion than the problem of the direction of time ... [T]he main problem with 'the problem of the direction of time' is to figure out exactly what the problem is or is supposed to be!"[2]

Before we start talking about the physics of time travel, let me say a few more words on time itself, in a way slightly less metaphysical that was the discussion in the previous chapter (which is why I'm writing this *here*, in a chapter with an increased emphasis on the analytical). When we speak of journeying to either the future or the past, we are implicitly making a distinction in the direction of the time traveler's trip. But does time actually have a *direction*? Is there an *arrow* that points the way? The answer seems obvious: *of course* time has a direction. After all, everybody 'knows' it flows from past to future. There is a curious language problem here, however, because we also like to say the present recedes into the past, which implies a 'flow' in the opposite direction, from future to past. Well, despite this snarled syntax, can we at least distinguish past from future, whichever way time flows?

---

[1] L. Sprague de Camp, "Some Curious Effects of Time Travel," in *Analog Readers' Choice*, Dial 1981.

[2] See note 54 in Chapter 2.

© Springer International Publishing AG 2017
P.J. Nahin, *Time Machine Tales*, Science and Fiction,
DOI 10.1007/978-3-319-48864-6_3

This would seem to be an important question to answer because for the phrases *flow of time* and *direction of time* to have any objective meaning at all, it must be somehow possible to identify a difference between past events and future ones. The special moment at which that distinction occurs is known as the *now* or the *present* and, as events make the transition associated with that distinctive difference, between past and future, we say that the *now* (the present) moves or flows. Philosophers—and science fiction writers and physicists, too, who after all are human beings with human senses like everybody else—call this common feeling that we all have, of the passage of time, the *psychological arrow of time*. One philosopher gave an amusing (tongue-in-cheek) gastronomical interpretation of the moving *now* as follows:

> "New slices of salami are continually being cut from a nonexistent chunk of salami called the future. The *present* is the slice on top of the pile. The past are the pieces beneath this, and even though they are not present they still continue to exist in the same way that the top slice of salami does. . . . This [concept] faces humiliation before the embarrassing question of how fast the pile of salami slices grows."[3]

The 'moving now' does present a problem for physicists because there is nothing in the laws of physics that marks the present moment as unique, and therefore nothing that reflects a 'flow' of time, nothing that models the reality of a 'moving now' becoming part of the past and the events of the future becoming, successively, the new 'now.' As a philosopher wrote long before time travel became a serious topic in the physics literature, "Talk of the flow of time or the advance of consciousness is a dangerous metaphor that must not be taken literally."[4]

What that philosopher may well have had in mind is that all events in the block universe simply have coordinates in spacetime, and there is nothing corresponding to 'have been' (past), 'are' (present), or 'will be' (future). There is no 'moving now' in the block universe except for its subjective presence in our conscious minds. All we can say from physics is that events are ordered in an earlier/later sequence, and in fact, even that relatively weak condition holds only for *causally* related events.[5] The relativistic, four-dimensional block universe view of spacetime that so many physicists (including Einstein) so dearly love seems to have no room for an objective theory of the *flow* of time. And yet, even for those same physicists, there is a powerful psychological sense that time *does* flow. But are they mistaken? It is a fact that, with not just a little irony, Gödel (the 'discoverer' of time travel) was convinced that the possibility of a block universe spacetime with CTLs/CTCs

---

[3]R. Gale, "Some Metaphysical Statements About Time," *Journal of Philosophy*, April 25, 1963, pp. 225-237. For many, this analogy may well bring to mind a pile of baloney rather than one of salami (and I think this was Gale's intention).

[4]J. J. C. Smart, "The Temporal Asymmetry of the World," *Analysis*, March 1954, pp. 79-83.

[5]Two events A and B are *non*-causally related if their separation in spacetime is such that a particle would have to travel at a *superluminal* speed (faster than light) to go from A to B. We'll discuss the physics of causally related events later in this chapter.

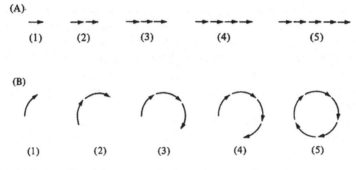

Fig. 3.1 Gödel's unreality of time argument

implies that the passage or flow of time *makes no sense!*[6] To see how Gödel arrived at such an astonishing conclusion, consider Fig. 3.1.

In part A of the figure[7] we see what most people who talk of a passage of time intuitively mean, as time progresses through a sequence of instants (shown as line of left-to-right arrows going from (1) to (2) to (3) and so on). At each stage the rightmost arrow is the present, and the arrows to the left of it (behind it) are the past, and the arrows to the right of the present are not shown because they are in the future and so don't exist yet. When you ask, at each step, which arrow is the *now*, the answer is clear.

Matters are dramatically different in part B of the figure, which shows the arrows forming a closed (circular) loop. Now there is no distinction between past and future, as each arrow is both ahead *and* behind any other arrow. In addition, there is no arrow that is uniquely the now. So, concluded Gödel, the passage of time can have no meaning in a temporal loop. As should come as no surprise, not everybody is convinced by this sort of argument.[8]

In principle, so it would seem, we can achieve perfect knowledge of what has happened but only imperfect prediction of what might happen. This observation seems to be at least a start at being able to tell past from future. And, in fact, the nature of the distinction between the two intervals of time seems obvious: we *remember* past events, but not future ones. As philosophers have so nicely put it, events in the past have formed *traces*, such as skulls, footprints in the sand, fossilized skeletons, surgical scars, photographs, taped recordings, carved stones,

---

[6]Gödel clearly states this in his 1949 philosophical essay (note 15 in the Introduction) concerning his 1949 technical paper (note 11 in the Introduction).

[7]This figure is based on the interpretation of Gödel's reasoning as presented by the philosopher Palle Yourgrau's 1991 book *The Disappearance of Time: Kurt Gödel and the idealistic tradition in philosophy* (Cambridge), which was expanded and reprinted a few years later under the new title *Gödel Meets Einstein: time travel in the Gödel universe*, Open Court 1999. Yourgrau later wrote a less technical version: *A World Without Time: the forgotten legacy of Gödel and Einstein*, Basic Books 2005.

[8]See, for example, S. Savitt, "Time Travel and Becoming," *The Monist*, July 2005, pp. 413-422.

and the like, whereas future events appear not to have formed traces. But is that necessarily so? Is it impossible for future events to create traces? The common-sense answer is *yes*, because of cause and effect, which dictates that there *must* be a temporal asymmetry in trace formation. That is, traces are the effects of *prior* causes. That line of reasoning leads us quickly to the fundamental issue of causation (which we've already encountered in Chap. 2), an issue that no discussion of time travel can avoid.

Part of the problem we have with backward time travel, and cause and effect, is as I've already mentioned, with language. The distinct and separate concepts of the temporal ordering of events, and of causality, have become merged in everyday thought. It is considered *obvious* to modern minds that if event A causes event B, then A *must* happen first. There is, however, at least one historical example of a similar merging of concepts that is parallel to our modern mixing of time order and causality—an example that shows how an issue can seem obvious and natural to the minds of one period of time, and yet to the minds of another period (our modern times) seem confused, odd, peculiar, even laughable.

As a *physicist* wrote in a paper[9] on advanced (that is, inverted causality) effects:

> Ancient Egypt was an essentially one-dimensional country strung out along the Nile, which flows from south to north. The winds were conveniently arranged to be predominantly northerly. To go north, a traveler could let his boat drift, while with a sail he could move south against the slow current. For this reason, in the writing of the ancient Egyptians, "go downstream (north)" was represented by a boat without sails, and "go upstream (south)" by a boat with sails. The words (and concepts) or north-south and up-downstream became merged. Since the Nile and its tributaries were the only rivers known to the ancient Egyptians, this caused no difficulties until they reached the Euphrates, which happened to flow from north to south. The resulting confusion in the ancient Egyptian mind is recorded for us to read today in their reference to "that inverted water which goes downstream (north) in going upstream (south)."

Often we can work our way free of the difficulties we create for ourselves with language, but only through common agreement. For example, the chairman of the board calls a meeting to order with mixed tenses by declaring "The meeting *will* take place *now*" and then saying at the end, "We will meet again *next* month, *same* time." We all know what these sentences mean, but only by our cultural heritage and not by the process of applying logic. The language problem causes similar difficulties for not only for fictional time travelers, but also for the physicists/philosophers who study the possibility of time machines. So—beware!

The idea of time flowing is a popular one, and it repeatedly appears in the time travel literature as the "river of time" or the "ocean of time." The deep psychological appeal of this sort of 'water language' has, not surprisingly, attracted the attention of philosophers. We can find one of the earliest expressions of the view in the *Meditations* of the second-century A.D. Roman emperor and Stoic philosopher Marcus Aurelius, who wrote: "Time is like a river made up of events which

---

[9]P. L. Csonka, "Advanced Effects in Particle Physics," *Physical Review*, April 1969, pp. 1266-1281.

happen, and a violent stream; for as soon as a thing has been, it is carried away, and another comes in its place, and this will be carried away, too." A most interesting essay on why such metaphors often seem so intuitively appropriate has been offered by one philosopher, who points out[10] that the seductiveness of the image of 'time as flowing water' is sufficiently great that you often find it in the scientific literature, too (Newton, you'll recall, wrote specifically in his *Principia* of time flowing). As for why such an image has such a powerful grip on our imaginations, I think we need look no further than to Kant. As he wrote (1781) in *Critique of Pure Reason*, "Time is nothing but the form of inner sense, that is, of the intuition of ourselves and of our inner state ... Because this inner intuition yields no shape, we endeavor to make up for this want by analogies." And what better than a rushing stream of water to represent our feeling of time rushing by?

Still, no matter how intuitive such water metaphors may be, they can still easily befuddle us as well. To quote our philosopher (note 10), "Time a river! A queer sort of river that. Of what sort of liquid does it consist? Is time a liquid? A very peculiar liquid indeed!" A classic paper by the philosopher Donald Williams (discussed in Chap. 2, note 119) expresses similar doubt about the water image of time. In the course of his writing, he presents a truly staggering collection of entertaining examples of 'time as metaphor,' of which I repeat just a few here: time flies, goes, marches, and rolls, as well as flows And then he offers *this* provocative imagery: the evolution of our lives is like "a moving picture film, unwinding from the dark reel of the future, projected briefly on the screen of the present, and rewound into the dark can of the past." Wow![11]

Returning to the water metaphor, the French astronomer Charles Nordmann (1881–1940) opened and closed his 1925 book *The Tyranny of Time* with following gloomy but all too true summary of the overwhelming sense we all have of the inexorable, one-way 'flow' of time. (The ellipses in what follows denote over 200 pages!) "Nothing can equal the bitter sweetness of dreaming on the banks of Time, that impalpable and fatal river strewn with dead leaves, our wistful hours carried downstream like rudderless wrecks ... In the eternal wave which rocks us, carries us along, and soon swallows us up, there is no rock to which we can fasten our frail barques; the very buoys we put out to measure our course are only floating mirages; and on the mysterious foundation of things our anchors slide along and fail to bite." A young person sees time, from Nordmann's perspective, as an ocean on which golden mornings arrive like waves from the future, whereas for an older person, liquid time is a nightmare flood, a swollen black torrent sweeping him first into the yawning abyss of the past and, ultimately and finally, into the eternal silence of the dark grave.

---

[10] J. J. C. Smart, "The River of Time," *Mind*, October 1949, pp. 483-494.

[11] And how about *this* image of time: Time is a snowball, with the center marking the beginning of the past, with ever new 'presents' accreting on the ever increasing surface as the snowball rolls down the hill of history!

The metaphor of time as a flowing river was ready-made for early science fiction writers, such as Caltech math professor Eric Temple Bell (1883–1960). His eventual novel *The Time Stream* began to appear in December 1931 as a serial in the science fiction pulp *Wonder Stories*, and Bell (writing as 'John Taine') made great use of the idea of time as a flowing stream, a stream in which one could swim into either the future or the past. There is strong evidence that Bell actually wrote the novel in July of 1921, but was unable to find a publisher for a decade, so odd did editors find the premise. By the time of its publication, others had beaten Bell into print.

The watery image of time had appeared a year earlier, for example, in a tale that played with the erosive nature of time in a dramatic way. As two time travelers speed into the future to rescue a friend, one of them describes the scene for us: "We huddled together in the whirling time girdling machine, cutting through the years as a ship's prow breasts surging waves. I could not help but think of the years as waves, beating in endless succession on the sands of eternity. They wore all away before them with pitiless attrition. Time seemed to eat all with dragon jaws."[12]

This image of time was taken a step further 3 years later in a story in which a large number of adventurers, from all across time, find themselves stranded at precisely the same place (in space and time). One of them offers his theory of what is behind this remarkable coincidence: They all have faulty time machines, like faulty boats, and all have hit the same snag on the 'river of time.' As he explains, "You may turn boats adrift on a river at many points, and they will all collect together at the same serious obstacle whether they have traveled a hundred or two miles. We are now at some period where the straight flow of time has been checked — perhaps it is even turning back on itself ... [We] have struck some barrier and been thrown up like so much jetsam."[13]

The 'flow' of time does have its critics, of course. The British-American philosopher Max Black (1909–1988) argued[14] that questions about the direction of time are meaningless because there can be no direction to something that (he asserted) does not flow. His reasoning was that if time does flow, then he ought to be entitled to ask *how fast* it flows. That requires, in turn, a *metatime* or *supertime* for measuring the flow rate of 'ordinary' time. But because supertime must flow, too, we would then need a super-supertime, and so off we trip into what would appear to be the black hole of a McTaggert-like infinite regress. The view, of an infinite regress of times, was forcefully rejected by another philosopher with

---

[12]E. A. Manley and W. Thode, "The Time Annihilator," *Wonder Stories*, November 1930. This is the same magazine that, months later, finally published Bell.

[13]J. Wyndham, "Wanderers of Time," *Wonder Stories*, March 1933. Notice again, that we have the same magazine (whose editor must have had a particular fancy for such tales).

[14]M. Black, "The 'Direction' of Time," *Analysis*, January 1959, pp. 54-63.

these sharp words: "the very idea of super (or hyper)-time is indeed repulsive in its redundancy and its aroma of dilettante physics."[15]

A hierarchy of hypertimes has not bothered other analysts, however, and an entire subfield of specialty among philosophers (and some physicists, too) in time analysis has developed in what is called *multidimensional time*. One practitioner in this specialty sarcastically rejected the infinite regress complaint as a valid objection—he called it "a crushing and unanswerable position" but actually meant just the opposite—and stated that it was not at all clear (at least, not to him) why supertime must flow.[16] After all, he argued, we measure the flow of a river with respect to its banks without requiring that the banks themselves flow. (That actually strikes me as being a point that deserves debating, but I have not been able to find any mention of it in the later philosophical literature.) The idea of multiple time dimensions is particularly attractive for one sort of time travel (we'll take it up at the end of this chapter), but it enjoys far more popularity among science fiction writers and philosophers than it does with physicists.

Professor Black's objection (note 14) to talk of time 'flowing' was based, at least in part, on the observation that there are uses of the word *direction* that are not directly tied to something flowing. For example, consider the statement 'He is facing in the direction of north.' Black argued that this is mere pointing, and it is not at all the same as *moving* north. He then dismissed the possibility of there being any meaning to the direction of time, writing that making an analogy of time "with a sign-post or an index finger is too far-fetched to be worth considering." This claim (which some may feel leans too much on grammar) is, of course, an affirmation of the myth-of-passage view made famous a few years earlier by Donald Williams (note 119 in Chap. 2).

Not just philosophers have rejected the idea of time flowing. In his 1966 novel *October the First Is Too Late*, which deals with a world in which different parts of Earth simultaneously experience different eras of the past (see *For Further Discussion* at the end of this chapter for more on what this might mean), the British cosmologist Fred Hoyle (1915–2001) calls the 'river of time' a "grotesque and absurd illusion," and a "bogus idea." Another fictional work that agrees with Hoyle's non-moving image of time is the 1979 *Roadmarks* by Roger Zelazny (1937–1995). In that novel we read of "the Road," along which story characters can travel but which doesn't itself move; exits from the Road lead to the various centuries (which sounds a lot like the Francis Bradley's 1883 book that may have given the block universe its name). *Roadmarks* is a clever bit of writing, with many allusions to the paradoxes of time travel, but its explanation of the Road's origin as having been constructed by *dragons* (!) greatly undermines its interest for physicists.

---

[15]D. Zeilicovici, "Temporal Becoming Minus the Moving-Now," *Nous*, September 1989, pp. 505-524.

[16]C. W. Webb, "Could Time Flow? If So, How Fast?" *Journal of Philosophy*, May 1960, pp. 357-365.

In a block universe spacetime, there is no flow of time, but one philosopher believed that to be simply because the block universe is *incomplete* in its representation of reality. Writing in 1925, Hans Reichenbach (1891–1953) asked "What does 'now' mean? Plato lived before me, and Napoleon IV will live after me. But which one of these three lives *now*? I understandably have a clear feeling that *I* live now. But does this assertion have an objective significance beyond my subjective experience?"[17] Reichenbach went on to answer his question in the affirmative, and to deduce that the block universe view is missing something: "In the condition of the world, a cross-section called the present is distinguished; the 'now' has objective significance. *Even when no human is alive any longer, there is a 'now'* [my emphasis] ... In the four-dimensional picture of the world, such as used by the theory of relativity, there is no such distinguished cross-section But this is due only to the fact that an essential content is omitted from this picture."

So, what *is* Reichenbach's 'missing essential content'? Feeling that the block universe is unacceptably fatalistic—in his words of ridicule, "the morrow has already occurred today in the same sense as yesterday"—he found his answer in the probabilistic theory of quantum mechanics. Classical physics argues that given total information about the state of the world *now*, one could in principle calculate perfectly the future or the past; one could both predict and retrodict. In contrast, quantum mechanics distinguishes past from future in a fundamental way.

Quantum mechanics does not deny that in principle we can know the past with exquisite accuracy, because each and every event leaves traces, evidence that is available to all with the means to find and decode them. But quantum mechanics also takes as truth that there is an unavoidable uncertainty to the future. The instant that this uncertainty is crystallized into fact was taken by Reichenbach to be the very definition of 'now.' The ever-increasing record of the past, in turn, defines (for Reichenbach) the *movement* of the 'now.' Reichenbach believed that with these observations he had at last captured the 'moving now' in mathematical theory, and that he had finally elevated the present from speculative psychology to solid physics, and that he had shown that the 'flow of time' is independent of the need for a conscious mind. However—

A later, powerful analysis[18] of the time-flow issue, combining philosophy with physics, comes down solidly in support of the *opposite* conclusion: it expresses the view that a 'moving now' *is* only in our minds and is *not* an intrinsic attribute of reality. The premise of that argument is that a mind-dependent flow of time is incompatible with what is called the *relativity of simultaneity* (to be discussed later in this chapter) which states that there is no universal cosmic-wide 'now' (this is a fundamental conclusion of special relativity). For example, it is meaningless to ask

---

[17]I've taken this quotation from A. Grünbaum, "Is There a 'Flow' of Time or Temporal Becoming?" in *Philosophical Problems of Space and Time*, Knopf 1963.

[18]L. R. Baker, "Temporal Becoming: The Argument from Physics," *Philosophical Forum*, Spring 1975, pp. 218-236.

what is happening on a planet in the Andromeda galaxy (two million light-years distant) *right now*.

Early science fiction stories are full of theories about the nature of 'now,' and the vast majority of them have no basis in scientific thought. Some of them are ingenious, however, and even though they are largely the pet ideas of the authors (and no one else's), perhaps they resulted in some young readers of the science fiction pulps of the 1930s and 1940s thinking about deeper matters than did the comic strips of "Buck Rogers," "The Lone Ranger," or "Terry and the Pirates." For example, according to one story, time is a wave and the 'moving now' we experience is carried on a crest of that wave. There are time waves both ahead and behind the crest we happen to be on (so we are told), and so each such crest carries a different 'now' for a different reality—hence the curious title.[19]

In another, more recent tale[20] about object duplication via time travel (which we'll discuss in Chap. 4, but you'll recall H. G. Wells was worried about this long ago), nine (!) copies of the same person from the year 2314 meet in 1870 to try and figure out what is going on. Part of their interesting discussion is the following analysis of the 'present':

"Gentlemen, I think I understand," said the first James Thomas."

"Eight faces turned toward him, and he felt as though he were looking into multiple mirrors."

"We hold that time is a single instant — the instant of the Present —which travels through Duration — do we not?"

"Eight heads nodded."

"We assume that time passes in a manner analogous to the stringing of an infinite number of beads. Each bead is the instant of Now when it is last on the chain. Beads are continually being added, and each one is the only Now until another is placed after it."

"Yes, that is my theory," said another James Thomas. "It can also be likened to the process of knitting. No matter how many stitches are knitted, there is only one last stitch, only one Now."

Einstein, too, was greatly bothered by the place of 'now' in time, perhaps even more than were James Thomas and his 'friends.' In an autobiographical essay, the philosopher Rudolf Carnap (1891–1970) recalled a conversation about this with Einstein in the early 1950s, at the Institute for Advanced Study in Princeton: "Once Einstein said that the problem of the Now worried him seriously. He explained that the experience of the Now means something special for man, something essentially different from the past and the future. That this experience cannot be grasped by

[19]R. Ray, "Today's Yesterday," *Wonder Stories*, January 1934.

[20]A. and P. Eisentein, "The Trouble With the Past," in *New Dimensions 1* (R. Silverberg, editor), Doubleday 1971.

science seemed to him a matter of painful but inevitable resignation. ... Einstein thought ... that there is something essential about the Now which is just outside the realm of science."[21]

3.2   The Arrows of Time

"On a microscopic level there is no preferred direction for time. The equations of motion don't give a damn whether time moves forward or backward."[22]

The central issue for philosophers of time (and for physicists, too, I think) is that of its reality (or not): is time *objective and something that really flows*, or is time simply a *mind-dependent illusion* and nothing more than an artifact of our incomplete perception of reality? As the previous section shows, there is little consensus on this issue. As a start on trying to get a handle on the matter, looking into a so-called 'arrow of time' may give us some guidance. I'll begin with the arrow I've already mentioned, the psychological arrow. As discussed before, this is the feeling we have of a 'moving now,' a feeling that has no appearance anywhere in physics. A 'moving now' simply has no place in any universe devoid of the physical processes in a brain that give rise to what we call consciousness—but that doesn't mean physicists don't wonder about the 'moving now' just as much as does everybody else (remember Einstein)! As one physicist wrote in a technical journal, "What does 'Now' mean? This question must surely be the starting point of any attempt at understanding the nature of time."[23]

Well, no matter whether time actually flows or not, most of us still believe we have had a past and hope we will have a future. Each of us thinks we can easily tell one from the other, too. We have, in fact, many not so subtle indications from our everyday lives of the obvious direction of time. Nearly all of these indications have the common theme of *irreversible change*. As the British mathematician J. J. Sylvester once put it, "The whirligig of time brings about its revenges."[24] The Roman poet Ovid, who died when Christ was a teenager, said the same in his *Metamorphoses* with the famous words "Time, the devourer of all things." The

---

[21]Quoted from *The Philosophy of Rudolp Carnap* (P. A. Schlipp, editor), The Library of Living Philosophers, Open Court 1963, pp. 37-38. For a view contrary to Einstein's, from another physicist, see K. B. M. Nor, "A Topological Explanation for Three Properties of Time," *Il Nuovo Cimento B*, January 1992, pp. 65-70, which claims to develop a geometrical explanation for the flow of time, and so (says Nor) there *is* an objective, mathematical reality to the 'moving now.'

[22]A science fiction character pretty accurately sums-up what a modern physicist would tell you today, in L. Eisenberg's story "The Time of His Life," *The Magazine of Fantasy & Science Fiction*, April 1968.

[23]J. P. Cullerne, "Free Will and the Resolution of Time Travel Paradoxes," *Contemporary Physics*, July-August 2001, pp. 243-245.

[24]J. J. Sylvester, "A Plea for the Mathematician," *Nature*, December 30, 1869, pp. 237-239.

image of time as devourer of all that is mortal was brilliantly presented by James Barrie in his *Peter Pan*, with the crocodile who had swallowed a ticking clock chasing Captain Hook all about Neverland.

No one yet has escaped the biological decay processes of time, and inanimate objects are no less immune to this aspect of time. Logs and cigarettes burn in the stove and ashtray, but they never unburn. Our cars rust but never 'unrust.' An explosion has never been seen to reverse itself, to form a dynamite stick or a bomb casing out of a collapsing fireball. Our world seems, indeed, literally to be built on an irreversible movement toward chaos, death, and decay. Lewis Carroll uses this observation in his *Through the Looking-Glass* when Alice tells Humpty Dumpty "one can't help growing older." And speaking of Humpty Dumpty, his famous fall provides a dramatic example of a one-way evolution from past to future; he wasn't at all convinced that Alice was correct but, once he had splattered, then

All the King's horses and all the King's men
Couldn't put Humpty Dumpty together again.

While we are on the subject of Mr. Dumpty, it is also worthwhile to note that nobody has ever figured out how to unscramble an egg. Why is that? One philosopher speculated that the answer is found in the "irreversible organic phenomena" taking place in our brains which results in our flow of consciousness always being in the same direction.[25]

More subtle than the undignified undoing of a prideful egg is the phenomenon of memory, which seems trivial only because most people have not thought very carefully about it. We remember the past while remembering nothing about the future. We might, in fact, be tempted to use the phenomenon of memory to answer the question of how to tell past from future. Anything you can remember *is* the past. But that is a circular definition, as discussed by Professor Smart (note 4) who observed that to ask why memory is always of the past "is as foolish as to ask why uncles are always male, never female." In *Through the Looking-Glass* the White Queen tells Alice that "it's a poor sort of memory that only works backward" but, except for the claims of clairvoyants, it seems that is the only sort of memory any of us has. Why is that so? Of course, that would *not* be the case for a time traveler while in the past. His personal past, which he would remember, would be the future for the world around him.

For physicists, the question of the direction of time is one of profound mystery. There seems, in fact, to be no fundamental reason why time should not be able to go from future to past—but then what would 'future' and 'past' *mean*—even though no one has ever observed time to do so. All the laws of classical physics, including general relativity, and quantum mechanics, too (except for the K-mesons mentioned in Chap. 1) involve time in such a way that they ignore its sign. In other words, replacing t with $-t$ results in a perfectly valid description of something that could actually happen. But not all such possibilities are observed to occur. Why not?

---

[25]H. Margenau, "Can Time Flow Backwards?" *Philosophy of Science*, April 1954, pp. 79-92.

In an unpublished paper written in 1949, while doing the work that would bring him a share of the 1965 Nobel prize in physics, Richard Feynman (1918–1988) wrote[26] "The relation of time in physics to that of gross experience has suffered many changes in the history of physics. The obvious difference of past and future does not appear in physical time for microscopic events . . . Einstein discovered that the present is not the same for all people [the relativity of simultaneity, to be discussed later in this chapter] . . . It may prove useful in physics to consider events in all of time at once and to imagine that we at each instant are only aware of those that lie behind us. The complete relation of this concept of physical time to the time of experience and causality is a physical problem which has not been worked out in detail. It may be that more problems and difficulties are produced than are solved by such a point of view."

Feynman did not elaborate on what he meant by the "problems and difficulties" with that point of view (which is clearly that of the block universe), but surely he had the logical paradoxes of time travel high on his list. As the Yale philosopher Henry Margenau wrote (note 25) in a tutorial on Feynman's work, "The theory of quantum electrodynamics developed by Feynman incorporates reversals in the course of time and thereby cherishes, in the minds of many, *an age-old phantasy* [my emphasis] of more than scientific appeal [which sounds like time travel to me]."

Because the individual classical equations of microscopic physics are time-reversible, the distinction between past and future for individual particles disappears. The equations are said to be symmetric with respect to time; the algebraic sign of *t* is irrelevant in the classical laws. It must be understood, however, that there *is* a crucial point to appreciate. When a physicist says *time reversal*, she is talking about a system evolving backward in forward time—that is, all the individual particle velocity vectors are instantly reversed at once. This is distinct from the time-reversed worlds of philosophers and science fiction writers (which we'll get into later in this chapter) in which time itself 'runs backwards.' The physicist's point of view is clearly expressed in an early essay by a chemist: "Every equation and every explanation used in physics must be compatible with the symmetry of time. Thus we can no longer regard effect as subsequent to cause. If we think of the present as pushed into existence by the past, we must in precisely the same sense think of it pulled into existence by the future."[27] More than three decades later, a mathematician and a physicist presented a similar statement: "In classical dynamics, the past completely determines the present, and therefore, by symmetry, the future also completely determines the present."[28]

[26]S. S. Schweber, "Feynman and the Visualization of Space-Time Processes," *Reviews of Modern Physics*, April 1986, pp. 449-508.

[27]G. N. Lewis, "The Symmetry of Time in Physics," *Science*, June 6, 1930, pp. 569-577.

[28]O. Penrose and I. C. Percival, "The Direction of Time," *Proceedings of the Physical Society* (London), March 1962, pp. 605-616.

Besides the physics, there is also an interesting theological connection to time reversal. As one philosopher put it, "If all the laws are time reversal invariant and so no irreversible processes occur in the physical Universe then there is no inherent, intrinsically meaningful difference between past and future . . . If this is actually the natural case, then all mankind's major religions which preach a creation of the Universe (by a supernatural agency) and imply, accordingly, a differentiation between the past and the future . . . would have to make appropriate adjustments."[29]

There are, as you might expect on such a controversial topic, dissenters to the view that the classical laws of physics are necessarily time-reversible. Dirac himself wrote that "I do not believe there is any need for physical laws to be invariant under time and space reflections, although all the exact laws of nature so far known do have this invariance."[30] Dirac did not, unfortunately, elaborate on just why he felt that way, but with the later discovery of K-mesons his position is seen to have been 'ahead of its time'! In a famous science fiction story[31] dealing with the direction of time, one character finally puts his finger on the real puzzle of the question of time: "How can a man live backward? You might as well ask the Universe to run in reverse entropy." That cogent question brings us, in fact, to the first scientific explanation developed to explain the observed asymmetric nature of time.

It was the Englishman A. S. Eddington (1882–1944) who gave the picturesque name, the *arrow of time*, to the observed asymmetric nature of time's direction from past to future. He was also one of the popularizers of an explanation for the arrow, using the famous second law of thermodynamics.[32] The second law of thermodynamics states that a measure of the internal randomness or disorder—what is called the *entropy*—of any closed system (that is, one free of external influences) continually evolves toward that of maximum disorder, toward the condition called *thermodynamic equilibrium*. Indeed, so striking is this increase in entropy S with time in a macroscopically large system that the increase in entropy has come to be thought of as actually *defining* the direction of time. Eddington, however, was not the originator of the entropy concept. The history of entropy can be traced back to before the turn of the century, to the great Austrian scientist Ludwig Boltzmann (1844–1906) and his famous H-theorem. The quantity H in that theorem is directly related to the more familiar entropy,[33] defined by Boltzmann in 1877.

---

[29]H. Mehlberg, "Philosophical Aspects of Physical Time," in *Basic Issues in the Philosophy of Time* (E. Freeman and W. Sellars, editors), Open Court 1971.

[30]P. A. M. Dirac, "Forms of Relativistic Dynamics," *Reviews of Modern Physics*, July 1949, pp. 392-399.

[31]A. Boucher, "The Chronokinesis of Jonathan Hull," *Astounding Science Fiction*, June 1946.

[32]A. S. Eddington, *The Nature of the Physical World*, Macmillan 1929.

[33]The H-theorem was a direct continuation of the work by the Scottish physicist James Clerk Maxwell (1831-1879) on the statistical properties of gas molecules (determining the probability density function of the molecules' speeds). In 1866 Maxwell found this function for the particular case of thermodynamic equilibrium. In 1872 Boltzmann found the differential-integral equation the function satisfies in general, even if the condition of thermodynamic equilibrium doesn't hold. From this Boltzmann was able to define a quantity H that he showed evolves in time such that

The entropy S of a system in a given state is proportional to W, which is the number of different possible ways the state can occur as a result of all possible variations of system's internal, microscopic structure. The calculation of W is usually quite complicated, but in various highly idealized systems it can be straightforward. Consider, for example, a vacuum cylinder with a thin membrane dividing the interior into halves. Suppose that we insert (to be specific) six molecules into the left half of the cylinder (and none into the right half). If we define the microscopic state of the system to be the number of molecules in the left half, then initially $W = 1$ because there is just one way to put all six molecules on the left side. This represents the state of *minimum* entropy, the state of maximum order that is most distant from thermodynamic equilibrium. If we now puncture the membrane then the molecules, once confined to the left side, are free to move about the entire cylinder. At any given instant we can imagine counting the number of molecules on the left side—suppose that at some particular instant we count five, with one molecule having moved to the right side. Then, $W = 6$, because there are six ways to pick the molecule that has moved from left to right, and so the entropy has increased.

We think of the thermodynamic equilibrium state as being the state with equal numbers of molecules in both halves of the cylinder, and that state has the *maximum* entropy. (Can you show that this state is associated with $W = 20$?) With such a small number of molecules, it is *not* clear that W (and so S) will *inexorably* increase with time; perhaps, after one of the six molecules has gone to the right, it then returns to the left side before any of its companions have joined it on the right. Such an event is called a *reversal*, and it will happen with some non-zero probability. But the more molecules there are in the cylinder (instead of six, make the number a million million million—still a small amount of gas in our everyday world, hardly enough to fill a sewing thimble), the more likely it becomes that the value of S *will* monotonically increase with time.

The steady increase in entropy is often observed in the everyday, large-scale world. A drop of ink in a glass of water spreads out in an expanding cloud, a cloud we *never* see collapse backward into an ink drop. A long rod of metal, initially hotter at one end than at the other, evolves toward a constant temperature along its entire length. We *never* see a uniformly warm rod spontaneously begin to cool at one end and grow hot at the other. A hot bath grows cold—nobody has *ever* seen a bath at room temperature suddenly, all by itself, begin to heat up and then boil in the middle of the tub while the edges freeze into ice chunks. In all of these cases, the end (future) state represents greater internal randomness or disorder than does the beginning (past) state.

---

solution to his differential-integral equation approaches Maxwell's equilibrium solution. The H-theorem says that H always *decreases* in systems not in equilibrium and is at a *minimum* in systems in equilibrium.

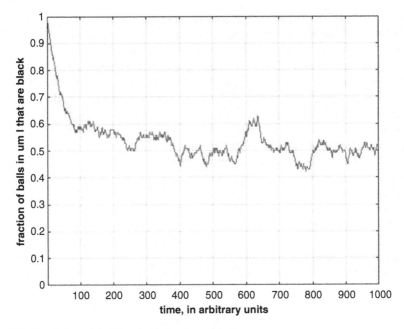

Fig. 3.2 Simulation of the Ehrenfest entropic gas clock

That is, low entropy *was* the past, and high entropy *will be* the future. The increase in entropy seems to define a direction to time, and so entropy has come to be called the *thermodynamic arrow of time*.

The first formal entropy model for the direction of time was put forth in a 1907 paper by the Austrian physicist Paul Ehrenfest (1880–1933) who was a friend of Einstein, and his Russian-born wife Tatyana (1880–1964), who was a skilled mathematician and her husband's occasional collaborator. In their paper the Ehrenfests developed one of the mainstays of physics, the so-called *entropic clock*. This clock, a statistical model based on the then new probability mathematics of Markov chains—after the Russian mathematician A. A. Markov (1856–1922)—describes how gases diffuse, and it is both a simple and a powerful concept. The Ehrenfest model is illustrated in Fig. 3.2, in a computer-generated plot based on a discussion by Princeton physicist John Wheeler of black hole fame (see Chap. 1).[34]

Imagine two urns, I and II, each containing *n* balls. Initially, at time $t = 0$, all of the balls in Urn I are black and all of the balls in Urn II are white. Then, at time $t = 1$ (in arbitrary units), a ball is selected at random from each urn and (instantaneously) placed in the other urn. This select-and-transfer process is repeated at times $t = 2$,

---

[34]J. A. Wheeler, "Frontiers of Time," in *Problems in the Foundations of Physics* (G. T. diFrancia, editor), *Proceedings of the International School of Physics* (Course 72), North-Holland 1979. See also W. J. Cocke, "Statistical Time Symmetry and Two-Time Boundary Conditions in Physics and Cosmology," *Physical Review*, August 25, 1967, pp. 1165-1170.

3, 4, ... At any given time each urn always contains n balls, but only at $t = 0$ are the colors of all the balls in a given urn necessarily the same. The phrase "selected at random" means (for example) that the probability of selecting a black ball from an urn containing b black balls is b/n. At any given time we completely describe the state of both urns by specifying the number of black balls in Urn I (or the number of white balls in Urn II, and so on). It is easy to write a computer simulation of this physical process,[35] and Fig. 3.2 shows how the fraction of black balls in Urn I evolves toward 0.5 as time increases. The plot is for $n = 100$ (200 balls total). The important observations are that (1) the evolution of the state of the system is toward 50 % black balls in Urn I (and this would be the case for 'almost all' sequences of random selections of the balls from the urns), and (2) the evolution is not monotonically decreasing from 100 % black balls to 50 % black balls, but rather has never-ending fluctuations about 50 % that may, in fact, be rather large in both amplitude and in duration.

There is a real puzzle with the entropic clock that may not be immediately apparent. The motion of each of the individual molecules is described by time-reversible physics, but when we average over 'many' molecules (assuming 200 molecules is 'many') we lose detailed information about the individual molecules. The puzzle is then how is it that by *reducing* our knowledge of a system, through statistical averaging, we then find it displaying a *new property*, that of asymmetric time evolution, that we didn't see before when we watched the individual molecules. And if that question isn't troublesome enough, we also have two additional puzzles called the 'reversibility' and the 'recurrence' paradoxes to consider as well.

The reversibility paradox is the question raised earlier: the classical equations of physics work just as well with time running in either direction, and so *why don't* things actually go 'backward'? This question, originally raised by the British mathematical physicist and engineer William Thomson (1824–1907)—better known as Lord Kelvin—in 1874, was brought to Boltzmann's attention in 1876 by the German physical chemist Johann Loschmidt (1821–1895), one of Boltzmann's professors at the University of Vienna. Boltzmann's answer to this apparent paradox was that it *is* imaginable that a world could run backward if initial conditions were suitable. For example, if all the velocity vectors of every particle in an equilibrium state were reversed, then the system *would* unwind backward in time toward its original non-equilibrium condition. That is, a system in thermodynamic equilibrium, the state of highest entropy, could evolve toward one of low entropy. Boltzmann even suggested that such might be the case for regions in our own universe, that there might actually be beings in a world somewhere 'out there' who

---

[35]I used MATLAB, and you can find the code — gasclock.m — in Appendix C, written in such a low-level way as to be virtually 100% transferable to just about any of the popular scientific programming languages, and easily executed on an inexpensive laptop. Note that there are no K-mesons in the code (!) and so, as stated in Chapter 1, they aren't responsible for the uni-directional time behavior depicted in the figure.

experience time running counter to our earthly experience. He said that in 1877, and it is a remarkable statement for a conservative nineteenth-century professor.[36] However, Boltzmann continued, from most given states there are vastly more ways for entropy to increase than there are for it to decrease, and that is why we see what we see, a continuous *increase* in entropy.[37]

To find a science fiction writer speculating on reversed time people is, of course, much less remarkable! One pulp story, in fact, presents a curious treatment of the nuances of reversed time in which people talk backward (along with a marvelous bathroom scene of a man un-washing his hands!). This tale[38] tells us a young physics teacher who is "twisted into a reversed Time Stream" by an electrical discharge. As he lives backward in time, he observes everybody about him appearing to run in reverse, but even more puzzling is that they have developed a "dreadful, granite-like hardness." We soon learn why:

"For a while he could not understand the impenetrable hardness of external objects which he had experienced; it seemed they ought rather to be of intangible transiency, much as a dream, since he was re-viewing the Past. But a moment's thought gave him the logical answer. The Past is definite, shaped, unalterable, as nothing else in Creation is. Therefore, to argue that he could move or alter any object here [the past] was to argue that he could change the whole history of the world or cosmos. Everything he saw about him had happened, and could not be changed in any way. On the other hand, he was fluid, movable, alterable, since *his* future still lay before him, even if it had been reversed; he was the intruder, the anomaly. In any clash between himself and the Past, the Past would prove irresistible every time."

This passage reflects the modern view that the past cannot be changed, but explains that view in a way different from that generally accepted today. Modern physicists and philosophers invoke consistency requirements (which we'll take up in the next chapter) to explain the 'solidity' of the past. The author of this story also failed to explain why his physics teacher had no trouble moving about through the air of the past, which apparently is not any more resistant to being displaced than were air molecules before time reversal occurred.

---

[36]The Austrian-British philosopher Karl Popper (1902-1994) called Boltzmann's willingness to consider the possibility that different regions of the universe could have different directions of time "staggering in its boldness and beauty," but when on to say that Boltzmann must be wrong because "it brands unidirectional change an illusion [which] makes the catastrophe of Hiroshima an illusion." That is an emotional argument, of course, and although one of great power, I fail to see how it is related to physics. See Volume 1 of *The Philosophy of Karl Popper* (P. A. Schlipp, editor), Open Court 1974, pp. 127-128.

[37]For more on Boltzmann's views on entropy, see the end of his letter "On Certain Questions of the Theory of Gases," *Nature*, February 28, 1895, pp. 413-415.

[38]C. F. Hall, "The Man Who Lived Backwards," *Tales of Wonder*, Summer 1938. The modern classic of a time-reversed world is Philip K. Dick's 1967 novel *Counter-Clock World*. We'll encounter another time-reversed world again in Chapter 4.

It didn't take long for science fiction writers to incorporate entropy as time's arrow into time travel. In one early tale there is the brief statement that entropy is behind the operation of its gadget.[39] And a few years later the inventor of a "warp gun" tells us that "The stupendous distortion of the warp may actually bring about a sort of kink in spacetime, and result in a reversal of entropy"[40] and, sure enough, when the gun is fired a woman, who is hit by the warp, ages 70 years in seconds (which is, of course, exactly the opposite of what we would expect from a "reversal of entropy"!). Just a year later, the story of a college student about to flunk his senior physics course appeared.[41] An examination is scheduled for the following day, but he needs a week and a half of study time. To his rescue comes ENTROPY, INC., a company that sells time by placing its clients inside a "time-cabinet" in which the local entropy is greatly accelerated. To someone looking through a window at the interior of the time-cabinet, the occupants would appear as characters in a speeded-up movie. Referring to Eddington by name, the author tells us that "entropy is what makes time irreversible — is what gives us the feeling of the flow of time."[42]

In a hilarious, melodramatic story featuring one of early science fiction's stereotypical 'mad scientists,' the entropic arrow of time is the scientific explanation for time travel.[43] There we read of Bryce Field, "a master-scientist, a demon, cruel, ruthless," who is rejected in love by the stupendously beautiful Lucy Grantham. Her lack of enthusiasm is perhaps understandable, as Bryce is described as having "a lean-jawed, sunken-eyed" appearance, along with "lank, untidy hair sprawled across his massive forehead." As Lucy tells him at one point, "I could never love you; you are too clever, too brilliantly scientific." After hearing that, it is no surprise that before we are more than a page or two into the tale that we learn Bryce has Lucy strapped to a steel table in an underground laboratory-in-a-cave. There he tells her of her fate: "You are going on a long journey, my dear. So long a journey that even I, master-scientist, do not know when it will end. A journey into the future — alone! ... You, Lucy, shall be the victim of entropy! ... I have discovered how to make a [globe] of non-time. Entropy will be halted ... You will be plunged into an eternal 'now.'"

And so the mad Doctor Field throws the switch on the wall of his "instrument-littered" cave on July 17, 1941, and Lucy remains "suspended" in time until the outside world reaches the date of August 9, 2450. That is the day she is at last dug-up from the cave by "big and muscular" engineer Clem Bradley and his "square-jawed" sidekick Buck Cardew, who uses a "warp in spacetime" to release Lucy from her "globe of non-time."

---

[39]M. J. Breuer, "The Time Valve," *Wonder Stories*, July 1930.

[40]F. B. Long, "Temporary Warp," *Astounding Stories*, August 1937.

[41]R. M. Farley, "Time for Sale," *Amazing Stories*, August 1938.

[42]Also citing Eddington was a tale by D. W. O'Brien, "The Man Who Lived Next Week," *Amazing Stories*, March 1941, which uses entropy to explain time travel. This curious story has the traveler arriving in the future with his clothing aged, which later 'de-ages' when the return trip is made!

[43]P. Cross, "Prisoner of Time," *Super Science Stories*, May 1942.

A few years later entropy was used in a similar but vastly more 'scientific' way. In that tale[44] we read of a scientist who has discovered "a field in which entropy was held level." As the reader is told, "An object in such a field could not experience any time flow — for it, time would not exist," since time flow is a *change* in entropy, and the 'change' of a level (or constant) field is zero. This interesting tale speculates on how such a field could have fantastic home uses ("Imagine cooking a chicken dinner, putting it in the field, and taking it out piping hot whenever needed, maybe twenty years hence!"). But its real use in the story is as a stasis generator for preserving fatally ill people until medical science has learned how to cure their diseases. This is, then, a high-tech method of suspended animation, of time travel into the future that is different from simply freezing (a clock in such a field would not age or measure the passage of personal time).

The gadget that does all this is called, somewhat sinisterly, the "Crypt," which we are told also makes a great bomb shelter, too, because "not even an atom bomb could penetrate a stasis field." The reason for that is intriguing: "The field requires a finite time in which to collapse — only there is no time in it." The interior of the Crypt is, quite literally, a frozen block of time more rigid and unyielding than the strongest steel.

As science fiction left the age of pulps and moved into the modern era, entropy continued to be useful a justification for time travel. Arthur C. Clarke used it,[45] as did Robert Silverberg. This last tale[46] is particularly interesting, as Silverberg pursued entropy beyond simply invoking it as a mere casual throwaway mention. When a newspaper from the future appears on people's doorsteps, the initial astonishment is replaced with puzzlement as the papers rapidly disintegrate. That is the result (we are told) of "entropic creep." The explanation continues, informing us that it is sort of like a strain in a geological fault (Silverberg has lived for decades in California, now and then a place of large to huge earthquakes, and it isn't surprising that he uses this particular imagery): "Entropy you know is the natural tendency of everything in nature to come apart at the seams as time goes along. These newspapers must be subject to unusually strong entropic strains because of their anomalous position out of their proper place in time."

Earlier I mentioned we had *two* puzzles associated with entropy; we've discussed 'reversibility,' so what's the other one, the 'recurrence paradox,' all about? The recurrence paradox is quite different from reversibility; it is based on a result established in 1890 by the great French mathematician Henri Poincaré (1854–1912). Motivated by the question of the stability of the motion of three masses governed by Newton's laws of mechanics (think, for example, of the Sun, the Earth, and the Moon), Poincaré showed that starting from almost any initial state, any fixed volume system with a finite amount of energy and a finite number of

---

[44]P. Anderson, "Time Heals," *Astounding Science Fiction*, October 1949.

[45]In, for example, his story of the tragic end of a geologist fifty million years in the past: "Time's Arrow," *Science Fantasy*, Summer 1950.

[46]"What We Learned from This Morning's Newspaper," *Infinity 4*, November 1972.

degrees of freedom will return *infinitely often and with arbitrarily little deviation* to almost every previous state. If you wait long enough, implies Poincaré's astonishing theorem, Pearl Harbor will happen again—and again, and again, and .... In 1896 the German mathematician Ernst Zermelo (1871–1953) used this result, which philosophers call the 'eternal return,' to claim that there could be no truly irreversible processes and thereby cast doubt on the idea that entropy *always* and *inexorably* increases.

Even for very small systems, however, such as a mere handful of molecules, the recurrence time is extremely large, and this was, in essence, Boltzmann's reply to Zermelo's concern. For example, if the gas-filled cylinder of our entropic clock has just 100 molecules (not the six I used in the earlier example), and if transitions from one side of the cylinder to the other side take place at the rate of one million per second, then the recurrence time has been calculated to be something like 30 million billion years![47] And for the universe itself, the recurrence time is simply incomprehensible. Mathematicians call 1 followed by a hundred zeros a *googol*, and the recurrence time in years for the universe has been estimated to be 1 followed by a googol of zeros (a so-called *googolplex* of years).[48]

Using a wonderful bit of imagery, one analyst wrote of the enormity of the recurrence time of a system considerably *less* complex than the universe this way: "If a man shuffled just a single pack of cards as rapidly as an individual molecule hits other molecules in air, and if a snail started to crawl around the universe ... at the rate of one centimeter *during the life of the sidereal system* [my emphasis], the snail would have got round the universe many millions of times before it would become at all likely that the man would have got the pack back to the original order."[49] If this is what it takes to get a pack of cards back to its initial state, then try to conceive of the time interval required to restore *the world* to December 7, 1941.[50]

The notion of eternal recurrence considerably predates Poincaré, and its scientific (as opposed to astrological) study can be traced back to the fourteenth century.[51] A 'more recent' claim for eternal recurrence, also based on scientific arguments (conservation of energy), can be found in many places in the writings of the German philosopher Friedrich Nietzsche (1844–1900)—see, for example, his *The Gay Science* (1882) and *Thus Spake Zarathustra* (1883)—again predating

---

[47]J. M. Blatt, "Time Reversal," *Scientific American*, August 1956.

[48]The googol is a gigantic number, far greater than the number of raindrops that have fallen on the Earth during its entire history. And the googolplex is light years beyond that.

[49]R. B. Braithwaite, "Professor Eddington's Gifford Lectures," *Mind*, October 1929, pp. 409-435.

[50]Pulp science fiction writers, of course, were not discouraged by such calculations, as they depended on the *certainty* of recurrence over *infinite* time. See, for example, S. G. Weinbaum, "The Circle of Zero," *Thrilling Wonder Stories*, August 1936, and L. D. Gunn, "The Time Twin," *Thrilling Wonder Stories*, August 1939.

[51]R. Small, "Incommensurability and Recurrence: From Oresme to Simmel," *Journal of the History of Ideas*, January-March 1991, pp. 121-137.

Poincaré. All of Nietzsche's arguments are flawed,[52] but they are rational, physical arguments, as opposed to arguments based on metaphysics or theology. In fiction, a glimmer of the idea of a repetition of human affairs preceded Poincaré by some years, too.[53]

An important caveat concerning recurrence is that we could never *know* of it because the state of all the historical records (geological, memories, books, photographs, and so on) would, as part of the physical state of the universe, also recur. And so those records could, up to the instant *before* the recurrence, contain no signature *of* the recurrence because the recurrence has not 'yet' happened![54] The 1993 movie *Groundhog Day* stumbles on this point, as it has a character (for some unexplained reason) live through the same day over-and-over *and he is aware he is doing that*. Indeed, he can change events within that time loop at will. It is interesting to note that one 'time loop' pulp science fiction tale[55] specifically avoided that error (and cited Nietzsche, to boot), and so demonstrated that pulp science fiction *could* have some philosophical merit to it.

While the enormous recurrence time for the universe may seem reason enough to reject the possibility of Pearl Harbor repeating, there are more fundamental reasons for such a rejection. For example, an *expanding* universe, such as the one we live in, violates the Poincaré theorem's assumed condition of a *fixed* volume system. As Professor Eddington put it in a 1934 lecture at Cornell University, "In an expanding space any particular congruence becomes more and more improbable. The expansion of the Universe creates new possibilities of distribution faster than the atoms can work through them, and there is no longer any likelihood of a particular distribution being repeated. If we continue shuffling a pack of cards we are bound sometime to bring them into their standard form — but not if the conditions are that every morning one more card is added to the pack."[56]

An even more direct way to escape Poincaré's theorem is to use a result from general relativity. Using Einstein's theory instead of the classical dynamics that Poincaré used, it has been shown (by Frank Tipler, the inventor of the rotating cylinder time machine spacetime that was mentioned in the previous chapters and which we'll revisit later in the book) that the recurrence theorem is simply no longer true.[57] As Tipler wrote, "In general relativity, singularities intervene to prevent recurrence. General relativistic Universes are thought to begin and end in singularities of infinite spacetime curvature [the Big Bang and the Big Crunch,

---

[52] J. Krueger, "Nietzschean Recurrence as a Cosmological Hypothesis," *Journal of the History of Philosophy*, October 1978, pp. 435-444.

[53] See "Human Repetends" by Marcus Clarke (1846-1881), a story originally published in 1872 and reprinted *Australian Science Fiction* (V. Ikin, editor), Academy Chicago 1984.

[54] For more on this point, see D. W. Theobald, "On the Recurrence of Things Past," *Mind*, January 1976, pp. 107-111.

[55] C. F. Ksanda, "Forever Is Today," *Thrilling Wonder Stories*, Summer 1946.

[56] A. S. Eddington, "The End of the World," in *New Pathways in Science*, Macmillan 1935.

[57] F. J. Tipler, "General Relativity and the Eternal Return," in *Essays in General Relativity* (F. J. Tipler, editor), Academic Press 1980.

respectively], and these singularities force time in general relativity to be linear rather than cyclic." A twist to this, however, is that in his analysis Tipler assumed that gravity is *always* attractive, and that the spacetime satisfies a special condition (called the *Cauchy condition* that we'll take-up later) that avoids backward causation. The first assumption is violated in wormhole time machine spacetimes, though, and the second is *by definition* violated in *any* spacetime that supports time travel! So, who knows . .... ?

Despite all of the previous discussion it is *not* true that the evolution of a system from past to future is *always* accompanied by an increase in entropy—that is, by an irreversible increase in some measure of the system's 'disorder.' Yes, it can be calculated that entropy is *very likely* to monotonically increase in systems of macroscopic size, but that is not the same as certainty. There *can* be fluctuations in the thermodynamic evolution of a system so as to have, at least for a while, a *decrease* in entropy (take another look at Fig. 3.2). All we can say, for sure, is that for macroscopically sized systems even very small fluctuations in increasing entropy are most improbable. To quote no less an authority than the combined genius of Gilbert and Sullivan (from their opera *H. M. S. Pinafore*), here's what we can honestly say of the possibility of failure in the supposed inexorable increase of entropy: "What, never?/No, never!/What, never?/Well, hardly ever." Still, for physicists, entropy is just too useful a concept to give up even though it does not *always* increase with increasing time for an isolated system.

Love it though they may, there are some puzzling aspects to entropy for physicists that remain to this day. For example, the idea that the universe began in some sort of Big Bang process 15 billion years or so ago is the generally accepted view today, The puzzle of that event, one that has been described as literally being a 'fireball explosion,' is that it must have been *fantastically* hot. This means that at the beginning (of everything) there was complete thermodynamic disorder, which from our earlier discussion means *maximum* entropy. Thus, we immediately have the question of how can the entropy of the universe be continuously increasing if it was as large as possible right from the start?[58]

One possible answer is that the proper model of the universe to use is the so-called *inflationary* universe. The inflationary model has a very high expansion rate for the early universe, much higher than the rate in the standard hot Big Bang model. In the standard model, the entropy puzzle occurs because of the ability of all particle processes to readjust rapidly to the ever-changing state of the universe; the so-called *relaxation times* of all particle processes were *short* compared to the expansion rate of the universe. That means that the actual entropy of the universe would, indeed, be the maximum possible at every instant (and so we have the entropy puzzle). In the inflationary model, however, the expansion rate of the early universe was temporarily so high that the relaxation times of particle processes

---

[58]It has been estimated that over the next 10^{116} years the entropy of the universe will increase by a factor in excess of 10^{14}. See S. Frautschi, "Entropy in an Expanding Universe," *Science*, August 13, 1982, pp. 593-599.

were *very long* compared to the expansion rate. That means the maximum *possible* entropy of the universe, at every instant, would greatly exceed the *actual* entropy. This 'entropy gap' is the cause, then, of the thermodynamic arrow of time, as the universe tries to 'catch-up' and reduce the resulting entropy deficiency.

There is also a philosophical problem with associating increasing entropy with the flow of time from the past into the future. Events in the past leave traces, artifacts taken to be ordered states—or at least more ordered than are their immediate surroundings. The classic example of this is a footprint in the sand, which is clearly a highly organized structure compared to the surrounding sandy beach. The footprint is the trace of a *past* event; such a trace was all the evidence, for example, that Robinson Crusoe needed to conclude that another human had walked that way. But now consider this famous counter-example,[59] that of a bombed city. Certainly there are traces aplenty of past bombing, and in fact one has to be careful not to trip over or to fall into them! The puzzle, of course, is in trying to argue that random bomb craters, strewn rubble, and crushed buildings, somehow constitute a more organized state (a 'footprint') than did the original city and its surrounding undamaged areas. This fuzziness was captured by one physicist who asked "If it were found that the entropy of the universe were decreasing, would one say that time was flowing backward, or would one say that it is a law of nature that entropy decreases with time?"[60]

For another example of the fuzziness of the relationship between entropy and time, consider the situation[61] of a cloud of non-colliding particles all initially moving toward each other. At first the radius of the smallest sphere that contains the cloud decreases with time but, eventually, as the particles move past another, the radius will grow without bound. Indeed, that inexorable increase of the radius could be taken as defining the direction of time that points toward the future. But in what sense is the *disorder* of the particle cloud increasing? After all, as the cloud expands it 'looks the same' at all times; only its scale (radius) changes. What has entropy to do with this expanding-into-the-future cloud? Perhaps nothing. Perhaps what is need is a *new* arrow of time.

So far we have looked in some detail at two arrows of time: the subjective, psychological feeling we have of time 'flowing,' which has no explanation in physics, and the thermodynamic, statistical quantity of entropy. A third arrow is the so-called *cosmological arrow* of the expansion of the universe. This arrow is not nearly as obvious as the first two. Only in the last century (since the 1920s), as a result of the American astronomer Edwin Hubble (1889–1953), has science become aware that the universe *is* expanding. An interesting speculation about the thermodynamic and cosmological arrows, one made numerous times, is that if the cosmological arrow should ever reverse—that is, if the universe should ever begin to

---

[59]See note 54 in Chapter 2.

[60]P. W. Bridgeman, *Reflections of a Physicist*, Philosophical Library 1955, p. 251.

[61]Taken from K. G. Denbigh, "The Many Faces of Irreversibility," *British Journal for the Philosophy of Science*, December 1989, pp. 501-518.

contract toward a Big Crunch—then the thermodynamic arrow would also reverse.[62] The reasoning is that the thermodynamic arrow *follows* the cosmological arrow in an *expanding* universe because that universe can continually 'swallow-up' ever more electromagnetic radiation as it is produced by any physical process. *If* the thermodynamic arrow continues to follow the direction of the cosmological arrow during a contraction, *then* the thermodynamic arrow would also reverse direction.

The usual objection to that suggestion is straightforward. If the direction of time did reverse, then we would see (so goes this argument) all sorts of odd events that would require enormously improbable physics, such as a shattered glass mirror reassembling itself. The error in that objection is subtle but equally simple. *It presupposes the retarded causality of our expanding universe.* In a contracting universe with a reversed thermodynamic arrow of time, however, there would be *advanced* causality, and thus there would be nothing at all improbable about such doings as self-assembling mirrors. As two physicists observed, "The mere reversal of the cosmological expansion will not of itself serve to reverse the direction of thermodynamic and electrodynamic processes, any more than the compression phase of a piston-and-cylinder cycle in a heat engine serves to reduce the entropy of the confined gas."[63]

Those same physicists go on to then mention Stephen Hawking's interest in the relationships among the various temporal arrows. At one time Hawking thought[64] he had discovered a connection between the thermodynamic and cosmological arrows, but then later abandoned that claim.[65] Hawking, in fact, has labeled his original claim "my greatest mistake in science," and has quite openly (and most entertainingly!) discussed his interest in the arrows of time.[66] Indeed, it was to be the subject of his doctoral dissertation but, as he wrote, "I ... needed something more definite, and less airy fairy than the arrow of time, for my PhD, and I therefore switched to singularities and black holes. They were a lot easier."

Yet another arrow of time is the *electromagnetic arrow*, which refers to the fact that radio waves are observed to only propagate into the future, and never into the past. This is a mysterious fact, because Maxwell's equations for the electromagnetic field, like all the other laws of physics, have no intrinsic time sense. The electromagnetic arrow will be discussed in some detail in Chap. 4.

---

[62]T. Gold, "The Arrow of Time," *American Journal of Physics*, June 1962, pp. 403-410.

[63]P. C. W. Davies and J. Twamley, "Time-Symmetric Cosmology and the Opacity of the Future Light Cone," *Classical and Quantum Gravity*, May 1993, pp. 931-945.

[64]S. Hawking, "Arrow of Time in Cosmology," *Physical Review D*, November 15, 1985, pp. 2489-2495. See also the next paper in the same journal, D. N. Page, "Will Entropy Decrease if the Universe Recollapses?," pp. 2496-2499.

[65]For *why* he abandoned that claim, see S. Hawking *et al.*, "Origin of Time Asymmetry," *Physical Review D*, June 15, 1993, pp. 5342-5356.

[66]S. W. Hawking, "The No Boundary Condition and the Arrow of Time,' in *Physical Origins of Time Asymmetry* (J. J. Halliwell *et al.*, editors), Cambridge University Press 1994.

3.3 Time Dilation

"Time as we know it is not universally absolute. The rate of its passage depends to a great extent upon the velocity of its observer with regard to some certain reference system. A moving clock will run slower with respect to a selected coordinate system than a stationary one."

—an early science fiction time traveler explains how his time machine works[67]

In this section I'll set the stage for the scientific basis of time travel to the future, as well as for time travel to the past via the warped spacetime called a wormhole. We start by imagining two horizontal, parallel mirrors, one positioned over the other and separated by distance d. The two mirrors are in the same frame of reference with an Observer; that is, the Observer is looking at two mirrors that are stationary with respect to him. Between the two mirrors we further imagine that a particle of light, a photon, is bouncing endlessly back and forth, up and down, in relentless reflection. This simple system is called a *photon clock*, or the Einstein-Langevin clock, after the French physicist Paul Langevin (1872–1946), and it has been part of physics for decades. We define the time required for the photon to travel from one mirror to the other as a *tick* in time, and so the return trip defines the clock's *tock*. The rate of timekeeping measured by the Observer, the time interval separating consecutive ticks, is obviously then given by

$$t' = 2\frac{d}{c}$$

where c is the speed of light.

Suppose we next imagine that the Observer and the photon clock move at constant speed v to the right across *our* line-of-sight. That is, we remain in the original frame while the photon clock and the Observer are now moving at speed v relative to us. This means the photon clock is in a different frame of reference from ours and so *we* do not see the photon bouncing up and down vertically, but rather *we* see the photon tracing out the triangular path shown in Fig. 3.3.

A round trip of the photon evidently now requires more time than before because the distance in the stationary frame (our frame) is greater than the round trip distance in the Observer's frame (moving with the photon clock, he still sees a round trip distance of d). In fact, if t is the time between consecutive ticks as seen by a stationary viewer (us), then the round trip path length of the photon that we see is

$$2\sqrt{d^2 + \left(\frac{vt}{2}\right)^2}$$

---

[67]F. J. Bridge, "Via the Time Accelerator," *Amazing Stories*, January 1931.

Fig. 3.3 The moving
(relative to us) photon clock

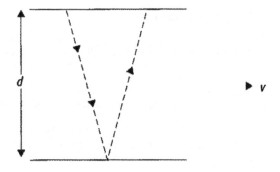

and so

$$t = \frac{2\sqrt{d^2 + \left(\frac{vt}{2}\right)^2}}{c}.$$

This can be easily and quickly manipulated algebraically and combined with the earlier expression for t', the tick interval for an Observer in the same frame as the photon clock, to give the tick interval for the moving clock *as measured by a stationary viewer* (us):

$$t = \frac{t'}{\sqrt{1 - \left(\frac{v}{c}\right)^2}}.$$

Notice that this reduces to $t = t'$ when $v = 0$—that is, when the photon clock is stationary with respect to us.

This last result is the famous Einstein time dilation formula, which shows that $t \geq t'$, and indeed that $t = \infty$ when $v = c$. That is, to us the moving photon clock appears to *run slow* compared to clocks in our stationary frame of reference and, at the speed of light, time *stands still*.[68] A curious anticipation of this association between light and timelessness can be found in a poem by the seventeenth-century poet Henry Vaughn who, in the opening words to his "The World"—which appeared in 1650 as part of his *Silex Scintillans* ("Sparking Flint")—wrote

"I saw Eternity the other night
Like a great *Ring* of pure and endless light,
All calm, as it was bright,
And round beneath it, Time in hours, days, years
Driven by the spheres
Like a vast shadow moved, in which the world
And all her train were hurled."

---

[68]For $v > c$ the time dilation formula says that time becomes imaginary, and this is one reason for claiming that $v > c$ is not possible. The time dilation formula has been experimentally verified: see H. E. Ives and G. R. Stilwell, "An Experimental Study of the Rate of a Moving Atomic Clock," *Journal of the Optical Society of America*, July 1938, pp. 215-226.

A similar modification in the length of a moving object (measured in the direction of motion) occurs when $v > 0$. While an Observer moving with the object will measure its length to be L', a stationary viewer will 'report' it to be contracted to the length

$$L = L' \sqrt{1 - \left(\frac{v}{c}\right)^2}.$$

This effect is called the *Lorentz-FitzGerald contraction*.[69] For 'everyday' objects and speeds the contraction effect is an extremely small one. For example, for a low-altitude satellite 100 m long, moving at 18,000 miles per hour (that is, at $v = 2.7 \times 10^{-5}c$), the contraction is less than 4×10^{-6} cm.

In the early days of science fiction the contraction effect was fascinating to readers, but authors often got it wrong. For example, in one story[70] of a runaway spaceship falling into the Sun, we read "When our racing [ship] was drawn from the Earth's gravity and fell at ever increasing speed toward the Sun it soon approached the speed of light. As we fell faster and faster our length in the direction of the Sun progressed into nothingness. Then — it reached the speed of light — passed it. Now — mind you this — when the [ship] attained the speed of light it was of a *minus* length." This author has managed to make four errors in three sentences!

The same author botched the Lorentz-FitzGerald contraction again 5 years later, and added yet more errors to his growing list. In that tale[71] there is an episode of faster-than-light radio communication along with a lengthy, unfortunate dissertation that actually denies special relativity's fundamental assertion that all inertial frames of references are indistinguishable from each other (two frames are *inertial* if they have no relative acceleration—I'll say more on this in the next section). And in yet another story[72] of high-speed space travel the author has the contraction working in the wrong direction—as the rocket ship moves faster and faster it gets longer and longer.

As bad as those errors are, first prize for mangling the laws of physics has to go to the story[73] of a near light-speed spaceship on its way to Alpha Centauri. The crew mutinies and puts the captain and first office 'overboard' (think *Mutiny on the Bounty*) with 6 months' worth of provisions. This happens at mid-voyage, about 2 light-years from both home and destination, so matters look grim. Indeed, the author tells his readers, several times, that things look *very* bad. But are they? With a stated speed of 162,000 miles per second, the time dilation factor is slightly more than 2 and so, because the space boat is traveling at $0.87c$, it will take a little more

---

[69]Named after the Dutch physicist H. A. Lorentz (1853-1928) and the Irish physicist G. F. FitzGerald (1851-1901).

[70]J. H. Haggard, "Faster Than Light," *Wonder Stories*, October 1930.

[71]J. H. Haggard, "Relativity to the Rescue," *Amazing Stories*, April 1935.

[72]D. Wandrei, "A Race Through Time," *Astounding Stories*, October 1933.

[73]N. Schachner, "Reverse Universe," *Astounding Stories*, June 1936.

Table 3.1 The Lorentz-
FitzGerald time slowing
factor

v	$\dfrac{1}{\sqrt{1-(\frac{v}{c})^2}}$
.1	1.005
.2	1.021
.5	1.155
.7	1.4
.9	2.294
.999	22.366
.9999	70.712

than 13 months of *space boat time* to complete the journey. If the men go on half-rations then it seems they *could* survive.

There is, of course, the problem of slowing down so as to arrive at Alpha Centauri at a reasonable speed, but that issue is ignored in the story. Instead, our attention is directed to the much more dramatic concern of a faster-than-light planet (don't ask!) colliding with the space boat and carrying the castaways onwards toward their destination. When this happens we read that time runs backwards (for what *really* occurs at superluminal speeds, keep reading this chapter) and, finally, in a repeat of an error I mentioned earlier, we are told that the Lorentz-FitzGerald contraction is negative for $v > c$.

The time-slowing (or size-shrinking) factor becomes pronounced only at values of v close to c, as shown in Table 3.1. For example, the last entry shows that a clock traveling at 99.99 % the speed of light will register the passage of 1 year while nearly 71 years pass on Earth. One science fiction writer got this dramatically wrong, even though he actually reproduced the Lorentz-FitzGerald equation in his story.[74] At one point he writes of the near light-speed rocket ship that stars in the tale, "If it [the ship's speed] was as slow as ninety-four percent [of the speed of light] ... for every moment ticked by the clocks of the [ship] hundreds passed on earth." In fact, the time dilation factor at that speed is 'only' 2.93.

One possible objection to time dilation is that the analysis done here has been for a *particular* clock. How do we know that another clock, one using wheels and pendulums, for example, instead of photons and mirrors, wouldn't be affected differently by motion? The answer comes from relativity itself, which says there is no way to detect uniform motion. If two clocks did behave differently, then this difference could be used as a motion detector. Since this is impossible within the framework of relativity, then *all* clocks, no matter what the details of their internal

---

[74]L. R. Hubbard, "To the Stars," *Astounding Science Fiction*, February and March 1950.

mechanisms may be (including the *biological* clocks of own bodies), must respond to motion just as does the photon clock.[75]

Time dilation can also be caused by gravity (it appears in the 2014 movie *Interstellar*), and that effect has been used to 'construct' a time machine from a wormhole (to be discussed later). You can get a qualitative understanding of how that happens by imagining a massive body (*massive*, to have a really big gravitational field) in space, on the surface of which is a hot object. That object emits electromagnetic radiation and, though it isn't essential to the following argument, further imagine that the temperature of the object is sufficiently high that some of the radiation (emitted by the very atoms of the object) is in the visible-light portion of the spectrum. Now, from elementary quantum theory we can also think of the object's atoms as emitting photons ('particles of light'), each of energy hf, where h is Planck's constant and f is the frequency in hertz (what used to be called 'cycles per second'). The higher the temperature, the higher the photon energy, and so the higher the frequency. In the visible spectrum, f is on the order of 10^{15} Hz, a frequency one *billion* times higher than commercial AM radio frequencies.

The radiating atoms can be thought of as tiny clocks, with alternate half-cycles of radiation being ticks and the half-cycles in-between being the tocks. The passage of time on the surface of the massive body can be measured by these atomic clocks in the hot, radiating object. To a distant observer, however, as she receives the photons from the hot object, the passage of surface time on the massive body will appear to occur at a reduced rate when compared with the photons emitted by her own identically hot object (her 'local' clock). That's because the radiation that arrives at the distant observer has traversed a gravitational field (a journey sometimes described as 'climbing out of a gravitational well') and so is down-shifted in frequency toward the red end of the visible spectrum. This effect is called either the *gravitational red shift* or the *gravitational time dilation* effect (or even the *Einstein shift*, because it was Einstein who predicted the effect in 1907).

You can 'understand' this dilation effect as follows. One can crudely think of a photon emitted by the hot object as something like a rock thrown upward. As the rock rises upward through the gravitational field, its *total* instantaneous energy is always constant, but the total, fixed energy is split between its kinetic and potential energies in an ever changing way. That is, as the rock rises, its kinetic energy continually decreases (the rock slows down), whereas its potential energy continually increases. A photon is not a rock, however, and it certainly can't slow down as

---

[75]Resistance to this conclusion persisted for years. See, for example, the letter "Relativity and Radio-activity," *Nature*, January 8, 1920, p. 468. The author of that letter wondered whether a clock based on radioactive decay might not somehow beat the 'conspiracy' of moving clocks running slow compared to stationary ones. And in a letter to *Science* (December 7, 1962, p. 1180), a reader objected to applying the laws of physics to biological systems, first asserting (incorrectly) that time dilation "has never been proved or disproved experimentally," and then "there is no known causal means by which greatly increased velocity could alter, without destroying the very biochemical basis of the life process, the metabolic changes which are responsible for the aging process."

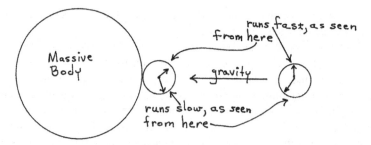

Fig. 3.4 Gravitational time dilation due to a massive body

it rises through a gravitational field (it *always* moves at the speed of light because the photon *is* light). The only way a photon can give up energy to balance the ever increasing potential energy (physicists will cringe at this, but read on) is to decrease its frequency. Hence, the red shift as seen by a distant observer of the photon, who thus sees time running slow on the massive body. Fig. 3.4 shows the case for a clock on a massive body, compared to a distant clock. Notice, carefully, the 'direction of gravity,' that is, the direction a small, unrestrained test mass will move.

A gravitational red shift in the *opposite* direction is nicely described in the famous 1966 science fiction story "Neutron Star" by Larry Niven. There a space traveler zooms down into a neutron star's intense gravity field (at half the speed of light!), passing within one mile of the star's surface. He reports what he observes in these dramatic words: "All around me were blue-white stars. Imagine light falling into a savagely steep gravitational well. It won't accelerate. Light can't move faster than light. But it can gain in energy by increasing its frequency. The light was falling on me, harder and harder, as I dropped." To Niven's intrepid spaceman, therefore, the passage of time on those distant blue-white stars appeared to be running *fast* compared to his wrist watch. This shows that the effect could equally well be called the 'gravitational *blue* shift.'

Notice that *gravitational*-induced time alterations do *not* have the symmetrical feature of motion-induced time dilations.[76] That is, for gravitational time dilations caused by photons either falling into or climbing out of gravity wells, observers at each end *agree* about whose clock is running slow, unlike in the motion-induced case where *each* of the relatively moving observers thinks it is the *other* observer's clock that is running *slow*.[77]

Now, comparing a massless photon to a rock which does have mass (and so potential energy), as each travels 'against' gravity, *is* straining the physics, with its one virtue being the provision of an initial plausibility argument. Gravitational time dilation is sufficiently important in the operation of wormhole time machines,

---

[76]Gravitational time dilation was experimentally observed in 1960, more than half a century after Einstein predicted it.

[77]A science fiction use of both the red and the blue gravitational shifts appears in the novel by J. P. Hogan, *Out of Time*, Bantam 1993.

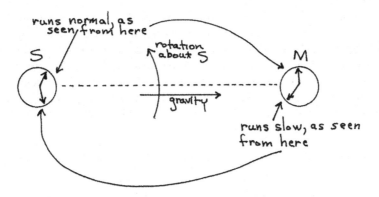

Fig. 3.5 Gravitational time dilation due to rotation

however, that perhaps another way to think about it that is more acceptable to hardcore physicists makes it worth another look. Imagine a turn-table disk that initially is *not* rotating (soon it *will* spin). On this disk imagine further that we fasten two clocks, one called S (for *stationary*) and one called M (for *moving*), as shown in Fig. 3.5. These two clocks are set to read the same time at some instant and then, thereafter, they tick-tock through time at precisely the same rate. We then start the disk rotating around a vertical axis through clock S. (Imagine S to be a *point* clock, and so it is *the* one point on the turn-table that remains at rest even as the disk rotates.) What happens to the time-keeping of S and M?

We can answer this question by using our earlier result concerning time dilation due to motion. Even with the disk now rotating, S appears stationary to an observer sitting on top of M, while M appears to be moving across the line of sight of an observer sitting on top of S. So, to the observer at S, clock M runs slow, while to the observer at M there is *no change* in the time keeping of clock S. This is a non-symmetrical outcome, and so should remind you of gravitational time dilation. This might be a puzzle to you, however, as we don't have a massive body in Fig. 3.5 to account for a gravity presence. This is where the genius of Einstein comes into play.

Anyone who has ever ridden on a merry-go-round knows there is an outward (pointing away from S) directed force called the *centrifugal force* that is 'trying' to toss you off the merry-go-round. Now, where there is a force there is an acceleration, and one of Einstein's starting points in his development of general relativity was to identify an acceleration, *whatever its origin*, with gravity. A massive body is, of course, one possible origin (the obvious one, in fact), but so is the rotation of the turn-table. So, we have the situation shown in Fig. 3.5, where now 'gravity' is directed as shown and, and as in Fig. 3.4, the direction of acceleration of gravity is toward the slow-running clock. Again, the direction of gravity is the direction a small, unrestrained test mass on the rotating disk will move. The further M is away from S, the greater the 'gravity' of the centrifugal acceleration and so the slower will M run as measured by the observer at S.

A 1968 story that uses the gravitational time dilation effect in a striking fashion tells of a starship's visit to a supernova, accompanied by a fantastic alien life-form—a ball of intelligent plasma named Lucifer that is telepathic.[78] While the ship stands off at a distance of 500 million kilometers, Lucifer will approach much closer to the *event horizon* (see the Glossary) of a black hole at the center of the supernova explosion and communicate its findings to a human telepath on the ship. A physicist in the crew is curious about one point, and asks the human telepath the following question:

"I have wondered about one item. Presumably Lucifer will go quite near the supernova. Can you still maintain contact with him? The time dilation effect, will that not change the frequency of his thoughts too much?"

Lucifer, in fact, dies in the black hole even as he saves the ship from destruction, and the human telepath will hear his death scream for the rest of her life. As the physicist later explains to the ship's captain, telepathy is instantaneous and has no limiting range (there is no known physical basis for believing any of this, but it is crucial for story effect):

> "Remember the time dilation. He fell from the sky and perished swiftly, yes.
> But in supernova time. Not the same as ours. To us, the final stellar collapse takes an infinite number of years. ... He will always be with her."[79]

3.4   The Lorentz Transformation

> "If only he'd paid more attention to mathematics in school."
>   —a science fiction time traveler laments missed opportunities[80]

In this (and the next) section the math gets about as 'deep' as it gets in this book, but to leave it out struck me as a cheat. *You* can skip part (or all) of the math and simply read the prose, but it seemed unfair for me to make that decision for you.

We begin by imagining two distinct frames of reference. One we take to be stationary, and the other as moving at a uniform speed v with the respect to the first. The moving frame is said to be *boosted* with respect to the stationary frame. We orient these two coordinate systems so that the motion occurs along just one axis (the x-axis, as shown in Fig. 3.6, where I am using primed variables for the moving frame). That is, the two frames have coincident x axes, and parallel y and z axes that

---

[78]Poul Anderson, "Kyrie," in *The Road to Science Fiction* (J. Gunn, editor), volume 3, New American Library 1979.

[79]A mathematical discussion of how signals take forever (even though they are emitted in a finite time interval) to travel from the event horizon of a black hole to a distant receiver can be found in James B. Hartle, *Gravity: an introduction to Einstein's general relativity*, Addison Wesley 2003, pp. 264-268.

[80]D. Knight, "Extempore," in *Far Out*, Simon and Schuster 1961. Similar words ("If only I had more mathematics") were spoken by Einstein the day before he died — see Walter Isaacson, *Einstein: his life and universe*, Simon & Schuster 2007, p. 542.

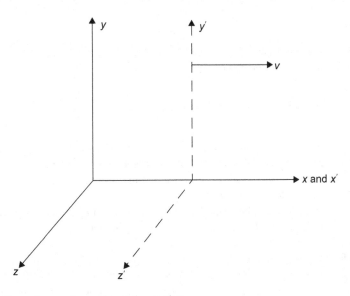

Fig. 3.6 Two reference frames in relative motion

are moving apart at the constant speed v. Let's also imagine that there is a clock at
the origin of each frame, and that at the instant the origins match the clocks are
synchronized; that is, $t = t' = 0$ is the instant the two coordinate systems coincide.

We further imagine that there is an observer at the origin in each frame. At some
arbitrary instant of time, each observer records the coordinates of the arbitrary point
P in space, as measured in his system. These observers could, for example, agree to
record the coordinates of P in their system when their clock reads 5 s. It seems
immediately obvious (as it was for Newton) that $t = t'$; that is, time runs at the same
rate in each frame and thus it makes sense to talk about "the same instant" at every
point in space. (After reading the previous section you know this *not* true, but
temporally forget that!) Thus, at this 'same instant' the stationary observer records
(x, y, z) and the moving observer records (x', y', z'). What are the relationships
between the primed and unprimed coordinates of P? That is, what mathematical
transformation converts from one frame to the other?

The answer seems obvious:

$$y' = y$$
$$z' = z$$
$$x' = x - vt.$$

This transformation, called the *Galilean transformation* after the Italian Galileo
Galilei (1564–1642), satisfies the relativity principle, which says that uniform
motion leaves the laws of physics unchanged. For example, in the stationary system
Newton's famous second law of motion for a constant mass m,

$$F = m\frac{d^2x}{dt^2}$$

becomes the identical-appearing form

$$F' = m\frac{d^2x'}{dt'^2}.$$

More precisely, *all* the laws of mechanics known to Newton are unchanged. Any frame of reference in which Newton's laws of mechanics hold true is said to be an *inertial* frame. Given one inertial frame, we can find infinitely many others simply by applying the Galilean transformation.

However, when the mathematical laws of electrodynamics were discovered by Maxwell in the nineteenth century, it was a shock to physicists to learn that the Galilean transformation does *not* leave Maxwell's equations unchanged in form; the transformed equations predict electromagnetic effects for the moving system that are *not* predicted to occur in the stationary system. This meant that there was theoretical support for the possibility that electromagnetic experiments might be devised to detect uniform motion, and this eventually led to the famous Michelson-Morley experiment of 1887. This experiment, sensitive enough to detect the motion of the Earth itself through space, failed to detect any such motion. The conclusion was clear: the new electromagnetic effects predicted by the Galilean transformation do not exist, and so the transformation must be wrong even though it works for the laws of mechanics. So—what is going on?

The answer is inspired, and again returns us to the cornerstone of relativity: the idea that the laws of physics, *all* the laws, should look the same to observers in uniform relative motion. That is, there is no special or preferred system of coordinates—*all* inertial systems are equivalent in physics. Evidence from an extremely broad variety of sensitive experiments had, by the end of the nineteenth century, convinced physicists that Maxwell's equations *are* correct. Thus, a new transformation was needed that leaves both the laws of mechanics *and* the laws of electrodynamics unchanged with uniform motion. But, a single transformation that works on Maxwell's equations and on the mechanical laws would therefore mean that Newton's mechanical laws as stated cannot be correct, and this was a breathtaking conclusion: Newton had been unchallenged for two centuries.

As it turns out, Newton's laws are *almost* right. The only correction required is that the mass of a moving body is not independent of motion, but rather varies as

$$m = \frac{m_0}{\sqrt{1 - (v/c)^2}}$$

where m_0 is the so-called *rest mass* when $v = 0$.[81] This result says that m is infinite at $v = c$ unless $m_0 = 0$ (as it is for a photon), which is another reason for the belief that accelerating a mass (such as a spaceship) up to the speed of light is impossible because it would require infinite energy (look back at note 68, too). With this modification, the transformation that leaves all the laws of physics unaltered in form by uniform motion is what is called the *Lorentz transformation* (after the same Lorentz the contraction effect is named for), who discovered it in 1904 by direct manipulation of Maxwell's equations:

$$y' = y$$
$$z' = z$$
$$x' = \frac{x - vt}{\sqrt{1 - (v/c)^2}}$$
$$t' = \frac{t - vx/c^2}{\sqrt{1 - (v/c)^2}}.$$

In 1905 Einstein discovered how to derive these equations from a fundamental reexamination of space and time without concerning oneself about the details of specific physical laws.

By simple algebraic manipulation, the transformation equations can be rewritten as

$$ct = \gamma ct' + \beta\gamma x'$$
$$x = \beta\gamma ct' + \gamma x'$$

where

$$\beta = \frac{v}{c}$$

and

$$\gamma = 1/\left(1 - \beta^2\right)$$

are dimensionless constants. In compact matrix form, the Lorentz transformation becomes

---

[81]This variation of mass with speed was experimentally observed in 1901.

$$\begin{bmatrix} ct \\ x \end{bmatrix} = \begin{bmatrix} \gamma & \beta\gamma \\ \beta\gamma & \gamma \end{bmatrix} \begin{bmatrix} ct' \\ x' \end{bmatrix}$$

and the symmetrical 2×2 matrix is called the Lorentz boost matrix or simply the *boost*. Notice that when $v = 0$ (zero boost) the boost matrix reduces to the identity matrix; that is, the two frames are one and the same with at most a shift in the location of the origins. Note, too, that $\beta = 0$ and $\gamma = 1$ for *any* v when c is infinite—that is, the boost matrix is again reduced to the identity matrix and the Lorentz transformation becomes the Galilean *if c* is infinite. But c is not infinite, and all the implications of special relativity are the direct result of the *finite* speed of light.

The Lorentz transformation contains two results I have mentioned earlier in the book. For example, in Chap. 2 it was mentioned that simultaneity is a relative concept in reference frames in relative motion. Let's see what the transformation says about that. Consider two events that occur specifically on the x-axis. They are simultaneous in the stationary system (at, say, time $t = T$) but are at different places (at, say, $x = X$ and $x = X + \Delta X$). Their occurrences in time for the moving observer are

$$t_1' = \frac{T - vX/c^2}{\sqrt{1 - (v/c)^2}}$$

and

$$t_2' = \frac{T - v(X + \Delta X)/c^2}{\sqrt{1 - (v/c)^2}}.$$

For the moving observer, therefore, the two events are *not* simultaneous, being separated in time by

$$t_1' - t_2' = \frac{v\Delta X/c^2}{\sqrt{1 - (v/c)^2}}.$$

Only if $\Delta X = 0$ (the two events occur at the same place) will $t_1' = t_2'$. That is, only if $\Delta X = 0$ are simultaneous events in one frame also simultaneous in another frame in relative motion.

And in the previous section we found that time runs slow in one frame as observed from another frame that is in relative motion. We can get this result from the t' equation of the Lorentz transformation by differentiating it with respect to t. Thus,

$$\frac{dt'}{dt} = \frac{1 - (v/c^2)\frac{dx}{dt}}{\sqrt{1 - (v/c)^2}}.$$

But since

$$\frac{dx}{dt} = v,$$

the speed of the moving frame as measured by the observer in the stationary frame, this gives

$$dt' = \sqrt{1 - (v/c)^2}\, dt$$

which is the same result we obtained by analyzing the photon clock.

The Lorentz transformation contains other interesting implications beyond these. For example, mention has been made several times to the 'relativity principle,' the belief that uniform motion has no observable effect on the forms of physical laws. But how do we know who is moving and who is stationary? After all, a system moving to the right past a stationary system could just as well be thought as the stationary system, while it's the *other* system that is moving to the *left* (at speed $-v$).

To study this question with the Lorentz transformation, we'll invert the transformation (that is, solve for the unprimed variables in terms of the primed ones). What we get back is just what you probably thought—the Lorentz transformation with v replaced by $-v$. That is, the Lorentz transformation is symmetrical, so two observers in different frames of reference each say it is the *other's* clock that is running slow! This follows immediately, in fact, from the original transformation written in matrix form. That is, multiplying through the earlier matrix equation by the inverse of the boost matrix, we get

$$\begin{bmatrix} ct' \\ x' \end{bmatrix} = \begin{bmatrix} \gamma & \beta\gamma \\ \beta\gamma & \gamma \end{bmatrix}^{-1} \begin{bmatrix} ct \\ x \end{bmatrix} = \begin{bmatrix} \gamma & -\beta\gamma \\ -\beta\gamma & \gamma \end{bmatrix} \begin{bmatrix} ct \\ x \end{bmatrix}.$$

The only difference between the original boost matrix and its inverse (which is, of course, the new boost matrix for the new interpretation of which frame is the moving one) is a change in sign for β, that is, in the sign of v. The inverse transformation is

$$y = y'$$

$$z = z'$$

$$x = \frac{x' + vt'}{\sqrt{1 - (v/c)^2}}$$

$$t = \frac{t' + vx'/c^2}{\sqrt{1 - (v/c)^2}}.$$

As a final example of what the Lorentz transformation tells us, consider the so-called *addition of velocities* problem. Suppose you are in a high-speed spaceship traveling past Earth at speed v. Earth is the stationary system (with the unprimed variables), and the spaceship is the moving system (with the primed variables). Assume the x and x′ axes are along the direction of motion. Imagine next that while standing in the nose of the spaceship, just as the spaceship passes Earth, you fire a gun in the direction of motion (away from the Earth), with the bullet exiting the gun with a muzzle speed of w. How fast is the bullet moving away from Earth? The common-sense answer in Galileo's time was $v+w$, but we now know that the Galilean transformation is wrong. What does the Lorentz transformation say?

Inside the spaceship, the position of the bullet at time t' after the gun is fired is

$$x' = wt'.$$

From the inverse Lorentz transformation, the location of the bullet earth's frame is

$$x = \frac{x' + vt'}{\sqrt{1 - (v/c)^2}} = \frac{w + v}{\sqrt{1 - (v/c)^2}}t'.$$

The transformation also tells us that (using $x' = wt'$)

$$t = \frac{t' + vx'/c^2}{\sqrt{1 - (v/c)^2}} = \frac{1 + wv/c^2}{\sqrt{1 - (v/c)^2}}t'.$$

Thus, the speed of the bullet in *Earth's frame* is

$$\frac{x}{t} = \frac{w + v}{1 + wv/c^2}.$$

Notice that for a low-speed bullet ($w \ll c$) this result[82] is close to $w + v$, but at high values for w the result is very much different. Indeed, suppose we don't fire a gun at all, but rather replace it with a flashlight. Now, instead of a bullet, we shoot *photons* at $w = c$. The Galilean transformation would (incorrectly) say that a stationary observer on Earth would see the photons moving away at speed $v+c$,

---

[82]This result was found by the French physicist Henri Poincaré (1854-1912) in June 1905, three months *before* the publication of Einstein's special theory of relativity which also contains the result. And it was Poincaré who first stated (in 1904) that "no velocity can surpass that of light, any more than any temperature could fall below the zero absolute."

which is a superluminal speed. The Lorentz transformation, however, says that the Earth observer would see a speed of

$$\frac{c+v}{1+cv/c^2} = \frac{c^2(c+v)}{c^2+cv} = \frac{c^2(c+v)}{c(c+v)} = c.$$

That is, no matter what the speed of the moving observer on the spaceship is, he sees the light from his flashlight traveling at the same speed as does the stationary observer back on Earth. This peculiar effect is unique to the speed of light ($w=c$). We've derived it here as a consequence of the Lorentz transformation, but in fact Einstein actually did things in reverse order. That is, he began by *postulating* the invariance of the speed of light[83] for all observers in uniform motion, combined that with the principle of relativity which says all physical laws look the same to those observers, and so derived the Lorentz transformation using no mathematics beyond high school algebra.

A mathematically elegant alternative derivation of the addition-of-velocities formula can be done by simply noticing that the condition of two successive boosts should be, itself, a boost. Thus, if we have a frame moving relative to a second frame (which is itself moving relative to a third frame), then the boost matrix of the first frame relative to the third frame is the product of the two individual boost matrices. That is,

$$\begin{bmatrix} \gamma_3 & \beta_3\gamma_3 \\ \beta_3\gamma_3 & \gamma_3 \end{bmatrix} = \begin{bmatrix} \gamma_2 & \beta_2\gamma_2 \\ \beta_2\gamma_2 & \gamma_2 \end{bmatrix} \begin{bmatrix} \gamma_1 & \beta_1\gamma_1 \\ \beta_1\gamma_1 & \gamma_1 \end{bmatrix}.$$

From this it is easy to show (if you know how to multiply matrices!) that

$$\beta_3 = \frac{\beta_1+\beta_2}{1+\beta_1\beta_2}.$$

Substitution of

$$\beta_1 = \frac{w}{c}, \beta_2 = \frac{v}{c}$$

immediately gives the addition-of-velocities formula.

A failure to understand the implications of the invariance of the speed of light resulted in two stupendous errors in the story "To the Stars," cited in note 74. At all times an officer stands watch on the bridge of a near light-speed rocket ship to be sure the ship doesn't accidently *reach* the speed of light. This is to be avoided (according to the author) because to reach the speed of light would cause the ship to

---

[83]Einstein's postulate was experimentally confirmed in 1932.

"hang there forever unmoving [in time] ... locked, protected and condemned to eternity by zero time." This horrible state is so easy to stumble into (the author was apparently unaware that it would require *infinite* energy) that occasionally the ship has to fire a 'check-blast' from its forward rocket tubes to slow down! Equally absurd is the means by which the development of this 'fatal' condition is detected: the nose of the ship mounts a forward-pointing light source (our earlier flashlight) and so, if the ship is getting too near the speed of light, it will start to *overtake* the photons emitted by that source!

The biggest puzzle of all, actually, is why the editor of *Astounding Science Fiction* let such a technically goofy story appear in a magazine recognized for its usual faithfulness to known science. Particularly so since the story appeared in 1950 and, as long ago as December 1937, none other than Isaac Asimov (then 17 years old) had written a letter to the editor[84] giving the proper interpretation of what happens when $v = c$. Here's what the young Asimov wrote (notice the early hint of his life-long pessimism concerning time travel to the past that appears near the end of his letter):

> "The effect on time of increasing speeds is ... well known. Relativity states that as speed approaches that of light, time slows up until at 186,000 miles a second, time (so to speak) stands still. This seems to refute statements found in so many astronomy books (and science fiction stories) that even at the speed of light it would take four years to reach the nearest star. No such thing! As time halts at the speed of light, a person traveling from Alpha Centauri to the solar system, or vice-versa, would not be aware of any lapse of time. In that sense the speed of light is infinite (as was thought in ancient times). This by the way offers an entirely scientific (if impractical) means of travel into the future. Say that someone wants to see how the world would look a hundred years from now. His procedure would be as follows: getting into his spaceship, he would proceed to a spot fifty light-years away at the speed of light. The journey would, for him, be practically instantaneous (due to the curious behavior of time at the speed of light). But fifty years would have elapsed on earth. He makes the return trip at the same speed. Another fifty years lapse on earth and he lands a hundred years after his time. With this system, however, it would be impossible to travel into the past, so I don't think it will ever be adopted."

Young Asimov missed an important detail concerning the reversal of the space-ship's direction of travel for the return to Earth, and I'll come back to it later in this chapter. But certainly he displayed a *far* better knowledge of the physics of time than did the author of "To the Stars" (who was, by the way, L. Ron Hubbard, a prolific writer of fantasy and science fiction before founding the Church of Scientology).

---

[84]The editor who bought "To the Stars" was the same editor editor of *Astounding* when Asimov's letter appeared, and so he was certainly aware of it.

3.5 Spacetime Diagrams, Light Cones, Metrics, and Invariant Intervals

"Come back when you know tensor calculus and I'll explain to you about n-dimensional forces and the warping of world-lines."
—a science fiction physicist's reply after being asked how his time machine works[85]

It is helpful in discussions about the spacetime of special relativity to use what are called *Minkowski spacetime diagrams*. These are plots of the spacetime coordinates of a particle; the resulting curve is called the *world line* of the particle. Such diagrams are four-dimensional—three space axes and one time axis—and hard to visualize, much less draw on a flat sheet of paper! The convention is to make do, whenever possible, with a simplified spacetime that has just one space axis (horizontal) and one time axis (vertical). As you'll recall from Chap. 1, physicists often call such a simplified diagram a *toy spacetime*.

So, for a particle at rest in some observer's frame of reference, its spacetime diagram for that observer is a *vertical* world line. If the particle is not at rest then its world line will tilt away from the vertical; the greater the speed the greater the deviation from the vertical. Accelerated particles will have world-lines that *curve* away from the vertical. Straight, uncurved world lines represent unaccelerated particles, that is, particles experiencing no forces and so in free fall. Such a world line is called a *geodesic*. In Fig. 3.7 the world lines for these various cases are shown on the same axes. It is assumed in the figure that all three particles are at $x = x_0$ when $t = 0$.

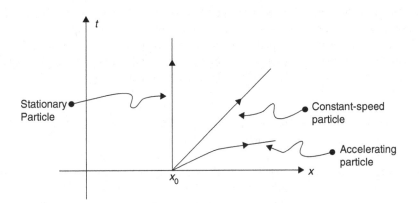

Fig. 3.7 World lines of three particles

---

[85]Poul Anderson, "The Little Monster," in *Science Fiction Adventure from WAY OUT* (R. Elwood, editor), Whitman 1973.

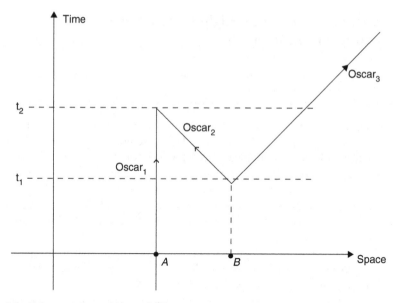

Fig. 3.8 A time traveler and his world line

Spacetime diagrams were embraced decades ago by philosophers looking for 'scientific' ways to support their position on time travel (whatever it might be), as opposed to the mere verbiage of traditional colleagues. A famous example of this is a 1962 paper by the Harvard philosophy professor Hillary Putnam (note 74 in Chap. 1). There we are asked to imagine the spacetime diagram of one Oscar Smith who, in Fig. 3.8, is at spatial location A next to his time machine. At time t_0 Oscar has not yet gotten into his time machine. A little later, at time t_1, we suddenly see not only Oscar at A but also *two more* Oscars who have appeared (apparently out of thin air) moving away from spatial location B! Between t_1 and t_2 we see the original Oscar at A and the two mysterious Oscars at B (for a total of three Oscars, labeled in the figure as $Oscar_1$, $Oscar_2$, and $Oscar_3$) move forward in time—but one of the new Oscars ($Oscar_2$) lives a decidedly odd existence in that his life seems to be running in reverse!

Eventually, at time t_2, the original $Oscar_1$ and the weird, reverse $Oscar_2$, merge and seemingly annihilate one another, vanishing into thin air to leave only a single Oscar ($Oscar_3$) for all time after t_2. Putnam argues that, although strange, what has just been described is still sensible and that, indeed, the very fact that we can draw the spacetime diagram of Fig. 3.8 supports the case for backward time travel. He claims this because although the spacetime diagram does show time increasing upward for all three Oscars (that is the time direction for an external observer) there is actually no 'spontaneous creation' or 'mutual annihilation' and all is sensible *if* $Oscar_2$ is understood actually to be a time traveler into the past with *his* time direction thus pointed opposite to that of the 'other two' Oscars. There is, of course, just *one* Oscar!

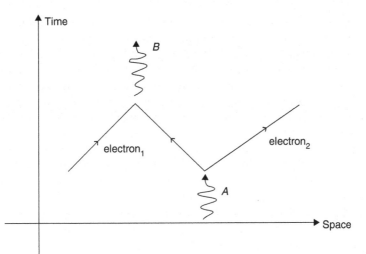

Fig. 3.9 Anti-matter via backward time travel

As mentioned back in Chap. 1, Putnam's suggestion was rebutted (see that chapter's note 76) by what is (in my opinion) an even less plausible mechanism for Putnam's kinked spacetime than is Putnam's invocation of time travel. That critic advocated, instead, an explanation based on matter transmitters and anti-matter, the latter an idea he credits to Feynman (who actually got it from Wheeler). In a paper by Feynman we do find the famous suggestion that a positron that appears to us to be moving forward in time is actually an electron traveling backward through time.[86] Logically, that greatly weakens the critic's view, because it puts him in the position of using anti-matter (explained in terms of backward time travel) to argue *against* backward time travel! But let's ignore that concern, give the critic the benefit of the doubt, and explore how anti-matter and backward time travel are imagined to be connected.

Feynman asks us to imagine the process shown in Fig. 3.9. Gamma ray A spontaneously creates an electron-positron pair, with electron$_2$ moving off to some distant region while the positron soon meets with electron$_1$, resulting in mutual annihilation and the production of gamma ray B. This description involves *three* particles, and each segment of the kinked line is a distinct particle. But Feynman said there is another way to look at this, a way that involves just *one* particle. According to Feynman, the kinked line in Fig. 3.9 (which should remind you of Oscar's kinked world line) is the world line of a single electron; the middle

---

[86]R. Feynman, "The Theory of Positrons," *Physical Review*, September 15, 1949, pp. 749-759. In a paper published the year before ("A Relativistic Cut-Off for Classical Electrodynamics," *Physical Review*, October 1948, pp. 939-946), he wrote "This idea that positrons might be electrons with the proper time reversed was suggested to me by Professor J. A. Wheeler." The identification of anti-matter with backward time travel occurred in science fiction (see note 124 in Chapter 1) almost simultaneously with Wheeler's speculation.

segment, that we call a positron, is just the electron traveling backward in time, and so we must reverse the arrowhead on it (indicating the opposite of the direction shown in Fig. 3.9).

There are two central questions at this point. First, why is a positron (with positive electric charge) moving forward in time mathematically (and physically) equivalent to a negatively charged electron moving backward in time? The answer is that the reversal in charge sign, which results from the reversal of the electron's proper time, follows from the TCP theorem that was mentioned in Chap. 2 (see note 38 there). And second, what causes the electron to suddenly move backward in time? Picturesquely, the electron is recoiling from the emitted burst of gamma ray B. Similarly, the absorption of the energy of gamma ray A by the electron that is recoiling backward in time causes a second recoil, giving the world line of what was originally called electron$_2$. This reinterpretation of a kinked spacetime diagram was described as follows (in Feynman's famous words): "It is as though a bombardier flying low over a road suddenly sees three roads and it is only when two of them come together and disappear that he realizes that he has simply passed over a long switchback in a single road."

In a later paper[87] Putnam's critic presented another line of attack against Putnam's interpretation of spacetime diagrams as lending support to time travel. There the critic observed that the presence of the time-reversed Oscar$_2$ shows that the "world of the Oscars" is not *temporally orientable*. A temporally orientable spacetime is one in which *every* point in it agrees with its local neighbors on the directions of past and future—a condition clearly *not* satisfied for the case of Oscar$_2$. As the critic pointed out, the first time travel spacetime discovered, the Gödel universe, *is* temporally orientable, and so in it the ambiguity of Oscar$_2$ (whether he is traveling backward in time as opposed to living forward 'in reverse') does not occur. That is, the critic agreed with Putnam's acceptance of the conceivability of time travel to the past, but not with his use of Feynman's concept of anti-matter as time-traveling matter. That critic wasn't alone in that opinion.

One physicist, for example, wrote of "Feynman's rather loose talk of particles 'traveling' backward ... in time,"[88] and the well-known philosopher John Earman declared "It is true that Feynman uses the slogan 'Positrons are electrons running backward in time,' but it is dangerous to draw conclusions from slogans."[89] I am not sure what Earman meant by "slogans": a careful reading of Feynman indicates that he actually took the matter quite seriously.[90] In his 1949 positron paper (note 86), for example, he wrote that "the idea that positrons can be represented as electrons

---

[87]See R. Weingard in note 114 in Chapter 1.

[88]H. Price, "The Asymmetry of Radiation: Reinterpreting the Wheeler-Feynman Argument," *Foundations of Physics*, August 1991, pp. 959-975.

[89]J. Earman, "On Going Backward in Time," *Philosophy of Science*, September 1967, pp. 211-222.

[90]Feynman declares the view of a positron as a time traveling electron to be of value, for example, in his famous book *Quantum Electrodynamics*, W. A. Benjamin 1961, p. 68.

with proper time reversed relative to true time has been discussed by the author and others,"[91] and also that "Previous results suggest waves propagating ... toward the past, and that such waves represent the propagation of a positron."

In any case, spacetime diagrams are highly useful in discussing time travel, but they do have some curious twists. In our everyday world, a path that joins two points on a surface with the *minimum* length is called a geodesic of that surface. As you'll see later in this chapter spacetime geodesics do indeed possess an extremal property, but rather than being a minimum it is a *maximum* property. Spacetime diagrams can be misleading on this matter, so it is important to remember that such diagrams are not a perfect representation of all the properties of a spacetime.

It is customary to draw spacetime diagrams with the speed of light as unity ($c = 1$). That is, a distance of 300,000 km on the space axis is represented by the same extension as is one second on the time axis. This means that the world line of a photon is tilted away from the vertical time axis by 45°. Because photons can travel in both space directions (to the left and to the right) in the two-dimensional spacetime we can draw on a piece of paper, and because the speed of light is the limiting speed of the universe, we can represent the collection of all possible world lines as those paths that *never* tilt more than 45° away from the vertical, which forms what is called a *light cone* in spacetime, as shown in Fig. 3.10 (which attempts to represent a *three*-dimensional spacetime, one with two space—imagine a y-axis, out of the paper, perpendicular to both the x and t axes—and one time dimension).

In Fig. 3.10 I have taken $x = y = 0$ at $t = 0$ for all the possible world lines involving speeds below the speed of light. Let's agree to call this spacetime point the Here-Now. Then, spacetime points in the upward half of the light cone are in the Future of Here-Now; similarly, spacetime points in the lower half of the light cone are in the Past of the Here-Now. We can draw a straight world line from the Here-Now to any point in the Future half-cone with a tilt of less than 45° away from the vertical, which means that a massive particle could travel from the Here-Now to that point at less than the speed of light. Similarly, a massive particle starting at any point in the Past half-cone could have reached the Here-Now by traveling at less than the speed of light. Such a world line is called *timelike* because its projection on the time axis is greater than its projection on a space axis—they are the world lines connecting spacetime points that are potentially causally linked. That is, an event at a spacetime point in the Past half-cone could have had an effect on the event at the Here-Now, even though its influence propagated at less than the speed of light. Also, an event at the Here-Now could potentially affect the event at any spacetime point in the Future half-cone of the Here-Now.

Any points in the regions of spacetime outside the Future and Past half-cones *cannot* be reached from the Here-Now except by world lines tilted more than 45°

---

[91]The "others" Feynman had in mind included, in particular, the eminent Swiss physicist Ernest C. G. Stükelberg (1905-1984), who in a 1942 article in the journal *Helvetica Physica Acta* also wrote of waves scattering backward in time.

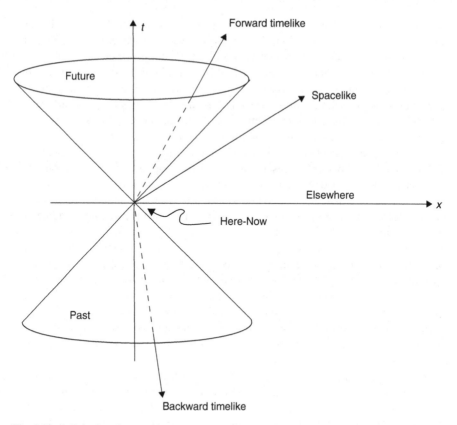

Fig. 3.10 A light cone in spacetime

away from the vertical. Such world lines, which represent travel at a speed faster than light, are called *spacelike* because their projections on a space axis is greater than their projections on the time axis. It is impossible for these world lines to connect causally linked events, and collectively they form the Elsewhere of the Here-Now. Notice that every point in spacetime has its own light cone. If A and B are causally linked, then if B is in the Future half-cone of A, then A is in the Past half-cone of B.

The imagery of the light cone is often useful in making seemingly quite abstract ideas appear transparent. For example, can an observer predict his own future from perfect knowledge of his own past? The easy answer is "No, because quantum uncertainties prohibit perfect knowledge of even the present, much less the past." But suppose we ignore quantum mechanics and limit our question to a universe that obeys only classical physics (which includes the special and general theories of relativity). Surprisingly (perhaps), the answer is *still* no. Having perfect knowledge of your own Past half-cone doesn't include knowing the *entire* past, so if you attempt to predict your own future (say, 1 min from now), there can be influences in Elsewhere that will arrive in the future (say, 59 s from now) about which you

Fig. 3.11 World line of a
particle traveling backward
in time from A to B

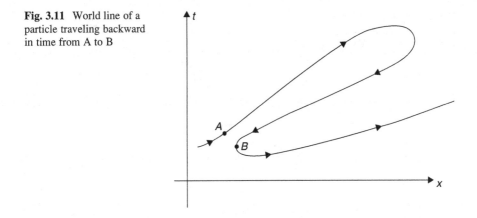

presently, by definition, cannot have any knowledge. And without that knowledge, you cannot predict. As one writer amusingly concludes a tutorial on this topic, "the prospect of predicting the future looks pretty bleak."[92]

A spacetime diagram does not always have to have future directed world lines. If a particle moves backward through time, assuming such a thing is possible, then the diagram can show this by having the world line double back on itself, as in Fig. 3.11 (in which the world line curves back and comes arbitrarily close to itself: it is the world line of a particle that visits itself in the past). Note that the world line in Fig. 3.11 does not actually touch or cross itself, because that would represent more than just a visit—it would represent a particle occupying the same spatial location at the same time as its earlier self. That would be catastrophic and, because it *did not* happen it *cannot* happen. Since the arrowheads on the world line always point in the direction of the local future of the particle, if the 'particle' is actually human then increasing memories are formed in the direction of the arrows. The time traveler at B has more memories than he does at A, even though A and B are nearly identical points in spacetime.

There is a problem with Fig. 3.11 that you may have caught. It is impossible to draw such a doubled-back world line in such a way that *at all places* it never tilts more than 45° from the vertical. That is, at least some portion of the world line will have

$$\left| \frac{dx}{dt} \right| > 1$$

which represents superluminal motion (we'll return to this in Chap. 5). One way to keep a bent-back world line always subluminal is to arrange for the light cones along the world line to be tilted relative to each other, as shown in Fig. 3.12, which

---

[92]M. Hogarth, "Predicting the Future in Relativistic Spacetimes," *Studies in the History and Philosophy of Science*, December 1993, pp. 721-739.

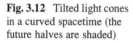

Fig. 3.12 Tilted light cones in a curved spacetime (the future halves are shaded)

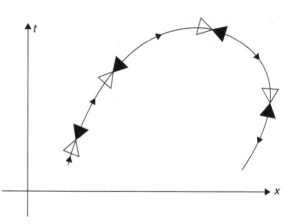

is possible only in a *curved* spacetime. This is an illustration of how general relativity *locally* obeys special relativity's demand that nothing travels faster than light and yet, *globally*, in curved spacetime, things are not so simple. In flat spacetimes all light cones are always 'aligned,' but in curved spacetimes they (generally) are not, and from that can come time travel to the past.[93]

Tilted light cones is the physics behind backward time travel around a rotating Tipler cylinder, for example, a particular time machine we'll discuss later in Chap. 5. Light cone tipping is, in fact, *essential* for time travel to the past. The mere presence of mass tips light cones ('warps spacetime'), but the effect is unnoticeable in everyday life on Earth. A truly enormous mass density is required to tip nearby light cones over so that their Future halves noticeably open up toward the massive body. If the massive body is additionally set to rotating, then a further consequence of Einstein's general theory is that the local light cones are tilted additionally in the direction of rotation. That is, the Future half-cones in spacetime open-up both toward the body and in the direction of the rotation.

It should now be clear that the only way a world line can bend back on itself for a close encounter visit is for both *x* and *t* to change. In other words, the world line of a particle that remains fixed in space and reverses just its time direction *runs into itself*. This is why the classic fictional time machine of H. G. Wells could not possibly work. A real time machine must move in space as well as through time, as does a Gödelian rocket (or, for that matter, as does the DeLorean time car in the *Back to the Future* films). The idea of warping world lines, to support time travel to the past, entered science fiction at an early date. For example, when the inventor of the time machine in a 1930s tale[94] is asked about the principle underlying his gadget, he replies "An electro-magnetic warping of the spacetime continuum.

---

[93]What is meant by a spacetime being *flat* will be formalized when we get to spacetime *metrics* later in this section. The Minkowski spacetime of special relativity is a flat spacetime, has no tilted light cones, and as such does *not* support time travel to the past.

[94]N. Schachner, "When the Future Dies," *Astounding Science Fiction*, June 1939.

The machine, if it works, will slide around the world-line of events and reappear at any specified time and place."

Using the Lorentz transformation equations from the previous section, we can establish quite general relationships between events in the Future, Past, and Elsewhere regions of spacetime. For example, (1) All events in the Future/Past for the Here-Now observer are in the Future/Past for any other nearby, relatively moving observer; (2) Any event in Elsewhere can appear to be simultaneous with the Here-Now for some observer and not simultaneous for another observer; and (3) The temporal ordering (the relations of *before* and *after*) of causally related events is the same for all observers. This is not so for events that are not causally related; if two events have a spacelike separation, then two observers can disagree over the temporal ordering of the events. This is, in fact, the basis for the two-wormhole and the cosmic string time machines, both of which will be discussed in Chap. 6. All these statements are easy to prove.

Consider, for example, statement (1). From the previous section we have (with $c = 1$)

$$t' = \frac{t - vx}{\sqrt{1 - v^2}} \text{ and } x' = \frac{x - vt}{\sqrt{1 - v^2}}$$

where t and x are the coordinates of some event A as measured by the observer in the stationary reference frame, and t' is the time measured by the observer in the reference frame moving at speed v. Thus

$$x'^2 - t'^2 = \frac{(x - vt)^2 - (t - vx)^2}{1 - v^2} = [\text{after a little algebra}] \ x^2 - t^2.$$

For the stationary observer the criterion for an event to be in the Future half-cone is $t > |x|$, that is, $t^2 > x^2$. Thus, $x^2 - t^2 < 0$ for all Future events. But the foregoing result then says $x'^2 - t'^2 < 0$, too, which is the moving observer's criterion for the event being in his Future half-cone. The same sort of argument shows that the two observers also agree on Past events.

Next, suppose that two events A and B occur such that the stationary observer measures them to be $\Delta T = t_B - t_A$ apart in time. Then, we can establish statement (2) by writing

$$t'_A = \frac{t_A - vx_A}{\sqrt{1 - v^2}} \text{ and } t'_B = \frac{(t_B + \Delta T) - vx_B}{\sqrt{1 - v^2}}$$

and so

$$\Delta T' = t'_B - t'_A = \frac{\Delta T + v(x_A - x_B)}{\sqrt{1 - v^2}}.$$

From this we have $\Delta T' = 0$ (that is, simultaneity) for the two events for the special observer moving at the speed

$$v = \frac{\Delta T}{x_B - x_A}$$

and this speed is less than the speed of light for the condition $x_B - x_A > \Delta T$. This is, of course, the condition for event B to be in the Elsewhere of event A. In fact, we can even have $\Delta T' < 0$ (with $\Delta T > 0$) for $v < 1$ in this case of spacelike separation of A and B. That is, a stationary observer and a sublight-speed moving observer can disagree about the temporal ordering of events with spacelike separation.

Similarly, for event B to be in the causal Future of event A, we have the condition $x_B - x_A < \Delta T$. Then,

$$\Delta T' = \frac{\Delta T - v(x_B - x_A)}{\sqrt{1 - v^2}} > \frac{\Delta T - v\Delta T}{\sqrt{1 - v^2}} = \Delta T \frac{1 - v}{\sqrt{1 - v^2}}.$$

Thus, $\Delta T > 0$ says $\Delta T' > 0$ for $v < 1$, and this establishes statement (3).

If we were drawing diagrams with both axes representing space (a plot of y versus x, for example), we would normally define a *distance metric* for the diagram using our everyday ideas about distance. That is, we could say that if we make differential movements of dx and dy along the two coordinate axes, then the differential distance ds is given by

$$(ds)^2 = (dx)^2 + (dy)^2.$$

This is, of course, just the Pythagorean theorem for the 'Euclidean' or 'as the crow flies' distance function. But it is not the only possible distance function. A distance function has several interesting mathematical properties,[95] but the one we are particularly interested in here is its invariance with respect to the coordinate system. For example, if we draw a line segment on a flat sheet of paper, the physical distance between its end-points does not depend on how we happen to select the x and y axes, a fact illustrated in Fig. 3.13 with the addition of a rotated and translated primed system. The coordinates for the endpoints A and B are obviously different in the two coordinate systems, but we still find that

$$(dx)^2 + (dy)^2 = (dx')^2 + (dy')^2.$$

---

[95]Mathematicians have defined the general properties of a distance function as follows: if A and B are any two points, and if $d(A,B)$ is the distance between A and B , then (1) $d(A,B) = d(B,A)$; (2) $d(A,B) = 0$ if and only if $A = B$; and (3) if C is any third point, then $d(A,B) \leq d(A,C) + d(C,B)$. The Pythagorean distance function possesses all three of these properties, but so do many other functions (for example, $ds = |dx| + |dy|$).

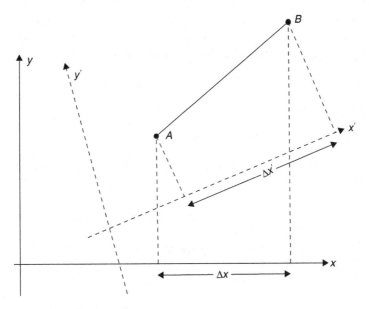

Fig. 3.13 Rotated/translated coordinate systems

We say that the Pythagorean distance function is *invariant*, that it is the same for all coordinates systems that are simply rotations and/or translations of each other.

We know that different observers, if in relative motion in the same spacetime, will see different space and time coordinates for the same event. Thus, it is natural to ask 'what is the metric for flat spacetime?' Is there, in fact, a metric that gives the same *distance* between two events for all observers? We might try to generalize in the obvious way from the Pythagorean theorem, and write

$$(ds)^2 = (dt)^2 + (dx)^2 + (dy)^2 + (dz)^2$$

where now all four dimensions are included. We would then ask ourselves whether it is true that

$$(ds)^2 = (ds')^2 = (dt')^2 + (dx')^2 + (dy')^2 + (dz')^2?$$

For our simple two-dimensional spacetime, this question reduces to asking whether

$$(dt)^2 + (dx)^2 = (dt')^2 + (dx')^2?$$

Using the Lorentz transformation equations from earlier, it is easy to discover that the answer is no. The 'natural' generalization of Pythagorean distance for flat, two-dimensional spacetime fails when four dimensions are included. So, what do we do now? Recalling the words of Professor Mundle from *Some First Words* (see note 23 there), we might wonder whether this difficulty could result from the fact that there is no fourth direction along which the time axis can point at right angles to the three space directions? At least, there is no *real* direction—but perhaps there is an imaginary one. Accordingly, with $= \sqrt{-1}$, let's try

$$(ds)^2 = (idt)^2 + (dx)^2 + (dy)^2 + (dz)^2 = -(dt)^2 + (dx)^2 + (dy)^2 + (dz)^2.$$

Using *imaginary time*, something that seems to be in the realm of science fiction, has resulted in a change in the sign of $(dt)^2$.

This is a crucial change, however, because this new metric *is* invariant. For a reason to be explained in the next section, I will use the negative of this metric (a choice that has no impact on the invariance property) and so write

$$(ds)^2 = (dt)^2 - (dx)^2 - (dy)^2 - (dz)^2.$$

For our simplified two-dimensional spacetime this reduces to

$$(ds)^2 = (dt)^2 - (dx)^2.$$

As before, the Lorentz transformation equations (with $c = 1$) are

$$t' = \frac{t - vx}{\sqrt{1 - v^2}} \quad \text{and} \quad x' = \frac{x - vt}{\sqrt{1 - v^2}}.$$

If we then calculate dx' and dt' from these equations, using

$$dx' = \frac{\partial x'}{\partial x}dx + \frac{\partial x'}{\partial t}dt, dt' = \frac{\partial t'}{\partial x}dx + \frac{\partial t'}{\partial t}dt$$

which are the fundamental relations for the *total* differential[96] of a function of two variables, and insert the results into $(dt')^2 - (dx')^2$, we quickly discover the invariance property of this quantity (a result we actually found earlier, in a different way, when we showed that observers in relative motion agree about what events are in the Future and what events are in the Past). Thus,

$$(dt')^2 - (dx')^2 = (dt)^2 - (dx)^2.$$

This quantity, on either side of the equality, is called the spacetime *interval* between the two events separated in flat spacetime by either dt, dx, dy, and dz, or by dt', dx', dy', and dz'. The observers in the unprimed and the primed systems see different individual space and time separations for two events, but they see the same

---

[96]The $\frac{\partial}{\partial x}$ and $\frac{\partial}{\partial t}$ symbols denote the *partial* derivatives with respect to x and t (see any good calculus book to brush-up on this). The rest of this chapter *will* have some more math in it, involving derivatives and even an integral or two, but nothing beyond freshman calculus. I've included it mostly for those who would feel cheated without *some* math!

interval. A single time coordinate, and three space coordinates, are said to form a *four-vector that is invariant under Lorentz transformation*. There are, in addition, other four-vectors that are also invariant under Lorentz transformation, such as the energy-momentum, velocity, and force four-vectors; all these quantities have invariants that are formed the same way, by taking the difference of the squares of the time components and the sum of the squares of the space components. The intrusion of the square root of minus one, in the time coordinate of the metric, seems a pretty clear indication that time *is* different from space.[97]

But still, the spatialization of time is nonetheless deeply embedded in Western culture. For example, when writing her popular 1978 book, drawing historical parallels between the fourteenth and twentieth centuries, Barbara Tuchman titled it *A Distant Mirror*, and not *An Old Mirror*. The mathematical mixing of space and time appeared quite early in science fiction, but often in comically mangled form. In one such tale, for example, as an evil scientist uses his time machine to transport captives into the past, he tells them, "We've got a longish journey before us, ten thousand years more, multiplied by the fourth power of two thousand miles."[98]

In general relativity, the metric of any four-dimensional spacetime has the structure of what mathematicians call a *symmetric quadratic Riemannian form*:

$$(ds)^2 = \sum_{i=1}^{4} \sum_{j=1}^{4} g_{ij}(dx_i)(dx_j), g_{ij} = g_{ji}$$

where $x_1 = t$, $x_2 = x$, $x_3 = y$, and $x_4 = z$, and the 16 g's are all functions of these four variables. (Because of the symmetry condition, only 10 of the g's are independent.) In this notation, a *flat* spacetime is mathematically characterized by $g_{11} = 1$, $g_{22} = g_{33} = g_{44} = -1$, $g_{ij} = 0$ for all $i \neq j$. Now, for a given spacetime, one can arbitrarily choose an infinity of coordinate systems. If just one of this infinity of systems is such that the ± 1, 0 values for the g's occur, then that spacetime is *globally* (that is, everywhere) flat. If no such coordinate system exists, then that spacetime is necessarily *curved*—I'll give you a more intuitive view of curvature in just a bit. (If this notation is extended to a fifth dimension[99] by including the additional coordinate x_5, then there are an additional eight off-diagonal g's and so

---

[97]Writing in 1972, one famous physicist said of his first encounter with the metric of special relativity (Minkowski spacetime), "Now, when I saw that minus sign [in $-(dt)^2$], it produced a tremendous effect on me. I immediately saw that here was something new." See P. A. M. Dirac, "Recollections of an Exciting Era," in *History of Twentieth Century Physics* (C. Weiner, editor), Academic Press 1977.

[98]V. Rousseau, "The Atom Smasher," *Astounding Stories*, May 1930.

[99]The fifth dimension was introduced in the 1920s by the German physicist Theodor Kaluza (1885-1954), but just what the nature of this fifth dimension might be remains a mystery. A few years after Kaluza, the Swedish physicist Oscar Klein (1894-1977) speculated that it might be a spatial dimension curled-up in a tiny circular path, so tiny that we don't notice it; the issue remains open. The idea of a fifth dimension appeared early in pulp science fiction, as in the January 1931 tale "The Fifth-Dimension Catapult," (*Astounding*) by Murray Leinster.

four new *independent* g's. These are just sufficient to describe the electromagnetic field, too, along with the gravitational field described by the other ten g's. That is what is meant by saying five-dimensional spacetime 'unifies' gravity with electromagnetism.) There is a g_{55}, too, which could be allowed to model a slowly varying gravitational constant, as suggested by Dirac in 1938.

The g functions are the components of the so-called *metric tensor of the second rank* (see the last discussion question for more on what this means) of that spacetime. The g's at each point in spacetime are related to the curvature of spacetime at that point, which in turn is dependent on the g's and on the energy density at that point. In fact, the ten equations for the g's, which *are* the famous Einstein gravitational partial differential field equations, are both nonlinear and coupled. That is, each g_{ij} is in general a nonlinear function of all the other g_{ik}, which accounts for the notorious difficulty[100] in finding analytical solutions to the field equations except in certain highly special cases, such as the spacetime of spinning spheres and rotating infinite cylinders.

What does it mean to 'solve' the field equations? It is useful to think of the equations schematically as follows:

local geometry of spacetime $\leftarrow$ local density, momentum and stress of the mass - energy of spacetime where the direction of the arrowhead means that the 'usual' practice is to assume the right-hand side (the so-called stress-energy tensor) as given, and then attempt to calculate the left-hand side. If the attempted calculation can be done, then one has solved the field equations for the spacetime geometry that is associated with the assumed mass-energy distribution.

Suppose, however, that we reverse the direction of the arrowhead. That is, suppose we assume a desired geometry. That is what Einstein did when he assumed the geometry of a *static* (non-expanding) universe and solved for the required mass-energy. What he found was just what had been observed by astronomers up to that time—a multitude of 'grains of matter' (what physicists call *dust*) plus the infamous cosmological constant. The constant, with its repulsive gravity, was needed to counteract the ordinary gravitational attraction of the stars that tends to pull them together. The later discovery that the universe is not static, but rather is expanding, rendered Einstein's solution moot.

Now let's go Einstein one better, and assume a spacetime geometry that contains closed timelike curves (a time machine, in other words) and then try to calculate the mass-energy distribution required by that spacetime. If that can be accomplished— and in fact the field equations themselves provide an algorithmic means of solution in this direction—then the physicist's work is done. The required mass-energy distribution requirements are put out to bid to 'spacetime engineers' and the lowest bidder 'simply' constructs that mass-energy distribution and so builds us our time machine! When the calculations are done, however, what has happened without

---

[100]An elementary, quite interesting discussion of the enormous computational complexity of the field equations is presented in Richard Pavelle and Paul S. Wang, "MACSYMA from F to G," *Journal of Symbolic Computation*, March 1985, pp. 69-100.

fail, at least up to now, is that the resulting mass-energy distribution comes out with an 'unphysical' nature, a technical way of saying our spacetime engineer wouldn't know *how* to assemble the required mass-energy distribution. What is meant by 'unphysical' will be explored, in the particular case of wormhole time machines, in Chap. 6.

The 16 g's for a four dimensional spacetime are often written in the form of a 4×4 matrix. In fact, the metric tensor is of the *second* rank precisely because a matrix has a *two*-dimensional form; scalars and vectors, which have forms of zero and one dimension, are tensors of rank zero and one, respectively. The collection of the algebraic signs of the main diagonal terms (the g_{ii}) is called the *signature* of the metric tensor. The signature of flat (Minkowski) spacetime is thus written as $[+, -, -, -]$; in more general (curved) spacetimes, this same signature is called *Lorentzian*. By contrast, the signature of a four-dimensional Euclidean space is $[+, +, +, +]$. This signature is called *Riemannian*. The geometry of flat, Minkowski spacetime is not Euclidean geometry because the spacetime signature of its metric has both plus *and* minus signs. As an uncurved spacetime, Minkowski spacetime *has no gravity*; to get gravity, we need a curved spacetime.

The idea of linking the curvature of a four-dimensional space to physical phenomena is the signature feature of general relativity, but it actually pre-dates Einstein by decades. It can be found, for example, in the work of the British mathematician William Kingdon Clifford (1845–1879), done in the 1860s and 1870s before Einstein's birth (in the year of Clifford's death). In his posthumously published book *The Common Sense of the Exact Sciences* (1885), Clifford wrote "We may conceive our space to have everywhere a nearly uniform curvature, but that slight variations of the curvature may occur from point to point and themselves vary with time. These variations of the curvature with time may produce effects which we not unnaturally attribute to physical causes independent of the curvature of our space. We may even go so far as to assign to this variation of the curvature what really happens in that phenomenon which we term the motion of matter." It isn't a long jump from "motion of matter" to gravity![101]

Not everyone enthusiastically embraced this new, radical view of nature. For example, the great Scottish mathematical physicist James Clerk Maxwell (1831–1879), of Maxwell's equations fame, who knew Clifford through their common membership in the London Mathematical Society, summarily dismissed this part of Clifford's work as simply the speculations of a "space crumbler." Decades later, Einstein faced the same rejection when the eminent British

---

[101]Clifford almost surely found inspiration in this part of his work from the even earlier efforts of the German mathematician Bernhard Riemann (1826-1866). See, for example, Clifford's translation of Riemann's famous 1854 lecture "On the Hypotheses Which Lie at the Bases of Geometry," *Nature*, May 1, 1873, pp. 14-17, and continued in the next issue (May 8, 1873, pp. 36-37).

astronomer Sir James Jeans declared "Einstein's crumbling of his four-dimensional space may ... be considered to be ... fictitious."[102]

Spacetime geometries are not easy concepts to grasp, and the metrics of curved spacetimes are even more complicated than is the metric of the flat spacetime of Minkowski's special relativity. As one paper so aptly put it, "Experience has taught us that the space in which we live has a geometry that is three-dimensional and Euclidean ... We are very much at home with [that] geometry ... But the geometric properties of a Minkowskian space are so alien to us that we may well despair of visualizing them, and a Riemannian [curved] space ... seems totally beyond comprehension."[103]

An important feature of Riemannian geometry is that although it is generally not globally flat, it is always *locally* flat. Thus any sufficiently small region in a curved Riemannian spacetime can be approximated, with arbitrarily small error, by a flat pseudo-Euclidean Minkowskian spacetime. That is, at each point in Riemannian spacetime, there is some *particular* inertial frame of reference in which special relativity is all there is to spacetime physics *at that point*. The particular inertial frame required is different, however, from point to point.

In a coordinate system different from rectangular, the flat Minkowskian metric can appear radically altered, but that is just an artifact of the mathematics and has no physical significance. For example, in spherical coordinates the Minkowskian metric becomes the equivalent

$$(ds)^2 = (dt)^2 - (dr)^2 - (rd\theta)^2 - (rsin(\theta)d\phi)^2$$

where ϕ is the azimuthal angle and θ is the angle measured from the polar axis. A related metric occurs in the theory of spherically symmetric, static (no time variation) time machine wormholes (discussed in Chap. 6), of the form

$$(ds)^2 = \left(e^{a(r)}dt\right)^2 - \frac{(dr)^2}{1 - \frac{b(r)}{r}} - (rd\theta)^2 - (rsin(\theta)d\phi)^2$$

where $a(r)$ is called the *redshift function* and $b(r)$ is called the *shape factor*. These two functions are nearly arbitrary, subject only to the constraints that both $b(r)/r$ and $a(r)$ vanish as r goes to infinity (r is the radial distance from the throat of the wormhole mouth). Indeed, as r increases, this curved wormhole spacetime metric

[102]Quoted in P. Kerszberg, "The Relativity of Rotation in the Early Foundations of General Relativity," *Studies in History and Philosophy of Science*, March 1987, pp. 53-79.

[103]R. W. Brehme and W. E. Moore, "Gravitational and Two-Dimensional Curved Surfaces," *American Journal of Physics*, July 1969, pp. 683-692.

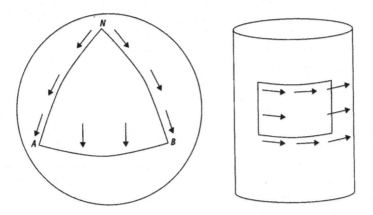

Fig. 3.14 The curvature of a space can be revealed by the process of parallel transport

reduces to that of flat Minkowskian spacetime, so this wormhole spacetime is said
to be *asymptotically flat*.[104]

An intuitive appreciation of 'flatness' can be realized in terms of what is called
the *parallel transport* of a vector around a closed path. In a curved space the vector
will experience a rotation, which will not occur in a flat space. Two examples of
parallel transport in ordinary three-dimensional space are shown in Fig. 3.14. The
spherical surface on the left is curved because if you slide the vector from N to A to
B and then back to N, all the while keeping it parallel to its immediately previous
orientation, then when it gets back to N the vector will point to B, *not* toward A as it
initially did. A similar trip on the cylindrical surface, however, results in zero
rotation of the vector. Thus, the cylindrical surface, despite superficial appearances,
is *not* curved.

Using the idea of the metric tensor, we can develop a more formal demonstration
that the surface of a sphere, unlike like that of a cylinder, is not flat. On the surface
of a sphere of radius a (*on the surface* $r = a$ everywhere, and so $dr = 0$), the measure
of the distance between two points (in spherical coordinates) is[105]

$$(ds)^2 = a^2 sin^2(\theta)(d\phi)^2 + a^2(d\theta)^2.$$

Writing $x_1 = \phi$ and $x_2 = \theta$ yields the more general form

---

[104]For wormhole spacetimes that are not asymptotically flat, see (for example) K. Narahara, *et al.*,
"Traversable Wormhole in the Expanding Universe," *Physics Letters B*, September 29, 1994,
pp. 319-323.

[105]Note *carefully* that this is a purely spatial problem, with no time, and we are taking all of the
metric coefficients as positive (unlike in the case of a spacetime metric).

$$(ds)^2 = g_{11}(dx_1)^2 + g_{12}(dx_1)(dx_2) + g_{21}(dx_2)(dx_1) + g_{22}(dx_2)^2$$

or, using the symmetry condition $g_{12} = g_{21}$, we have

$$(ds)^2 = g_{11}(dx_1)^2 + 2g_{12}(dx_1)(dx_2) + g_{22}(dx_2)^2.$$

From this we immediately have $g_{11} = a^2 sin^2(\theta)$, $g_{22} = a^2$, and $g_{12} = g_{21} = 0$.

Now, suppose we ask whether it is possible to find some new coordinate system (with variables x_1' and x_2') in which the invariant $(ds)^2$ is given by the flat Euclidean metric $(dx_1')^2 + (dx_2')^2$. In such a coordinate system (if it exists) we would have the 'flatness conditions' of $g_{11}' = g_{22}' = 1$ and $g_{12}' = g_{21}' = 0$. With such a change of coordinates, each of our original ϕ, θ coordinates would generally be a function of both of the new coordinates—that is, $\phi = \phi(x_1', x_2')$ and $\theta = \theta(x_1', x_2')$. Thus, writing the total differential of a function of two variables (see note 96 again), we have

$$d\phi = \frac{\partial \phi}{\partial x_1'} dx_1' + \frac{\partial \phi}{\partial x_2'} dx_2'$$

$$d\theta = \frac{\partial \theta}{\partial x_1'} dx_1' + \frac{\partial \theta}{\partial x_2'} dx_2'.$$

Substituting these expressions into the above expression for $(ds)^2$ on the surface of a sphere, and collecting terms, we arrive at

$$\begin{aligned}
(ds)^2 &= a^2 \left\{ \left[sin^2(\theta) \left(\frac{\partial \phi}{\partial x_1'} \right)^2 + \left(\frac{\partial \theta}{\partial x_1'} \right)^2 \right] (dx_1')^2 + \left[sin^2(\theta) \left(\frac{\partial \phi}{\partial x_2'} \right)^2 + \left(\frac{\partial \theta}{\partial x_2'} \right)^2 \right] (dx_2')^2 \right. \\
&\quad \left. + 2 \left[sin^2(\theta) \frac{\partial \phi}{\partial x_1'} \frac{\partial \phi}{\partial x_2'} + \frac{\partial \theta}{\partial x_1'} \frac{\partial \theta}{\partial x_2'} \right] (dx_1')(dx_2') \right\} \\
&= g_{11}'(dx_1')^2 + 2g_{12}'(dx_1')(dx_2') + g_{22}'(dx_2')^2
\end{aligned}$$

We can now immediately write down each of the g' and, if we demand that they satisfy the 'flatness conditions,' then we have the following three statements:

$$sin^2(\theta) \left(\frac{\partial \phi}{\partial x_1'} \right)^2 + \left(\frac{\partial \theta}{\partial x_1'} \right)^2 = \frac{1}{a^2}$$

$$sin^2(\theta) \left(\frac{\partial \phi}{\partial x_2'} \right)^2 + \left(\frac{\partial \theta}{\partial x_2'} \right)^2 = \frac{1}{a^2}$$

$$sin^2(\theta) \frac{\partial \phi}{\partial x_1'} \frac{\partial \phi}{\partial x_2'} + \frac{\partial \theta}{\partial x_1'} \frac{\partial \theta}{\partial x_2'} = 0.$$

For a *globally* flat surface, that is, a surface that is flat *everywhere*, these three statements must hold in particular at the poles of the sphere. That is, at $\theta = 0°$

and at $\theta = 180°$. At both of these points $\sin(\theta) = 0$, and so at the poles the three statements reduce to

$$\left(\frac{\partial\theta}{\partial x_1'}\right)^2 = \frac{1}{a^2}$$

$$\left(\frac{\partial\theta}{\partial x_2'}\right)^2 = \frac{1}{a^2}$$

$$\frac{\partial\theta}{\partial x_1'}\frac{\partial\theta}{\partial x_2'} = 0.$$

But the third statement is incompatible with the first two and, because of that incompatibility, there is no primed coordinate system in which the g' coefficients in the metric are those of a globally flat metric. Thus, unlike the surface of a cylinder, the surface of a sphere is not flat but rather is curved. This almost surely comes as no surprise to you, but now you can *prove* it!

3.6   Proper Time and the Twin Paradox in Time Travel to the Future

"[The] equations of duo-quadrant lineations [have] been substantiated ... Our fourth-angle deviation from the six conceivable electronic dimensions did the trick all right. I went forward in Time."[106]

The spacetime interval of the previous section has an important interpretation that leads to one of the more dazzling results of special relativity—time travel into the future. First, recall the flat spacetime metric

$$(ds)^2 = (dt)^2 - (dx)^2 - (dy)^2 - (dz)^2$$

in which the use of unprimed variables indicates that the measurements on the space and time coordinates of a moving particle are made with respect to a stationary observer's frame of reference. Now, suppose that the space and time coordinates of a moving particle are made with respect to the particle instead. Then, using primed variables for measurements made in this new frame of reference, $dx' = dy' = dz' = 0$ because the particle is always at the origin (by definition)! Recalling the invariance of the spacetime interval for all observers, we conclude that

---

[106] A science fiction scientist babbles incoherent nonsense, not special relativity, about how to travel into the future, in a story by J. H. Haggard, "He Who Masters Time," *Thrilling Wonder Stories*, February 1937.

$$(ds')^2 = (ds)^2 = (dt')^2.$$

That is, the spacetime interval between two events is the time lapse measured by a clock attached to a particle that moves from one event to the other. This time is called *proper time*, which gets its name from the idea that it belongs to (is the *property* of) the moving particle. This is the technical reason for taking (as we did in the previous section) $(ds)^2 = (dt)^2 - (dx)^2$ rather than $(ds)^2 = (dx)^2 - (dt)^2$. The first choice avoids the somewhat awkward result of an imaginary proper time.

Next, we'll adopt what has come to be called the *clock hypothesis*, which states that an accelerated clock runs at the same instantaneous rate as an unaccelerated clock that is moving alongside at the instantaneously same speed. As we showed earlier (take a look back at Sect. 3.3), if the accelerated clock's instantaneous speed is v, then its rate of time keeping (dt') is related to that of the 'stationary' (unaccelerated) clock (dt) as

$$dt' = \sqrt{1 - \left(\frac{v}{c}\right)^2}\, dt$$

where c is the speed of light. The clock hypothesis is generally assumed to be true. Einstein, himself, in his famous 1905 special relativity paper, specifically took the rate of a clock's timekeeping to be velocity-dependent only. When asked during an interview decades later whether it is permissible to use special relativity in situations involving acceleration, Einstein replied "Oh, yes, that is all right as long as gravity does not enter; in all other cases, special relativity is applicable. Although, perhaps the general theory approach might be better, it is not necessary."[107]

The clock hypothesis has long had experimental verification. For example, in one experiment the time keeping of accelerated atomic clocks was determined to be given precisely by the time dilation formula of special relativity, even when their direct mechanical acceleration (the centripetal acceleration produced by a rapidly spinning disk) exceeded 66,000 gees![108] And even more impressive are the time dilation results of a later experiment in which the time keeping of near light-speed charged particles, orbiting in a magnetic field, was in excellent agreement with the time dilation formula, even as accelerations well in excess of 10^{15} gees were reached![109]

The total elapsed time between two events A and B, as measured by the proper time of an accelerated clock making the journey, is, therefore, given by

[107]R. S. Shankland, "Conversations with Albert Einstein," *American Journal of Physics*, January 1963, pp. 47-57.

[108]H. J. Hay, *et al.*, "Measurement of the Red Shift in an Accelerated System Using the Mössbauer Effect in Fe^{57}," *Physical Review Letters*, February 15, 1960, pp. 165-166.

[109]J. Bailey, *et al.*, "Measurements of Relativistic Time Dilation for Positive and Negative Muons in a Circular Orbit," *Nature*, July 29, 1977, pp. 301-305.

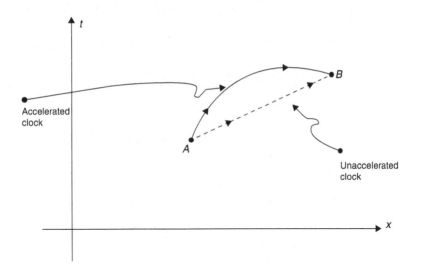

Fig. 3.15 World lines of two clocks (one accelerated and one unaccelerated)

$$t' = \int dt' = \int_{t_A}^{t_B} \sqrt{1 - \left(\frac{v}{c}\right)^2} \, dt < t_B - t_A \text{ if } v \neq 0$$

where $t_B - t_A$ is the elapsed time between A and B as measured by the unaccelerated clock. The inequality results because, for $v \neq 0$, the integrand is always less than 1. Now, we know that in a spacetime diagram the world line of the unaccelerated clock is a straight line, whereas the world line of the accelerated clock is a curved line. Thus, using Fig. 3.15, combined with the inequality $t' < t_B - t_A$, we have the following central result: the world line of *maximum* proper time is the one that *looks the shortest*, that is, the straight (or free-falling geodesic) world line. In the spacetime diagram the curved line looks longer, but in fact any curved line will have a smaller proper time than does the straight world line. This is a dramatic example of how Minkowskian *spacetime* geometry differs from Euclidean *space* geometry; in the latter geometry, there is no longest path between two points.

From this, we can now understand the famous paradox of the twins, Bob and Bill. Bill remains on Earth, but Bob gets into a rocket ship and goes on a high speed trip out into space. Eventually he brings his ship to a stop, turns around, and returns to Earth. The world lines of Bob and Bill are initially together, then they diverge as Bob goes on his trip, and then they come together again at the end of Bob's trip, as shown in Fig. 3.16. The details of Bob's trip are not important for a general statement of the paradox (although in just a bit I *will* present the details for one possible trip). All we need observe for now is that Bill's world line from A to B is straight, whereas Bob's is curved. Bill's body (that is, his local clock) will therefore measure a greater proper time than will Bob's; that is, Bob will be younger than his stay-at-home twin! Equivalently, upon his return Bob will hear his Earthbound

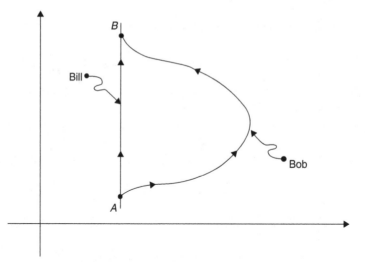

Fig. 3.16 World lines of unaccelerated (Bill) and accelerated (Bob) twins

brother declare the date to be further in the future than Bob's trip lasted (according to Bob). Bob will therefore conclude that he has traveled into the future.[110]

This situation is called a paradox, not because of the time travel aspect (there are no logical paradoxes associated with travel into the future), but rather because it seems to violate the very spirit of special relativity. That is, from Bill's point of view, Bob at first travels away and then returns. But one can argue that from Bob's point of view it is Bill who first recedes and then returns. So why is it *Bob* who is the younger, and not Bill? Long after Einstein's 1905 publication of special relativity, this point remained a great puzzle for many. For example, in the 1923 Presidential Address to the Eastern Division of the American Philosophical Association, we read this very objection to the twin paradox. The conclusion by the speaker is that such a thing "could happen only in a universe in which all squares were round and the *principio contradiction* had been put to sleep."[111]

So, what's the answer, why is it *Bob* who is the younger of the twins? The classic physics answer is that the two points of view are actually *not* symmetrical, that there is a definite asymmetry between Bill and Bob. After all, it is Bob who feels the acceleration from the rocket's engines as he blasts off from Earth—it is Bob, not

---

[110]The twin paradox is hinted at in Einstein's 1905 paper, but it is in a 1911 address to the International Congress of Philosophy in Bologna, by the French physicist Paul Langevin (1872-1946), that a human space traveler is first introduced (in a cannonball moving at near light-speed, an idea motivated by Langevin's reading of Jules Verne's 1872 novel *From the Earth to the Moon*). The writer Pierre Boulle proudly mentioned this contribution by his fellow Frenchman in the time travel story "Time Out of Mind" (you can find it in Boulle's collection *Time Out of Mind*, Vanguard Press 1966).

[111]W. P. Montague, "The Einstein Theory and a Possible Alternative," *The Philosophical Review*, March 1924, pp. 143-170.

Bill, who feels *force*—whereas Bill feels nothing unusual as he remains on Earth. The more fundamental physics answer, however, is that Bob's world line in spacetime is curved, whereas Bill's is straight.

In an open, spatially unbounded, flat spacetime, *curved* is indeed synonymous with *accelerated*, but this need not be so in a *closed*, flat spacetime. In an open, flat spacetime, the only way two world lines can diverge in the past and then meet again in the future is for at least one of them to curve, but in a closed but still flat spacetime, it is possible for two straight world lines to meet more than once. For example, Fig. 3.17 shows a simple two-dimensional spacetime that is the surface of a cylinder (which you'll recall we argued earlier is flat), rather than an infinite flat plane. The two world lines in that figure are *both* straight: to visualize this, imagine cutting the cylinder open along the (vertical) time dimension, and then flattening it out. Bob's world line, however, looks longer in the spacetime diagram, so Bob's proper time will be less than Bill's when they meet again, even though now *neither* of them has experienced any acceleration.[112] This is simply an interesting mathematical exercise, however, and as far as is known the spacetime we live in is not cylindrical and so Bob's trip into the future will require an accelerating rocketship.[113]

If we specify the details of Bob's trip, we can then precisely calculate the difference in elapsed time for the twins. In an analysis that dates back to 1962, the German astrophysicist Sebastian von Hoerner (1919–2003) did that for the following trip[114]: To begin, Bob gets into his rocket ship at time $t = t' = 0$ (t is time measured on Earth by Bill, and t' is time measured by Bob in his rocket). The Bill and Bob synchronize their clocks at the instant of departure. Bob's trip is to be made in comfort, and so his rocket accelerates at a constant rate (a one gee acceleration, for example, would be equivalent to Earth's gravity, and Bob would feel right at home). This is of practical importance, obviously, because we do not want the experienced acceleration to be incompatible with the physical survival of Bob. Bob

---

[112]See C. H. Brans and D. R. Stewart, "Unaccelerated-Returning Twin Paradox in Flat Space-Time," *Physical Review D*, September 15, 1973, pp. 1662-1666. For a similar treatment, this time by a mathematician, see Jeffrey R. Weeks, "The Twin Paradox in a Closed Universe," *American Mathematical Monthly*, August-September 2001, pp. 585-589.

[113]In Chapter 6 we'll discuss the idea of traveling into the past by moving faster than light (*superluminal* motion). A treatment of such travel, in Bob's cylindrical spacetime, is by S. K. Blau, "Would a Topology Change Allow Ms. Bright to Travel Backward in Time?" *American Journal of Physics*, March 1998, pp. 179-185, which answers that question in its last line: "Ms. Bright cannot [return] 'the previous night' and alter history," a conclusion that no doubt met with Hawking's approval. The 'Ms. Bright' in the title is the heroine of a 1923 limerick that you can find quoted in the first *For Further Discussion* of Chapter 6.

[114]Originally appearing in the journal *Science*, under the title of "The General Limits of Space Travel," von Hoerner's analysis was reprinted in the classic anthology *Interstellar Communication* (A. G. W. Cameron, editor), W. A. Benjamin 1963. The arithmetic was, alas, just a bit sloppy (the final formulas, fortunately, are correct), and many of von Hoerner's numerical evaluations are incorrect. Later, the British mathematician Leslie Marder cleaned-up the analysis in his beautiful little book on the twin paradox, *Time and the Space-Traveler*, George Allen & Unwin, Ltd. 1971.

Fig. 3.17 Unaccelerated
twin paradox in a
cylindrical spacetime

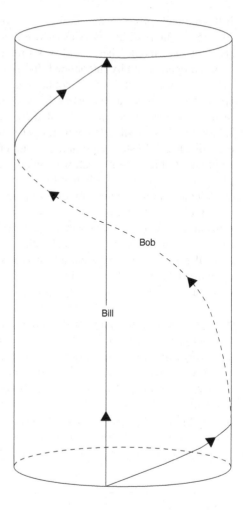

travels this way for a time interval of T (as measured by Bill on Earth) and T'
(as measured by Bob in his rocket). At that time the rocket is traveling at its
maximum speed. Bob then turns off the rearward engine and turns on a forward-
mounted engine so as to experience a constant deceleration. Floor and ceiling
interchange, but Bob always weighs the same. If he does this for the same time
interval T (as measured by Bill on Earth) and T' (as measured by Bob) as for the
initial acceleration phase of the trip, the rocket will be brought to rest with respect to
Earth. At that time, $2T$ (as measured by Bill on Earth) and $2T'$ (as measured by Bob),
the rocket is at its maximum distance from Earth. Bob then returns to Earth, using
the same acceleration/deceleration process, and Bob arrives back home, gently,
with a final speed of zero with respect to Earth (ignoring, of course, all the
navigational problems due to the motion of Earth during the trip).

Assuming the clock hypothesis is true (not all physicists believe this,[115] but remember, we have Einstein[116] on our side with this!), the result is that if a is the constant acceleration of the rocket (as experienced by Bob), and c is the speed of light, then the relationship between the *total roundtrip time* as measured by Bill ($4T$) and as measured by Bob ($4T'$), is given by

$$T = \frac{c}{a}\sinh\left(\frac{a}{c}T'\right).$$

The difference in what each twin believes to be the date can, in fact, be truly astonishing. For example, if $a = 1$ gee and Bob travels (by his clock) for $4T' = 20$ years, then $4T = 339$ years.[117] Of course, there will actually be no disagreement between Bill and Bob over the date upon Bob's return because Bill will be long dead!

Is it likely that such a time trip[118] will someday be made into the future? It's just my opinion, but I suspect not. Bob will be traveling at virtually the speed of light at maximum speed ($0.9993\,c$, to be precise, at which point he will be 84 light-years from Earth), and to zip through space at such a speed would result in a very high rate of collision with stray hydrogen atoms (about one per cubic centimeter). The result of those energetic collisions would be the intense blasting of Bob with a lethal dose of gamma radiation. And, as von Hoerner showed in his original analysis (note 114), the energy required by Bob's rocket would be simply mind-boggling.

Professor Schild (see note 116), on the other hand, seems to have been less bothered by such considerations when he wrote, at the beginning of a prose discussion of the twin paradox, "I have no doubt that if our technology should ever advance to the stage where large-scale twin effects become noticeable with our unaided senses, then [people] will have no difficulty in adjusting their concepts of time until the new phenomena see quite natural."[119]

I'll end this chapter on time with the observation that it is the distinction between the proper time of Bob on a rocket ship, and the time of those who are not fellow

---

[115]Consider, for example, this remark by 1965 Nobel physics laureate Julian Schwinger (1918-1994) about the twin paradox in his 1986 book *Einstein's Legacy*: "The observer on the spaceship ... is *not* in uniform, unaccelerated motion ... The special theory of relativity does not apply to such an accelerated observer." Schwinger was wrong in this conclusion (see the next note).

[116]As a physicist wrote on this point after Einstein's death, "A good many physicists believe that [the twin] paradox can only be resolved by the general theory of relativity. ... However, they are quite wrong. The twin effect ... is one of special relativity." See A. Schild, "The Clock Paradox in Relativity Theory," *American Mathematical Monthly*, January 1959, pp. 1-18. Alfred Schild (1921-1977) was professor of physics at the University of Texas, and a recognized expert in the general theory.

[117]In using the T, T' formula, one has to be careful to use MKS units, that is, length and time measured in meters and seconds, respectively. Thus, $a = 1$ gee $= 9.81$ meters/second^2 and $c = 186,210$ miles/second $= 2.997 \times 10^8$ meters/second.

[118]For a science fiction use of such a trip, see Robert Heinlein's 1956 novel *Time for the Stars*.

[119]A. Schild, "Time," *Texas Quarterly*, Autumn 1960, pp. 42-62.

space travelers, that eliminates the occasional suggestion of simply *freezing* one's way via suspended animation to the future (as in the 1992 film *Forever Young*). That is, suppose you manage to talk a friend into climbing into a freezer (please don't actually try this!); at the moment he gets in, his wrist watch agrees with yours (both are powered and maintained at a constant temperature by 100-year nuclear batteries). Years later, when you thaw your friend out, you'll find that your watches still agree. But when Bob returns from his rocket trip, his watch will not agree with yours, but instead will be far behind. Bob is a true time traveler, but your frozen/ thawed friend is not.[120]

A provocative illustration of the distinction between the proper time of a time traveler and of those who are not time travelers was given by a philosopher.[121] He begins his analysis by asking what appears to be a question with an obvious answer. Suppose, he says, it is [2018] and you suddenly wake up in a hospital and are told that you have been in a coma for the past 2 weeks. You are also told that you were in an auto accident 2 weeks ago, that you suffered temporary neural damage, and that the eventual reversal of such damage always, at some time within 4 weeks after the damage occurs, causes a day of excruciating pain *if you are conscious at the time*. Would you prefer for the day of damage reversal to be in the past 2 weeks (when you were in the coma) or in the next 2 weeks? The answer seems clear. After all, if the day that damage reversal occurs has already happened, then you simply slept through it and missed the pain. To prefer the day of pain to be in the future (when you will be awake) seems absurd. Now, let's add time travel to the equation.

All is as before, but now you immediately leave the hospital upon regaining consciousness to take a trip back to 1892, where you will stay for 2 weeks. Again, it seems clear that you would prefer to have had the day of pain in the past 2 weeks (in 2018), not in the next 2 weeks (in 1892). Note that the *next* is a reference to your proper time, because whereas 1892 is the global past, it is *your* personal future. Thus, now your preference would be to have the day of pain in the *recent personal past* of 2018, not the *distant global* past of 1892. Now, let's put another time travel twist to this story.

All is as before in the original tale, except now you are told in the hospital that the auto accident happened just after you made a time trip to 2092: as you walked out of your time machine in 2092, you were hit by a car. The 2 weeks you were in a coma were in 2092, before you were judged fit enough (although still unconscious) to make the time journey back to 2018. When would you now prefer to have the day of pain? Clearly, as always, in your *personal past*, which is now the *global future*.

Time *is* different for those who time travel and those who don't!

---

[120]In his 1956 novel *The Door Into Summer*, the always ingenious Robert Heinlein used *both* ideas, with the cold-storage method of reaching the future combined with a true time machine to allow his hero to return to his 'present' (the future's past).

[121]A. Gallois, "Asymmetry in Attitudes About the Nature of Time," *Philosophical Studies*, October 1994, pp. 51-59.

3.7 For Further Discussion

An interesting *theological* analysis of Feynman's idea of an electron traveling backward in time, expressed as thinking of the electron as being in 'one of God's films played in reverse,' is given by J. W. Smith, "Time Travel and Backward Causation," *Cogito* 1985, pp. 57–67. Wondering what it would be like for an electron to travel backwards in time, Smith's answer was: "Consider an electron e_1. At time t_0 it is at (x_0, y_0, z_0). At time t_1 it is at (x_1, y_1, z_1). If the direction of time for the electron was reversed, then the electron would be observed on the 'film of the world' to travel back along the same path as it did before, i.e., back to (x_0, y_0, z_0). If God stopped the 'film of the world' and examined the charge of e_1, then He would find that it was *negative*, not *positive*. Hence the electron traveling backwards in time is simply that: an electron traveling backwards in time, it is not a positron. Time reversal does not result in a reversal of charge. Thus, the Stückelberg-Feynman position is incorrect ...." Discuss this in terms of the TCP theorem. Is there a conflict? When God stops the 'film of the world,' does the electron even *have* an arrow of time?

All the modern, major religions of the world are in agreement on these two points: (1) God created the universe and (2) At some time in the past, the universe came into existence. This does raise the question of what was God doing *before* He created the universe (see note 7 in the Introduction). In his *Confessions*, Saint Augustine comments on the conundrum that (1) and (2) are possibly in conflict. After all, if God *created* the universe then, given any time t in the finite past, He must have been doing something *before* time t, which means that for any time t in the finite past the universe already existed. Thus, the universe had no instant of creation in the finite past and so had no first moment of existence—which implies (2) is false. However, like any good philosopher of religion, Augustine not only provided this theological puzzle, but also a way to wiggle free of it. His suggested counter is the assertion that time is *itself* a creation of God, that is, He made time as *part* of creating the universe. Thus, there was no time *before* He created the universe, and so the very question of 'what was He doing before He made the universe' has no meaning. What do you think of Augustine's two arguments?

The Australian philosopher J. J. C. Smart (1920–2012) invoked five dimensional spacetime in a way very different from that of including an 'eternity' axis for God's temporal time (see note 33 in "Some First Words"). What Smart argued for, instead, was *multiple* four-dimensional worlds existing together without conflict, just as an infinity of two-dimensional worlds can exist without conflict in a three dimensional space. As he wrote, "The reason why there could be two totally disparate space-times is simply the quite obvious one that two totally disparate four-dimensional spaces can exist within a suitable five-dimensional space. There is no difficulty in mathematical inconceivability here. Now let one of these four-spaces be our own space-time world, and let the other four-space be more or less similar, in accordance with whatever story you wish to tell about it." This idea had, in fact, been around long before Smart's 1967 paper. In his 1898 Presidential Address to the American Mathematical Society, for example, Simon Newcomb declared "Add a fourth dimension of space, and there is room for an indefinite number of Universes, all alongside of each other, as there is for an indefinite number of sheets of paper when we pile them upon each other." Newcomb's idea appealed to H. G. Wells' fancy so much that he built two novels, *The Wonderful Visit* and *Men Like Gods*, around it. In the first novel there is explicit mention of multiple worlds "lying somewhere close together, unsuspecting, as near as page to page in a book," and the second one speaks of one parallel universe being rotated into another. John Cramer (a University of Washington physicist) repeated Newcomb's and Wells' parallel universes/pages-of-a-book/rotation imagery almost word for word in his 1991 novel *Twistor*.

Read these three novels, and then discuss. In particular, is Wells consistent in his presentations (as a novelist there is, of course, no reason he should be!)?

Imagine the following two events, **A** and **B**, in Minkowski spacetime: **A** is the emission of a photon, and **B** is the absorption of that photon. What is the spacetime interval between **A** and **B**? It might seem that we need to know more about the precise spatial and temporal coordinates of **A** and **B**, but in fact the interval is *always zero* for any two events connected by light (by photons). To see this, write the flat Minkowski metric as

$$\left(\frac{ds}{dt}\right)^2 = 1 - \left(\frac{dx}{dt}\right)^2$$

(continued)

and so, since $(dx/dt)^2 = 1$ (because photons—by definition!—travel at the speed of light) we have $(ds)^2 = 0$. The world line of any photon has what is called a *null* interval, as do all world lines on the surface of a light cone. Timelike world lines (in the interior of light cones, with $(dx/dt)^2 < 1$) have positive intervals, that is, $(ds)^2 > 0$. Spacelike world lines (in the exterior of light cones, in Elsewhere, with $(dx/dt)^2 > 1$) have negative intervals, that is, $(ds)^2 < 0$. This is one of the significant differences between *distances* in space (which are never negative) and *intervals* in spacetime. The American chemist G. N. Lewis (1875–1946)—see note 60 in Chap. 2—constructed a romantic illustration of this when he wrote "Any pair of points [in spacetime] which are separated by zero distance [interval] are in *virtual contact*. In other words, I may say that my eye touches a star, not in the same sense as when I say that my hand touches a pen, but in an equally physical sense." (See Lewis' paper "Light Waves and Light Corpuscles," *Nature*, February 13, 1926, pp. 236–238; Lewis was the originator of the term *photon* for a particle of light.) To understand what Lewis was getting at, you must understand that in spacetime we can have the interval between **A** and **B** as zero, and the interval between **B** and **C** as zero, but the interval between **A** and **C** may *not* be zero. To convince yourself of this, suppose the Minkowski spacetime coordinates (x, t) of **A**, **B**, and **C** are (1,3), (2,2) and (1,1), respectively. Show that $(ds_{AC})^2 = 4$, while $(ds_{AB})^2 = (ds_{BC})^2 = 0$. (Hint: draw a diagram of this spacetime and simply plug the coordinates of **A**, **B**, and **C** into the metric.)

The use of the time dilation effect of high-speed space travel was used by science fiction writer Donald Wandrei (1908–1987) in his tale "A Race Through Time," *Astounding Stories*, October 1933. Initially set in 1950, this is the story of two scientists, one evil (of course!) and the other good (of course!), who develop quite different methods for travel into future times. The evil one does it with a drug that slows the metabolic processes of the body, while the good one builds an atomic-powered rocket in his home workshop! The evil scientist kidnaps the good one's girlfriend, seals the two of them inside a crystal dome, and then injects her and himself with his drug. He has arranged matters so that they will emerge from the dome in the year one million A.D. Learning what has happened, the good scientist rushes to the dome finds he can't break in, but sees an indicator dial pointing at 1,000,000. (The evil scientist has conveniently provided the dial, as well as having made the dome transparent, much as modern-day movie criminals always include a count-down clock with glowing red digits on their bombs so the hero always knows just how much time is left to disarm the bomb.)

(continued)

Returning to his rocket, the good scientist decides that he, too, will travel into the far future using time dilation. So, off he goes, on a trip like Bob's in Sect. 3.6. The story ends with an ironic twist—the good scientist thought the '1,000,000' he saw through the dome meant *one million years in the future beyond 1950*, and so he arrives back on Earth nearly 2000 years *after* his girlfriend and the evil scientist emerged from the dome. (They are, of course, long dead when the good scientist returns.)

On the outward leg of the good scientist's rocket flight we read of his "frightful speed — now thousands of light years per Earth second." Discuss this in terms of relativity theory. That is, does the rocket actually travel thousands of light years in one second of elapsed time back on Earth?

The American physicist Robert Forward (1932–2002) was, in addition to being an expert in general relativity, also a quite inventive science fiction writer. In his short story "Twin Paradox" (*Analog Science Fiction*, August 1983), for example, he used *biology* to give a surprising, ironic twist to the classic physics puzzle. The story flips the asymmetric aging of the twins by imagining that, just after the traveling twin's departure, the secret of immortality is discovered. The treatment has to be administered no later than at a certain age, however, and upon his return to Earth the traveling twin is just a bit too old for it to work. He thus becomes the last person to die of old age! In this tale, the details of the traveling twin's trip are somewhat different from Bob's trip. Read "Twin Paradox" (you can also find the story in Forward's 1995 book *Indistinguishable from Magic*), and summarize how the traveling twin's trip is accomplished.

In the mystical 1920 novel *A Voyage to Arcturus*, by the Scottish writer David Lindsay (1876–1945), we read of a spaceship that travels to Arcturus (the brightest star in the constellation Boötes, 36 light years from Earth) in just 19 h of proper time. The technical details of the trip are not explained in the novel, so assume they are the same as in Bob's trip. That is, $2T'$ $= 68,400$ s as measured by a clock on the spaceship. The distance traveled in this time is

$$2\frac{c\left\{\sqrt{a^2 T^2 + c^2} - c\right\}}{a}$$

(continued)

where c is the speed of light, a is the constant acceleration/deceleration of the rocket (as measured on the spaceship), and T is how long an *Earth*-based clock says the trip takes. Use this formula, and the one relating T and T' in the text, to calculate

1. The value of the constant acceleration/deceleration, a (does the result seem reasonable to you?) and
2. The value of T (the length of time that passes on Earth as 19 h pass on the spaceship).

You don't have to discuss general relativity to encounter tensors of the second rank (as is the metric tensor). Electrical engineers run into such a thing, for example, when studying the lowly Ohm's law! In a copper wire, that law says the current density (the vector $\mathbf{J}$, in units of amperes per square meter) at any point is related to the electric field (the vector $\mathbf{E}$, in units of volts per meter) at that point by the *scalar* σ as follows: $\mathbf{J} = \sigma\mathbf{E}$, where σ (called the *conductivity*) is a single number. This says, in rectangular coordinates, that the x-component of $\mathbf{J}$ depends *only* on the x-component of $\mathbf{E}$, and similarly for the y-component and z-components of $\mathbf{J}$ and $\mathbf{E}$. More generally, however, each component of $\mathbf{J}$ depends on *all* of the components of $\mathbf{E}$ (as in certain crystalline structures), and so we have the equations

$$J_x = \sigma_{11}E_x + \sigma_{12}E_y + \sigma_{13}E_z$$
$$J_y = \sigma_{21}E_x + \sigma_{22}E_y + \sigma_{23}E_z$$
$$J_z = \sigma_{31}E_x + \sigma_{32}E_y + \sigma_{33}E_z$$

or, in matrix form

$$\mathbf{J} = \begin{bmatrix} J_x \\ J_y \\ J_z \end{bmatrix} = \begin{bmatrix} \sigma_{11} & \sigma_{12} & \sigma_{13} \\ \sigma_{21} & \sigma_{22} & \sigma_{23} \\ \sigma_{31} & \sigma_{32} & \sigma_{33} \end{bmatrix} \begin{bmatrix} E_x \\ E_y \\ E_z \end{bmatrix} = \sigma\mathbf{E}.$$

So $\mathbf{J}$ is now related to $\mathbf{E}$ by σ, a 3×3 *matrix* (9 numbers, instead of just 1). The matrix σ is, in fact, a tensor of rank 2. This tensor is in the three-dimensional space of the copper wire, while the 4×4 metric tensor matrix is in a four-dimensional spacetime, but both are tensors of rank 2. The number of numbers in a tensor of rank n, in a space of dimension d, is d^n. In general relativity, tensors of higher rank than 2 are required. For example, the

(continued)

curvature tensor has rank 4, and so in a four-dimensional spacetime it is described by $4^4 = 256$ numbers. This goes a long way in explaining why, in general relativity, computing often involves a *lot* of arithmetic! The primary characteristic of tensors of any rank is that they are invariant under a change in coordinate systems (choosing a coordinate system is an arbitrary matter, made mostly for human convenience, about which Nature is indifferent). Read more about tensors, and write an essay on how they behave under a change in coordinates. A good place to start, at the level of this book, is Lillian Lieber's *The Einstein Theory of Relativity: a trip to the fourth dimension*, Paul Dry Books 2008 (an updated version of the original 1945 book).

Chapter 4
Philosophers, Physicists, and the Time Travel Paradoxes

"He felt the intellectual desperation of any honest philosopher. He knew that he had about as much chance of understanding such problems as a collie has of understanding how dog food gets into cans." [1]

4.1 Paradoxes and Their First Appearance in Science Fiction

"There's a lot we don't know about time travel. How do you expect logic to hold when paradoxes hold, too?" "Does that mean you don't know?" "Yes." [2]

More than 30 years ago Quentin Smith, a philosopher who believes in a finite length to the past, wrote a refutation to those who believe in an infinite past and, while that paper [3] has nothing to do with the paradoxes of time travel, in the course of presenting his reasoning he included the following curious passage:

"Why does the sun arise in the morning and not at some other time? Why do the hands of a properly functioning clock point to 12:00 at noon and midnight and not at other times? Why does the death of a person occur at a later time than his birth? The answer in all these cases is: Because by the very nature of these events they could not occur at other times. It belongs to the very nature of the sun's rising that it occur in the morning and not in the afternoon or evening. It belongs to the very nature of a properly functioning clock to point to 12:00 at noon and midnight and not at other times. And it belongs to the very nature of death to occur at a time later than a person's birth."

But what of a time traveler born in 1980 who, in 2018, enters her time machine, pushes a few buttons, and then boldly steps out into the Cretaceous period seventy million years earlier—and is promptly eaten for lunch by a passing *Tyrannosaurus*

---

[1] A time traveler admits (to himself) how perplexed he is by paradoxes in Robert Heinlein's classic tale "By His Bootstraps," *Astounding Science Fiction*, October 1941.

[2] Excerpt from a conversation between two paradox-puzzled time travelers in Larry Niven's story "Bird in the Hand," *The Magazine of Fantasy & Science Fiction*, October 1970.

[3] Q. Smith, "Kant and the Beginning of the World," *New Scholasticism*, Summer 1985, pp. 339–346.

© Springer International Publishing AG 2017
P.J. Nahin, *Time Machine Tales*, Science and Fiction,
DOI 10.1007/978-3-319-48864-6_4

rex? Perhaps Smith himself would say that there is no contradiction between this and his third claim because, in the time traveler's *proper* time, her spectacular death does indeed come after her birth. For many, however, for a time traveler to die before her mother is born *is* a paradox, plain and simple, say what you will about proper time.

One science fiction view of time travel paradoxes is that nature would be so disrupted by them that, should one occur, the universe would be torn apart. In one story, for example, a paradox is on the verge of happening through the use of a Tipler-cylinder time machine (mentioned back in Chap. 1, and which we'll discuss in more detail in the next chapter). In response to this 'threat to common sense,' the universe 'decides' to avoid the paradox by simply eliminating the perpetrators via a local nova![4] Niven wasn't the first to use this idea. A famous story[5] by L. Sprague de Camp (1907–2000), written two decades before, had already put forth the suggestion that nature will take all required corrective action to avoid paradoxes.

In De Camp's tale we read of two big-game hunter guides who use "Professor Prochuska's time machine at Washington University"—built with the aid of a "cool thirty million" dollar grant from the Rockefeller Foundation—to operate a safari-for-hire business that transports hunters back to the late Mesozoic era. When a disgruntled client tries to go back to the day *before* a previous trip to shoot the guides (who had displeased him, or rather *would* displease him the next day), we learn just how nasty De Camp thought Mother Nature would be to avoid a paradox. (After all, the guides had *not been* shot during the safari, so they *could not be* shot.[6]): "The instant James started [to ambush the guides] the space-time forces snapped him forward to the present to prevent a paradox. And the violence of the passage practically tore him to bits [making his body look] as though it had been pulverized and every blood vessel burst, so it was hardly more than a slimy mass of pink protoplasm."

And even earlier we have a famous tale[7] that was discussed in Chap. 1, by Fredric Brown (1906–1972), a master of the special category of science fiction called the "short-short," in which everything happens in 500 words or less. As you might expect, the oddities of time travel were natural attractions for Brown's quirky talent. You'll recall that in this story the inventor of the first time machine demonstrates it to two colleagues by sending a brass cube 5 min into the future. After being placed in the machine, the cube vanishes and then 5 min later reappears. No paradoxes with that—it is a trip into the *past* that has the potential for deadly repercussions. We learn just how deadly when the inventor next declares that at

[4]L. Niven, "Rotating Cylinders and the Possibility of Global Causality Violation," *Analog Science Fiction*, August 1977. Niven took this title from a physics paper with that title, authored by Tipler, that had appeared three years earlier in *Physical Review D* (April 15, 1974, pp. 2203–2206).

[5]L. Sprague de Camp, "A Gun for Dinosaur," *Galaxy Science Fiction*, March 1956.

[6]This is a statement of the belief that the past cannot be changed, an idea we will examine later in this chapter.

[7]Look back at note 93 in Chap. 1.

three o'clock he will again place the cube in the time machine but, until then, he will hold the cube in his hand. Thus, he says, at 5 min before three the cube will vanish from his hand and immediately appear in the time machine because 5 min after that, at three o'clock, he will send it 5 min into the past. And indeed, at 5 min before three the cube does simultaneously vanish from his hand and appear in the time machine. Then, slightly before three, as the three men stand pondering what has happened, one of the observers asks what will happen if the inventor does *not* send the cube back at three o'clock? "Wouldn't there be a paradox of some sort involved?" he wonders.[8] His curiosity aroused, the inventor can't resist the experiment—and the universe promptly vanishes.

Time travel, of course, is full of potential paradoxes. A paradox, according to the usual dictionary definition, is something that appears to contain contradictory or incompatible parts, thus reducing the whole to seeming nonsense. And yet, truth is also evident in the whole. The history of science and mathematics has left a long trail of paradoxes, and those that involve time travel are merely among the most recent. Not all of the puzzles of time travel involve physics or logic, however. As one philosopher observed, "Doubtless time travel will raise a host of legal difficulties, e.g., should a time traveler who punches his younger self (or vice versa) be charged with assault? Should the time traveler who murders someone and then flees into the past for sanctuary be tried in the past for his crime committed in the future? If he marries in the past can he be tried for bigamy even though his other wife will not be born for almost 5000 years? Etc., etc. I leave such questions for lawyers and writers of ethics textbooks to solve."[9]

One way early science fiction writers had of responding to the puzzle of time travel paradoxes was to just give up and to concede that the logical puzzles are overwhelming. In one tale, for example, the inventor of the Chronoscope (a gadget that can only *view* the past) explains, "There is no time travel machine. Such a thing is a logical impossibility, treated seriously only by half-cracked writers of fantasy. Such a machine would lead at once into a hopeless paradox."[10] Equally concerned about time travel paradoxes was the pulp science fiction time traveler who told his partner, just before their first trip in time, that "I'm not sure any more about getting back. There're some unpredictable terms in the time-travel equation—paradoxes. Maybe we *won't* get back."[11]

---

[8]This is what is called a *bilking paradox*, and such paradoxes will be discussed later in this chapter. Brown gave this story a lot of thought. At one point in the tale one of the colleagues, puzzled by how the inventor will be able to place the cube into the time machine at three if it has already vanished from his hand and appeared in the machine, asks "How can you place it there, then?" Replies the inventor, "It will, as my hand approaches, vanish from the [machine] and appear in my hand to be placed there."

[9]L. Dwyer, "Time Travel and Some Alleged Logical Asymmetries Between Past and Future," *Canadian Journal of Philosophy*, March 1978, pp. 15–38.

[10]M. Jameson, "Dead End," *Thrilling Wonder Stories*, March 1941.

[11]E. Binder, "The Time Cheaters," *Thrilling Wonder Stories*, March 1940.

Fig. 4.1 One way pulp science fiction avoided paradoxes was to use a 'time viewer' (like the chronoscope in "Dead End") as in "The Time Eliminator" (*Amazing Stories*, December 1926). This illustration from the story (authored by somebody who used only the initials K.A.W.) by Frank R. Paul (©1926 by Experimenter Publishing Co.) shows the inventor demonstrating his gadget to his future wife and father-in-law. Able to look back in time, the screen is displaying scenes from the older man's courtship of his wife, decades in the past. Reprinted by permission of the Ackerman Science Fiction Agency, 2495 Glendower Ave., Hollywood, CA 90027 for the Estate

But are there *really* paradoxes? Or is it true, as the extraordinary boy-prodigy who invented a time machine exclaimed (when his teacher asserted that some questions could never be answered because "Nature is full of paradoxes"), "Ah, Professor, what nonsense! Nature is harmonious; it is we who bring the paradoxes into it."[12] Saying the same are two physicists, in a paper on the circular orbits of photons around black holes: "There are no paradoxes in physics, but only in our

---

[12]V. Grigoriev, "Vanya," in *Last Door to Aiya* (M. Ginsburg, editor), S. G. Phillips 1968.

attempts to understand physical ideas by using inadequate reasoning or false intuition."[13] And as the time traveler in an early pulp story[14] casually declares to a friend, after an astonishing adventure in the year A.D. 1,001,930, "Paradoxical? My dear fellow, the Einstein Theory is full of apparent paradoxes, yet to him who understands it there is no inconsistency whatever. Give me another cigarette, will you, Frank?"

It is in *Amazing Stories* that we find the first non-fictional speculations about time travel in a pulp magazine—and certainly long before any *physics* journal would touch the subject! Publisher and editor Hugo Gernsback started those speculations by reprinting Wells' *Time Machine*, which in turn sparked a fair number of readers' letters that were printed in the magazine's "Discussions" section. Typical of the less interesting is the following comment from a letter in the July 1927 issue: "In the 'Time Machine' I found something amiss. How could one travel to the future in a machine when the beings of the future have not yet materialized?" (We answered that question in the previous chapter with the twin paradox.) Far more interesting was this letter, in the same issue:

"How about this 'Time Machine'? Let's suppose our inventor starts a 'Time Voyage' backward to about A.D. 1900, at which time he was a schoolboy ... His watch ticks forward although the clock on the laboratory wall goes backward. Now we are in June 1900, and he stops the machine, gets out and attends the graduating exercises of the class of 1900 of which he was a member. Will there be another 'he' on the stage? Of course, because he *did* graduate in 1900 ... Should he go up and shake hands with this 'alter ego'? Will there be two physically distinct but characteristically identical persons? Alas! No! He can't go up and shake hands with himself because ... this voyage back through time only duplicates actual past conditions and in 1900 this strange 'other he' did *not* appear suddenly in quaint ultra-new fashions and congratulate the graduate. How could they both be wearing the same watch they got from Aunt Lucy on their seventh birthday, the same watch in two different places at the same time. Boy! Page Einstein! No, he cannot be there because he wasn't there in 1900 (except in the person of the graduate) ... The journey backward must cease on the year of his birth. If he could pass that year it would certainly be an effect before a cause ... Suppose for instance in the graduating exercise above, the inventor should decide to shoot his former self ... He couldn't do it because if he did the inventor would have been cut off before he began to invent and he would never have gotten around to make the voyage, thus rendering it impossible for him to be there taking a shot at himself, so that as a matter of fact he *would* be there and *could* take a shot—help, help, I'm on a vicious circle merry-go-round ... Now as to trips into the future, I could probably think up some humorous adventures wherein [the inventor] digs up his own skeleton and finds by the process of actual examination that he must expect to have his leg amputated because the skeleton presents positive proof that this was done."[15]

---

[13]M. A. Abramowicz and J. P. Lasota, "On Traveling Round Without Feeling It and Uncurving Curves," *American Journal of Physics*, October 1986, pp. 936–939.

[14]F. J. Bridge, "Via the Time Accelerator," *Amazing Stories*, January 1931.

[15]This story idea (the letter was signed only with the initials T.J.D.) may well have been the inspiration for R. Rocklynne, "Time Wants a Skeleton," *Astounding Science Fiction*, June 1941. Not all fans agreed with T.J.D. A few years later, for example, a teenager named P. Schuyler Miller (1912–1974), who would author several time travel classics himself, wrote a letter to the editor of *Astounding Stories* (June 1931) stating "there is nothing in physics ... to prevent yourself from

All of the ingenious puzzles in this letter intrigued Gernsback, and may have, in fact, been the cause of his featuring a new, original time machine story[16] in the same issue. It was the tale of a scientist who transports an entire ship at sea 14,000 years back in time and causes it to hover over lost Atlantis! That story provoked a sharp letter from a reader who claimed its logic had a fatal flaw: the story's author indicated that the Atlantians observed the time travelers when, 'of course' (asserted the reader), the time travelers must actually have been invisible. The reader explained his reasoning as follows, beginning by defining A as one of the Atlantians.

> "Now A lived his life, thousands of years ago, and died. All right, now let us pass on in time 14,000 years. Now, back we come in time when A is again living his life. Lo and behold, this time A sees before he dies a strange phenomenon in the sky! He sees the shipload of people observing him. And yet these people are necessarily observing him during his one and only lifetime, wherein he certainly did not, could not, have observed them."

Gernsback printed this letter in his September 1927 editorial "The Mystery of Time," and concluded by saying "I do . . . agree . . . that the inhabitants of Atlantis would probably not have seen the . . . travelers in time." Other readers felt this way, too, because after Gernsback published yet another time machine tale[17] in 1927, the same invisibility argument appeared again in the magazine's "Discussions" column.

Two years later an amateurishly written tale[18] appeared in which a man travels in time from 1928 to 2930 with the aid of an "astounding machine based on advanced electro-physics and the non-Euclidean theory of hyperspace." The purpose of that story was two-fold: to present several of the classic paradoxes of time travel, and then to make the claim that although the simple minds of twentieth-century people cannot understand the explanations of the paradoxes (possibly explaining why the author offers none!), the paradoxes are all trivial to the scientists of the thirtieth century. Despite this shortcoming (as well as some pretty awful dialog) the story nonetheless still managed to entertain readers with the sheer mystery of the paradoxes. Letters poured into the magazine from young fans, all demanding more time travel fiction.

So, that same year Gernsback responded with a story[19] that plays with the question of the role of time travelers in the past. (That question was clearly 'in

---

going into the past . . . and shaking hands with yourself or killing yourself." That did, however, provoke the following harsh reply from another, more skeptical reader (in the December 1933 issue): "P. S. Miller once wrote that time traveling is not incompatible with any laws of physics . . . 'he don't know from nothing.'"

[16]C. B. White, "The Lost Continent," *Amazing Stories*, July 1927.

[17]F. Flagg, "The Machine Man of Ardathia," *Amazing Stories*, November 1927.

[18]C. Cloukey, "Paradox," *Amazing Stories Quarterly*, Summer 1929.

[19]H. F. Kirkham, "The Time Oscillator," *Science Wonder Stories*, December 1929.

the air,' as Gernsback's old magazine *Amazing*[20] simultaneously published a story[21] addressing this same puzzle of time travel.) Could time travelers actually participate in events ("mix into the affairs of the period," in Gernsback's words), or would they just be unseen observers? This question, obviously inspired by the earlier discussion in *Amazing Stories*, intrigued Gernsback as much as it did his readers and so, along with Kirkham's story, he printed a challenge titled "The Question of Time-Traveling" (see note 18 in the Introduction):

> "In presenting this story to our readers, we do so with an idea of bringing on a discussion as to time traveling in general. The question in brief is as follows: Can a time traveler, going back in time—whether ten years or ten million years—partake in the life of that time and mingle in with its people; or must he remain suspended in his own time-dimension, a spectator who merely looks on but is powerless to do more? Interesting problems would seem to arise, of which only one need be mentioned: Suppose I can travel back into time, let me say 200 years; and I visit the homestead of my great great great grandfather, and am able to take part in the life of his time. I am thus enabled to shoot him, while he is still a young man and as yet unmarried. From this it will be noted that I could have prevented my own birth; because the line of propagation would have ceased right there. Consequently, it would seem that the idea of time traveling into a past where the time traveler can freely participate in activities of a former age, becomes an absurdity. The editor wishes to receive letters from our readers on this point: the best of which will be published in a special section."

Gernsback's challenge did not go unnoticed and, over the next year or so, he published a large number of reader responses in the magazine's letters column.

Indeed, a few months after issuing the challenge, in his introduction to another time travel tale,[22] Gernsback wrote that ever since the publication of Kirkham's tale "there has been a great controversy among our readers as to the possibility of time flying and the conditions under which it may be done." Most of those letters are interesting if not particularly profound, with one exception. That was a letter written by a 14 year old boy in San Francisco, and its appearance in the February 1931 issue of *Science Wonder* may well have served as the inspiration for several of the classic time travel tales published during the next 20 years:

> "Some time ago you asked us (the readers) what our opinions on time traveling were. Although a bit late, I am now going to voice four opinions …
>
> (1) Now, in the first place if time traveling were a possibility there would be no need for some scientist getting a headache trying to invent an instrument or 'Time-Machine' to 'go back and kill grandpa' (in answer to the age-old argument of preventing your birth by killing your grandparents I would say: 'who the heck would want to kill his grandpa or gandma!'[23]) I figure it out thusly: A man takes a time machine, and travels into the

---

[20]By this time Gernsback had lost control of *Amazing*, and *Science Wonder* was his come-back as a publisher of pulp 'scientifiction.'

[21]E. L. Rementer, "The Time Deflector," *Amazing Stories*, December 1929. Gernsback may well have been the editor, before he lost *Amazing*, who bought this story, and the magazine's new management simply used what remained in inventory.

[22]F. Flagg, "An Adventure in Time," *Science Wonder Stories*, April 1930.

[23]Look back at note 26 in "Some First Words."

future from where he sends it (under automatic control) to the past so that he may find it and travel into the future and send it back to himself again. Hence the time machine was never invented, but!—from whence did the machine *come*?

(2) Another impossibility that might result could be: A man travels a few years into the future and sees himself killed in some unpleasant manner,—so—after returning to his correct time he commits suicide in order to avert death in the more terrible way which he was destined to. Therefore how could he have seen himself killed in an entirely different manner than really was the case?

(3) Another thing that might corrupt the laws of nature would be to: Travel into the future; find out how some ingenious invention of the time worked; return to your right time; build a machine, or whatever it may be, similar to the one you had recently learned the workings of; and use it until the time you saw it arrives, and then if your past self saw it as you did, he would take it and claim it to be an invention of his (your) own, as you did. Then—who really *did* invent the consarn thing?

(4) Here's the last knock on time traveling: What if a man were to travel back a few years and marry his mother, thereby resulting in his being his own 'father'?"

Jim H. Nicholson

Gernsback's reply, immediately following this letter, was favorable, opening with "Young Mr. Nicholson does present some of the more humorous [?] aspects of time traveling. Logically we are compelled to admit that he is right—that if people could go back into the past or into the future and partake of the life in those periods, they could disrupt the normal course of events."[24] Nicholson's letter *is* ingenious, and it anticipated the central ideas of a number of science fiction tales yet to be written.[25] However, as you'll see as you read the rest of this chapter, contrary to Gernsback's view Nicholson's comments are *not* logically correct.[26]

---

[24]Despite these words, Gernsback apparently hadn't given up entirely on the 'invisibility of time travelers' view, as he had only a few months earlier published another such tale: R. A. Palmer, "The Time Ray of Jandra," *Wonder Stories*, June 1930. In this story (one either silly or hilarious, take your pick) a time traveler moves into the future by means of a 'time ray.' Unfortunately, the ray works differently on the various chemical elements, and not at all on either hydrogen or oxygen. Thus the time traveler—or at least *much* of him—and his machine do vanish into the future, but left behind are "several gallons of water spilled on the floor." (The human body is about 60 % H_2O.)

[25]For example, Nicholson's item (2) is a precise plot outline for L. Raphael, "The Man Who Saw Through Time," *Fantastic Adventures*, September 1941, and a version of item (4) is in Robert Heinlein's famous "All You Zombies—," *Magazine of Fantasy & Science Fiction*, March 1959.

[26]One cannot, however, fault the imaginative powers of James Nicholson (1916–1972). He eventually became President of American International Films, the company that made such science fiction 'classics' as *Attack of the Crab Monsters* (1957), the 1963 X *(The Man with the X-Ray Eyes)*, and *The Time Travelers* (1964).

4.2   Changing the Past and the Grandfather Paradox

"I'm not kidding you at all Phil," Barney insisted. "I have produced a workable Time Machine, and I am going to use it to go back and kill my grandfather."[27]

As mentioned in "Some First Words," physicists and philosophers often have quite different approaches to time travel (see note 26 there again). A vivid illustration of that difference is found in a philosophical paper[28] that, after acknowledging the apparent restrictions of the grandfather paradox, turns its attention to a matter that almost surely is beyond the power of physics to study—namely, the nature of the *conversation* between a time traveler and his/her younger self. A physicist, on the other hand, views the restrictions as *the whole point*, because the central question that hovers over *any* discussion of time travel is that of 'changing the past.' As one science fiction fan summed it up in a letter to the editor of *Astounding Stories* (August 1931): "It is said that the past cannot be changed, and that any effort to do so would be useless. In my belief, no matter where or when a man goes in the past, if he appears in a year or day that has already gone by, *he is changing the past.* Then there should be no room for doubt: time traveling is impossible. It will never be done." And certainly killing your grandfather in the past (when he is still a baby) would qualify as a change to the past!

The idea of 'changing the past' occurred to the minds of philosophers long before it did in those of science fiction writers or physicists. Four centuries before Christ, the question had already been asked-and-answered by Aristotle. In his *Nicomachean Ethics*, in fact, we find him declaring that the Greek poet Agathon had known the answer a century earlier, and he quotes the poet as saying "Forever God lacks this one thing alone, To make a deed that has been done undone."

Agathon and Aristotle aside, some medieval theologians argued passionately that the past *could* be changed (but only by God). The eleventh-century Italian cleric Peter Damian (who became a Christian saint) is a famous exponent of that radical view.[29] Writing in his *De Omnipotentia Dei* ("On the Divine Omnipotence

---

[27]The opening line to F. M. Busby, "A Gun for Grandfather," *Future Science Fiction*, September 1957.

[28]Jiri Benovsky, "Endurance and Time Travel," *Kriterion—Journal of Philosophy*, 2011, pp. 65–72.

[29]R. P. McArthur and M. P. Slattery, "Peter Damian and Undoing the Past," *Philosophical Studies*, February 1974, pp. 137–141; P. Remnant, "Peter Damian: Could God Change the Past?" *Canadian Journal of Philosophy*, June 1978, pp. 259–268; R. Gaskin, "Peter Damian on Divine Power and the Contingency of the Past," *The British Journal for the History of Philosophy*, September 1997, pp. 229–247.

in Remaking What Has Been Destroyed and Undoing What Has Been Done"),[30] Damian made it clear that he believed nothing could withstand the power of God, not even the past. Ralph Waldo Emerson's poem "The Past" ("All is now secure and fast, Not the gods can shake the Past") would have been blasphemy for Damian. The following words from Damian testify to the strength of his commitment to a belief in the possibility of changing the past: "Just as we can duly say 'God was able to make it so [that] Rome, before it had been founded, should not have been founded,' in the same way we can equally and suitably say, 'God can make it so that Rome, even after it was founded, should not have been founded.'"[31]

Two centuries after Damian, Aquinas argued the contrary view, that changing the past is *not* within God's power. Whereas Damian felt it impossible to deny any act to God, Aquinas took the far more moderate position that part of God's law is that there be no contradictions in the world (this is, in fact, the modern view of time travel physicists) and that certainly God would be bound by his own law. As he wrote, "It is best to say that what involves contradiction cannot be done rather than that God cannot do it." In his *Paradise Lost*, John Milton's God is constrained even more: he is free to act or not, but if he does freely decide to act, it can only be to 'do right.' That might seem to preclude causing contradiction, as in changing the past, but perhaps not. Milton's contemporary, Thomas Hobbes, declared that there is no *a priori* standard of goodness, and thus (for Hobbes) there are no constraints on God's powers. For Hobbes, therefore, it would seem God could change the past. Theological changing of the past, as you might expect, leads to all sorts of mind-boggling puzzles. Because of such puzzles, theology would certainly be influenced by time travel, but just as certainly theological reasoning will not illuminate the puzzles of the time travel paradoxes.

The question of the immutability of past events is of special interest to theologians because it is directly related to the question of free will versus fatalism. That is, are humans the creators of the future, or are they mere fated puppets of destiny? One way theology gets involved with the issue of 'changing the past' is via what is called the *retroactive petitionary prayer*. (An 'ordinary' petitionary prayer, like the Lord's Prayer in Matthew 6 and Luke 11, asks for something in the present or the future.) Examples of retroactive prayers include that of the surgical patient who prays, just before an exploratory operation, for his tumor to be non-malignant, and

---

[30]This work is in the form of a letter to his friend Desiderius (who later became Pope Victor III), in which Damian rebutted Desiderius' defense of St. Jerome's claim that "while God can do all things, he cannot cause a virgin to be restored after she has fallen." Desiderius thought the reason God could not restore virgins is that he does not want to, to which Damian replied that this meant God is unable to do whatever he does not want to do, but this meant that God would then be less powerful than men, who are able to do things they don't want to do (such as go without food for a month). This is a good example of the dangers involved when getting into debates with theologians.

[31]The Argentinian writer Jorge Luis Borges (1899–1986) was so inspired by Damian's view that the past could be changed that he wrote a short story based on it (see "The Other Death," originally published in *The New Yorker*, November 2, 1968) and put a character in it named after Damian.

that of the soldier's wife who prays that her husband was not among those killed in yesterday's battle. Such prayers are for a happy outcome to an event that is over and done with at the time of the prayer. One might accept the rationality of praying about the future ("Please, God, let me survive tomorrow's battle and I'll be good for the rest of my life") but are prayers about the past even sensible? (The three major monotheistic religions of the world—Christianity, Judaism, and Islam—say *yes*.)

In an appendix titled "Special Providences" in his book Miracles, C. S. Lewis answers that question as follows:

> "When we are praying about the result, say, of a battle or a medical consultation, the thought will often cross our minds that (if only we knew it) the event is already decided one way or the other. I believe this to be no good reason for ceasing our prayers. The event certainly has been decided—in a sense it was decided 'before all worlds.' But one of the things taken into account in deciding it, and therefore one of the things that really causes it to happen, may be this very prayer that we are now offering. Thus, shocking as it may sound, I conclude that we can at noon become part causes of an event occurring at ten A.M. (Some scientists would find this easier than popular thought does.)"

Here we see Lewis, a prominent lay theologian, arguing for the present influencing (but not changing) the past. What can we make of that? Was Lewis arguing for backward causation, the close relative of time travel? I think perhaps so; the final two sentences in the above excerpt makes it plausible that he may have held that view. It is a view that does find much support in the block universe interpretation of Minkowskian spacetime. Lewis never mentions the block concept by name, but it is clear that he believed in the idea of God being able to see all of reality at once. Lewis believed, therefore, that God knows of a petitionary prayer before it is made; or, even stronger, if God is not a temporal being but rather is eternal and knows time 'all at once,' then God knows the prayer and the event being prayed about 'at the same time.'

Lewis did make it clear that he believed it is a sin to pray for something *known* not to have occurred—for example, to pray for the safety of someone known to have been killed yesterday. As he wrote, "The known event states God's will. It is psychologically impossible to pray for what we know to be unobtainable, and if it were possible, the prayer would sin against the duty of submission to God's known will." Taking a less judgmental position (but essentially agreeing with Lewis) were two philosophers who, writing of the battle of Waterloo, said "for one who knows the outcome of the battle more than a hundred and 50 years ago, [a retroactive petitionary] prayer is pointless and in that sense absurd. But a prayer prayed in ignorance of the outcome of the past event is not pointless in that way." Further, in support of backward causation, they also wrote that "to pray in 1980 that Napoleon lose at Waterloo" is logical because "why should your prayer not be efficacious in bringing about Napoleon's defeat?"[32] Disagreeing, however, was another philosopher who bluntly declared "A prayer for something to have happened is simply an absurdity, regardless of the utterer's knowledge or ignorance of how things went."[33]

---

[32]E. Stump and N. Kretzmann, "Eternity," *Journal of Philosophy*, August 1981, pp. 429–458.

[33]P. Geach, *God and the Soul*, Routledge & Kegan Paul 1969.

There have been all sorts of opinions expressed through the ages in reaction to the idea of affecting the past via retroactive petitionary prayers,[34] and on the role of backward causation. The British philosopher Michael Dummett (1925–2011), in particular, discussed Lewis' concept of such prayers with great sympathy,[35] and backward causation allows one to both explain them as well as retaining free will. That is, it is not God's foreknowledge that causes our later actions (forcing our behavior and so turning us into automatons), but rather it is our later *freely-chosen* actions that causes God's foreknowledge! While such theological speculations are interesting, in the end they are simply positions of faith, about which mathematical physics has nothing to say.

Eventually, of course, others besides philosophers and theologians began to ponder the questions raised by 'changing the past.' In a January 1963 personal letter to the editor of *The Magazine of Fantasy and Science Fiction*, Robert Heinlein wrote[36] "Mark Twain invented the time-travel story; 6 years later H. G. Wells perfected it *and its paradoxes* [my emphasis]. Between them they left little for latecomers to do." How a man as widely read as was Heinlein, who had authored some of the best short time travel stories ever written, could have written such an erroneous sentence is a mystery to me. *A Connecticut Yankee in King Arthur's Court* and *The Time Machine*, certainly both works of genius, are *not* pioneers in paradox. And Heinlein's own contributions are proof enough that there was a lot left to do with time travel, well after 1900.

The very first story to be written that even hints at the particular time travel paradox of changing the past seems to be by the Unitarian minister Edward Everett Hale (1822–1909), best known today as the author of the 1863 story "The Man Without a Country." Hale wrote "Hands Off" in 1881, and published it anonymously in *Harper's New Monthly Magazine* with the express purpose of stirring up some theological debate (which apparently it didn't). He certainly had no idea that he would come to be recognized by literary scholars as a pioneer in the yet-to-be invented genre of science fiction.

Hale's story opens with the mysterious words "I was in another stage of existence. I was free from the limits of Time, and in new relations to space." These words are spoken by an unnamed narrator, who seems to have just died, and who finds himself, in his new 'form,' observing "some twenty or thirty thousand solar systems" while in the company of "a Mentor [probably an angel] so loving and patient." Under the guidance of this Mentor, in an attempt to 'improve' history, the narrator alters the Biblical account of Joseph and his imprisonment in Egypt. At first, subsequent history *is* better, but then humanity

---

[34]A summary of those opinions can be found in G. Brown, "Praying About the Past," *Philosophical Quarterly*, January 1985, pp. 83–86. Debate continues on the retroactive prayer into the 21st century: see, for example, K. Timpe, "Prayers for the Past," September 2005, pp. 305–322, and T. J. Mawson, "Praying for Known Outcomes," March 2007, pp. 71–87, both in *Religious Studies*.

[35]M. Dummett, "Bringing About the Past," *Philosophical Review*, July 1964, pp. 338–359.

[36]Reprinted in the posthumously published *Grumbles from the Grave* (edited by Heinlein's widow, Virginia Heinlein), Del Rey 1990.

sinks into irreversible depravity. In the end the narrator watches the last handful of humans kill each other in a particularly symbolic place for the Christian world: "The last of these human brutes all lay stark dead on the one side and on the other side of the grim rock of Calvary!" There would be no Crucifixion and Resurrection for the salvation of humankind, which naturally greatly disturbs the narrator. But the Mentor calms him, saying "Do not be disturbed, you have done nothing." It has, you see, just been an experimental world, an alternate Universe, and the narrator has learned the lesson of "Hands Off" the past.

Hale's story is a better Sunday sermon than it is a change-the-past time travel tale, and the device of experimenting on a not-really-real Earth is disappointing from a modern science fiction point of view. But Hale's story almost certainly did have an immediate (if indirect) impact. There is no absolute proof, but with its appearance in a national magazine, it seems quite likely that "Hands Off" was read by Edward Page Mitchell (1852–1927), an editor on a daily New York newspaper, the *Sun*. I write that because, just 6 months after Hale's story appeared, the *Sun*, in its issue of September 18, 1881, printed Mitchell's "The Clock That Went Backward." That tale, published anonymously, used a machine (the clock) for time travel,[37] as well as incorporated the idea of time travel involving paradoxes. The story predates Wells' *Time Machine* by 14 years, and Wells' novel did not include a paradoxical element.[38]

There are, however, two *hints* at paradox in Wells' novel. In the opening, during the dinner party at which the Time Traveller tries to convince his friends of the possibility of a time machine, one of them observes that "It would be remarkably convenient for the historian. One might travel back and verify the accepted account of the Battle of Hastings, for instance." To that another guest replies, "Don't you think you would attract attention? Our ancestors had no great tolerance for anachronisms." The second hint occurs when the incredulous Editor, astonished at the disheveled appearance of the Time Traveller upon his return from the future, wonders "What *was* this time traveling? A man couldn't cover himself with dust by rolling in a paradox, could he?"

What might happen if time travelers *could* change the past? This question is nicely illustrated in one novel[39] where a time traveler finds himself stranded in the London of 1810. Despite his predicament, he takes solace with "I could invent things—the light bulb, the internal combustion engine, ..., flush toilets ... " But then he thinks better of doing any of that: "no, better not to do anything to change

---

[37]See Jan Pinkerton, "Backward Time Travel, Alternate Universes, and Edward Everett Hale," *Extrapolation*, Summer 1979, pp. 168–175. The time machine in Mitchell's story is more fantasy than anything else. It is simply stated that if the clock runs backward, then it travels backward in time.

[38]Wells' failure to use paradox in his famous novel surprises most modern readers and, in fact, one of the first reviewers specifically criticized him for this lapse. See the 1895 review of *The Time Machine* that appeared in *Pall Mall Magazine*, by Israel Zangwill, reprinted in Parrinder's book (note 1 in the Introduction).

[39]T. Powers, *The Anubis Gates*, Ace 1983, a work with equal shares of physics and magic.

the course of recorded history—such tampering might cancel the trip I got here by, or even the circumstances under which my mother and father met. I'll have to be careful."

This is really just a more recent treatment of the change-the-past paradox that was already well established in early science fiction. In a story[40] published a half-century before, we find the paradox explicitly stated, along with a possible solution that is similar to the kind of explanations that have appeared in the philosophical literature[41]:

> "Suppose you landed in your own past?," queried Eric."
> Dow smiled.
> "The eternal question," he said. "The inevitable objection to the very idea of time travel. Well, you never did, did you? You know it never happened!"

But, suppose you *could* land in your own past. What then?

One famous story[42] that considered this question *embraced* the idea of changing the past. In it a client on a dinosaur hunting safari fails to follow the instructions of his guide to do *nothing* in the past except shoot a dinosaur that is about to dies for "other reasons" anyway—alas, he accidently kills a butterfly. This results in enormous changes in history, as indicated by the 'before' and 'after' versions of the time machine company's ad:

before
TIME SAFARI, INC.
SAFARIS TO ANY YEAR IN THE PAST.
YOU NAME THE ANIMAL.
WE TAKE YOU THERE.
YOU SHOOT IT.
after
TYME SEFARI INC.
SEFARIS TU ANY YEER EN THE PAST.
YU NAIM THE ANIMALL.
WE TAEK YU THAIR.
YU SHOOT ITT.

Bradbury describes the death of the butterfly as having started the knocking "down [of] a line of small dominoes and then big dominoes and then gigantic dominoes, all down the years across Time." This is, of course, a somewhat unconvincing argument. After all, previous dinosaurs, when shot, must have fallen to the ground and flattened a lot of butterflies! With such threats for every decision, no matter how seemingly innocent, hanging over the head of a time traveler, it

---

[40]C. L. Moore, "Tryst in Time," *Astounding Stories*, December 1936.

[41]See, for example, P. J. Riggs, "The Principal Paradox of Time Travel," *Ratio*, April 1997, pp. 48–64. The 'principal paradox' is that time travel is inherently contradictory because it permits the possibility of traveling to an earlier time to prevent the trip. The grandfather paradox is a special case of this. For more discussion, see T. Chambers, "Time Travel: How Not to Defuse the Principal Paradox," *Ratio*, September 1999, pp. 296–301.

[42]R. Bradbury, "A Sound of Thunder," *Collier's*, June 1952.

would take a brave soul to do much more, while in the past, than just stand still and breathe.

Equally grim is the tale[43] that takes a different view by denying the past can be changed. There we read of a doomed hero who journeys back to 1865 to save Lincoln from Booth, but his "time-distorter" is quickly taken away from him by suspicious guards. Its internal workings tick, you see, and they think he is an assassin with a clock bomb. They destroy it, haul him away to his fate, and Lincoln goes on to meet his. In the same spirit (but even more shocking) is the result of a time traveler's intentional tampering with the past in David Gerrold's 1973 novel *The Man Who Folded Himself*. That traveler experiments with "making things different" and, in his words, "Once I created a world where Jesus Christ ... went out into the desert to fast and never came back. The twentieth century I returned to was—different. Alien."

With such a stupendous power to alter reality, assuming the past *can* be changed, perhaps one might imagine prospective time travelers to the past being required to first file Historical Impact Statements![44] Not all would receive permission. In one classic tale,[45] for example, we read of a time traveler who takes a rifle and 5000 rounds of explosive bullets back to Golgotha. His intention—to be history's first Rambo by picking off any Roman soldier who gets within a hundred yards of Jesus! As outrageous as this concept is (but who among those now reading this won't admit to at least a momentary thrill at the idea and, perhaps, even a secret willingness to do it themselves, if they could), it isn't the story's peak. That comes when the reader is reminded that it was Christ's *desire* to die on the Cross, that he *had* to die for our sins; to prevent that from happening would change all of history for the last 2000 years. What, then, should the time traveler's colleagues do when they discover his plan? Should they stop him or not? What might happen if they do interfere? Of course, if the time traveler is 'now' in the past, isn't it already 'too late' to stop him? Oh, the conundrums of time travel and changing the past!

The classic change-the-past paradox is, of course, the grandfather paradox. A famous story[46] pushed this paradox to its logical limit to illustrate its supposed dangers. Having traveled to Greece in the fifth century B.C., the traveler suddenly realizes (with just a little exaggeration): "Ninety-five generations back you'd have more grandfathers than there are people on Earth, or stars in the Galaxy! You're kin to everyone ... You as much as take a poke at anyone, and the odds are you won't even get to be a twinkle in your daddy's eye."

---

[43]R. Silverberg, "The Assassin," *Imaginative Tales*, July 1957.

[44]See, for example, the novel by C. L. Harness, *Krono*, Franklin Watts 1988.

[45]A. Porges, "The Rescuer," *Analog Science Fiction*, July 1962.

[46]P. S. Miller, "Status Quondam," *New Tales of Space and Time*, November 1951. This is the Miller I mentioned back in note 15.

Fig. 4.2 Illustrator Jack Binder (1902–1986) was the author of a continuing series called "IF—
…" in *Thrilling Wonder Stories*. In each issue, the ellipses would be replaced with some phrase
such as "the Sun exploded!," "there was another ice age!," or "there was no friction!." The
installment shown here (and in Fig. 4.3) appeared in December 1938 and asserted that the past
could be changed by a time traveler

Fig. 4.3 (Fig. 4.2 continued). Illustration for "IF—You Were Stranded in Time!" ©1938 by Better Publications, Inc.; Reprinted by permission of the Ackerman Science Fiction Agency, 2495 Glendower Ave., Hollywood, CA 90027 for the Estate

Even earlier than Miller's tale is an equally famous one[47] that illustrates the same point in a graphic way. In A.D. 452 a time traveler shoots and kills one of Attila's Huns (who would have been his great-grandfather many times over); the result is that 50,000 of the Hun's descendants vanish! So dramatic did the readers of this story find the concept that author repeated the idea the very next year.[48] Twenty years later another writer topped these tales by having a time traveler accidently kill the original 'intelligent baboon' in the ancient past, thereby wiping out the entire human species![49]

Science fiction writers have been as puzzled by the grandfather paradox[50] as have been nearly everybody else. As the inventor of the first time machine says in one tale,[51] "I have devised a method [for travel] into the distant past. The paradox is immediately pointed out—suppose [the time traveler] should kill an ancestor or otherwise change history? I do not claim to be able to explain how this apparent paradox is overcome in time travel; all I know is that time travel *is* possible. Undoubtedly, better minds than mine will one day resolve that paradox, but until then we shall continue to utilize time travel, paradox or not."

Some may feel it overly dramatic that the classic time travel paradox has such a murderous form, but that is its historical origin in science fiction (not in either physics or philosophy). We can find the grandfather paradox discussed as if already well-known in a letter to the editor at *Astounding Stories* (January 1933). The author of that letter wrote "Why pick on grandfather? It seems that the only way to prove that time travel is impossible is to cite a case of killing one's own grandfather. This incessant murdering of harmless ancestors must stop. Let's see some wide-awake fan make up some other method of disproving the theory." As we proceed, you'll see just how clever some of those who responded to that writer's plea have been but, even today, as the grandfather paradox stands revealed as a red herring, it is preeminent in most people's imaginations. If a solution to the grandfather paradox puzzle escaped an early science fiction writer, then he would generally just mysteriously mention it and then hasten on to other matters. For example, in one story[52] the following exchange between the stock pulp-fiction characters of a young hero and a brilliant old scientist occurs:

---

[47]N. Schachner, "Ancestral Voices," *Astounding Stories*, December 1933.

[48]N. Schachner, "The Time Imposter," *Astounding Stories*, March 1934.

[49]C. Dye, "Time Goes to Now," *Science Fiction Quarterly*, May 1953.

[50]The 'paradox' is that, assuming you do arrive in the past with a working gun, why *can't* you kill your grandfather? After all, you *must* fail in that quest because otherwise you wouldn't be there from the future to even try. But *why* must you fail? It is, of course, not actually necessary to try to kill your grandfather to run into this paradoxical situation—just go back in time to any moment in the past and try to kill yourself! You won't succeed (if the past is unchangeable), but *why not*? (To argue 'because the past is unchangeable' is to beg the question. We need more insight than that.)

[51]M. Reynolds and F. Brown, "Dark Interlude," *Galaxy Science Fiction*, January 1951.

[52]C. South, "The Time Mirror," *Amazing Stories*, December 1942.

"You mean that time travel really is possible? That men can be transported into the future or the past—?"

The other held up a restraining hand. "Yes. Time travel *is* possible . . ."

"But professor! Think of what you're saying! You're telling me that I could go back and murder my own grandfather. That I could prevent myself from being born—?"

Again the elder man sighed. "I was afraid of this," he said. "I knew you could not understand." He hesitated. Then: "At any rate, take my word for it that time travel is possible. Also, I assure you that there are any number of perfectly sound theoretical and practical reasons why you never could hope to murder your grandparents."

We are, however, not told just what those reasons might be.

Even when all has been said about the impossibility of changing the past, and even when they are finally willing to concede that point, most people still cannot help wondering *why* the time traveler can't kill his grandfather. There the time traveler is, after all, just two feet away from the nasty young codger (I assume he is nasty to make the whole unpleasant business of murder as palatable as possible), with a perfectly functioning and well-oiled revolver in his hand, cocked and loaded with powerful, factory-fresh ammunition that even Dirty Harry would find excessive. What can possibly prevent the time traveler from simply raising his arm and doing the deed? Indeed, the artwork (reproduced at the end of the Introduction) accompanying one 1944 story shows this act in detail, including the smoking gun in the hand of the time traveler who has just taken a shot at grandpop. And if that still leaves open the remote possibility of an aiming error through nervousness, then why can't a suicidal time traveler just wrap his body in factory-fresh dynamite and blow-up granddad—along with himself and everything else within a hundred feet?

I'll argue in this book that killing your grandfather in the past, before he sets *you* in motion, is *logically* impossible. The laws of physics will then faithfully do their duty. No one will ever find an unfinished note in the empty laboratory of a missing traveler who, skeptical of the grandfather paradox, has written "To prove the falsity of the grandfather paradox, I will take my time machine back 50 years and kill my grandf. . .." Nor will any time traveler have to be concerned about the twist in one tale, which opens with the inventor of a time machine is showing the gadget to three friends. One of them later steals the machine to go back 60 years to kill his grandfather—and the story closes with a *near* repeat of the opening, with the inventor showing the gadget to *two* friends.[53]

Invoking logic in this way, in the context of time travel to the past, was discussed in the philosophical literature nearly half a century ago: "If we assume that it is impossible for [a time traveler to kill his younger self], some people are inclined to ask such questions as this: 'But how can the laws of logic prevent him from killing his younger self? Do they cause his finger to slip on the trigger or the bullet to fly apart in mid-air?' The implication of such questions is that the laws of logic cannot prevent such actions. But such questions are like asking: 'How do the laws of logic

[53]F. Brown, "First Time Machine," *Honeymoon in Hell*, Bantam 1958.

prevent the geometer from trisecting the angle or squaring the circle? Do they, for example, cause his ruler to slip as a crucial moment every time he tries it?'"[54].

A similar point was made later by another philosopher: "Surely it is not an impairment of 'freedom of action' ... that, e.g., you cannot push another person harder than he/she pushes you. Just as one would explain this is the case by reference to Newton's third law ('to every action there is an equal and opposite reaction'), one could explain the impossibility of [causing a paradox] by reference to the laws which imply such a impossibility. If this explanation is taken to be unsatisfactory, it would seem that one is saddled with a general problem concerning the reconciliation of physics and 'freedom,' and not with a specific argument against [paradoxes]."[55]

The grandfather paradox unquestionably nags at all students of time travel. As a character in one story declares, "The resolution of [the grandfather paradox] is the key to time,"[56] and some incorrectly believe it remains unresolved.[57] The paradox *is* undeniably troublesome: as one philosopher put it, the apparent possibility of a time traveler being able to do away with both his grandfather and himself gives "rise to such puzzles that we are forced to question its [time travel's] intelligibility."[58] In the next section we'll explore how to answer this concern.

4.3 Changing Versus Affecting the Past

> "The past—it's pretty damn solid, Phil. It's a little like a compost pile—fairly soft near the surface but packed hard further down, with all that Time piled on top of it."[59]
>   —one 'explanation,' perhaps, for the unchangeability of the past

The common belief today, among physicists and philosophers alike, is that given any consistent description of reality it is simply impossible for a time traveler to kill himself as a baby. As one philosopher put it, "Autoinfanticide is *metaphysically impossible* [my emphasis]. This metaphysical impossibility is philosophically intriguing because unlike most impossible events, we can vividly picture how it might look. Time travel itself seems possible, and for those who arrive in the past

---

[54]J. W. Meiland, "A Two-Dimensional Passage Model of Time for Time Travel," *Philosophical Studies*, November 1974, pp. 153–173. Science fiction had already considered time travel suicide in, for example, K. Neville, "Mission," *Fantasy and Science Fiction*, April 1953.

[55]F. Arntzenius, "Causal Paradoxes in Special Relativity," *British Journal for the Philosophy of Science*, June 1990, pp. 223–243.

[56]P. Worth, "Typewriter from the Future," *Amazing Stories*," February 1950. See also note 106 in Chap. 1.

[57]For example, in J. H. Schmidt, "Newcomb's Paradox Realized with Backward Causation," *British Journal for the Philosophy of Science*, March 1998, pp. 67–87, we read that "there are as yet no generally accepted solutions" to the grandfather paradox.

[58]S. Gorovitz, "Leaving the Past Alone," *Philosophical Review*, July 1964, pp. 360–371.

[59]F. M. Busby, "A Gun for Grandfather," *Future Science Fiction*, Fall 1957.

with proper equipment and training, the actual infanticide should not be difficult."[60] And so the grandfather paradox lives on, bedeviling both physicists and philosophers alike.

Indeed, one physicist described the time travel paradoxes as "the most controversial issue related to time machines."[61] As he argued, "These paradoxes seem to be something inherent to time machines (their main attribute, perhaps), so it is reasonable to assume that if there exists a universal law prohibiting the time machines, it must have something to do with the paradoxes. And on the other hand, be the problem of the paradoxes satisfactorily solved there probably would be no need to look for such a law, the (supposed) paradoxicalness of the time machines being traditionally the main objection against them."

One of the persistent stumbling blocks to removing the confusion of the paradoxes is a failure to distinguish between *affecting* and *changing* the past. One philosopher wrote this, in a somewhat bungled attempt to explain what is meant by affecting the past: "Nothing anyone can do now can make it not have rained yesterday"[62] if, in fact, it *did* rain yesterday. This is correct, but it is *not* what is meant by *affecting* the past. Rather, if the reason it *did* rain yesterday is because a time traveler from the future seeded the clouds, then that time traveler affected the past. Making it not to have rained yesterday would be to *change* the past.

A classic[63] by Isaac Asimov illustrates what is meant by a time traveler affecting the past. An idealistic physics professor, convinced that the world's political problems are the result of the comparative newness of scientific thought and tradition, tries to change the past (and thus the present) by sending a Greek translation of a modern chemistry text back 2000 years to the Hellenic days of Leucippus, Lucretius, and Democritus. He dies in the attempt but succeeds in the transmission. When a government investigator—called in because the professor drained an entire nuclear power reactor to energize his time machine!—discovers that the transmission takes a day to travel back a hundred years (a little gimmick with no foundation in physics, but simply something Asimov needed for the story), he fears 'our' world will vanish in 20 days, to be replaced by a 'new' one. In the end, however, he learns you can't change the past. As one of the late professor's colleagues tells the investigator, "While you are right that any change in the course of past events, however trifling, would have incalculable consequences ... I must point out that are nevertheless wrong in your final conclusion. *Because THIS is the world in which the Greek chemistry text WAS sent back.*"

[60]David Horacek, "Time Travel in Indeterministic Worlds," *The Monist*, July 2005, pp. 423–436.

[61]S. Krasnikov, "Time Travel Paradox," *Physical Review D*, February 14, 2002, pp. 064013–1 to -8.

[62]R. G. Swinburne, "Affecting the Past," *Philosophical Quarterly*, October 1966, pp. 341–347.

[63]"The Red Queen's Race," *Astounding Science Fiction*, January 1949.

Another good illustration from science fiction of affecting the past can be found years before Asimov's. In that story,[64] a time traveler leaves the Chicago of 1942 for the year 3000. Much later, in the year 2564, another time traveler interested in history journeys back to 2253 in an attempt to learn the cause of the great Chicago explosion of that year. The explosion was centered on the site of an ancient laboratory once used by a scientist who mysteriously vanished in 1942. The second time traveler begins his trip into the past on the same spot, with plans to go back to the day before the explosion. At the story's end, we learn that the disaster was the result of the two time travelers colliding as both 'passed through' 2253. The backward traveling historian, therefore, by pushing a button in 2564, is the cause of an event that happened 311 years earlier.[65]

To give a science fiction example of changing the past, it would be hard to do better than with a story that appeared a few years after Asimov's The central character is a researcher in time travel who has concluded that what is wrong with the world can be traced back to the scientific method getting off to a late start (this story was almost surely written as a result of Asimov's tale!): the time traveler thinks he can correct matters by visiting 340 B.C. and educating Aristotle on the proper scientific attitude. (Aristotle believed that observing the world was inferior to pure thinking about how the things, in his opinion, *ought* to work.) This the time traveler does, with utterly disastrous consequences. He returns to the present to find a scientifically retarded world that makes him a slave. In his cell he writes on a wall the bitter lesson he has learned too late: "Leave Well Enough Alone."[66]

It is the fear of time travelers from the future attempting to alter the past that has led some philosophers (and not just a few physicists) to assert that time travel is impossible, because it would mean what they feel to be impossible might happen: changing the past. One philosopher, however, argued long ago that such a worry is unwarranted. As he wrote at the end of an essay (a polemic *against* the concept of four-dimensional spacetime, and so *against* the idea of time travel), "Squandering vast sums on foolish enterprises is an everyday occurrence. [For example], will the U.S. time explorer get back and eliminate Lenin before his Russian rival gets back even earlier and eliminates George Washington? ... If such spectacular folly once gets under way because governments have been convinced of some nonsensical theory, a logician will not ... lose any sleep about who is going to succeed."[67]

---

[64]O. Saari, "The Time Bender," *Astounding Stories*, August 1937.

[65]This story describes something a bit more than 'simply' affecting the past; it has a *causal loop* in it. The time traveling historian makes his trip *because* of an event in the past that his trip causes. Such paradoxes will be the subject of the next section.

[66]L. Sprague de Camp, "Aristotle and the Gun," *Astounding Science Fiction*, February 1958. Asimov and de Camp were close friends, and their two stories with similar premises are clearly the result of a bit of friendly rivalry.

[67]P. Geach, "Some Problems About Time," in *Studies in the Philosophy of Thought and Action* (P. F. Strawson, editor), Oxford 1968.

And long before that essay (with its correct conclusion reached through faulty reasoning) was penned, we learn the same lesson (in a 1923 tale by the English novelist May Sinclair (1863–1946)) as we follow a woman right into hell after her death; she ends up there because of an immoral life. She then wanders through time, into her past, but finds that she can change nothing. As she is told, "You think the past affects the future. Has it never struck you that the future may affect the past? ... You *were* what you *were to be*."[68] This last line, from a *non*-science fiction story, is consistent with the modern view held by physicists of time travel. You cannot travel anywhere into the past unless you've already been there, and when you do make the trip you will do what you've already done there. You could not, as does the time traveler in one tale,[69] change the course of history by revealing twentieth-century physics in the eighteenth century. That does not mean you would necessarily be ineffectual during your stay in the past, however (certainly it doesn't mean, as Hugo Gernsback thought, that you'd be *invisible*!) Not being able to change the past is not equivalent to being unable to *influence* or *affect* what happened in the past, and science fiction writers have used this distinction to good effect, as did Asimov (note 63) and de Camp (note 66).

Robert Heinlein was a science fiction writer who clearly understood time travel paradoxes, both what they mean and, at least as important, what they do not mean. In his 1964 cold-war novel *Farnham's Freehold*, for example—the story of a family that is literally blasted twenty-one centuries into the future when their bomb shelter receives a direct hit from a Soviet nuclear warhead—we find following exchange as two of the characters are about to return to their original time via time machine:

> "The way I see it, there are no paradoxes in time travel, there can't be. If we are going to make this jump, then we already did; that's what happened. And if it doesn't work, then it's because it didn't happen."
>
> "But it hasn't happened yet. Therefore, you are saying it didn't happen, so it can't happen. That's what I said."
>
> "No, no! We don't know whether it has already happened or not. If it did, it will. If it didn't, it won't."

Modern philosophers, and many physicists, too, as well, who have examined the concept of time travel in depth, agree with Heinlein's character and, indeed, it is now common practice to invoke the so-called *principle of self-consistency*—generally attributed to the Russian astrophysicist Igor Novikov (see note 117 in Chap. 1) because he and his colleagues did not simply invoke it, but rather were able to *derive* it from the *principle of least action*, a concept held by many to be at

---

[68]M. Sinclair, "Where Their Fire Is Not Quenched," in *After the Darkness Falls* (B. Karloff, editor), World Publishing 1946.

[69]D. Beason, "Ben Franklin's Laser," *Analog*, December 1990.

the highest level of importance in physics[70]: all that is required, argued Novikov, in *any* physical process (including time travel), is that a *logical* consistency exist between events.[71] In his book *Evolution of the Universe* (originally published in Russian in 1979), Novikov wrote "The closure of time curves does not necessarily imply a violation of causality, since the events along such a closed line may be all 'self-adjusted'—they all affect one another through the closed cycle and follow on another in a self-consistent way." He later repeated that view in one of the first time machine papers in the physics literature.[72]

In fact, despite the attachment of Novikov's name to the principle of self-consistency, it was actually around in physics *decades* earlier; it has been traced back to as far as 1903![73] And at least an intuitive understanding of the principle can be found in the mainstream literature from nearly as long ago. For example, in Lord Dunsany's short 1928 play *The Jest of Hahalaba* (the inspiration for the 1944 film *It Happened Tomorrow*), a man obtains (via supernatural means) a copy of tomorrow's newspaper. In it he reads his own obituary, which so shocks him that he promptly expires—thus explaining the obituary notice.

The principle of self-consistency has been in science fiction long before Novikov, too. An example is the 1941 story "Time Wants a Skeleton" (see note 15). In it one character, after puzzling over a time travel paradox, realizes that "Future and present demanded co-operation, if there was to be a logical future!" And a nice lecture on the principle (that pre-dates by 3 years the dialogue quoted earlier from Heinlein's *Farnham's Freehold*) is given by a character that is particularly interesting because it was published, not in a specialty science fiction magazine catering to an audience with 'genre knowledge' of time travel, but rather in an icon of general American culture.[74]

Not all science fiction writers, however, have understood the requirement for consistency around a loop in time. In one tale, for example, a man meets the

---

[70]Like just about everything concerning time travel, however, not *all* think this. For example, the great German physicist Max Planck (1858–1947), the 1918 Nobel physics laureate, said (in 1922): "Physics hence is inclined to view the principle of least action more as a formal and accidental curiosity than as a pillar of physical knowledge." Still, he *did* also declare that he thought it unlikely "the dominance of such a simple law could be a mere accident." Quoted from Marc Lange, "Conservation Laws in Scientific Explanations: Constraints or Coincidences," *Philosophy of Science*, July 2011, pp. 333–352.

[71]See A. Carlini, *et al.*, "Time Machines: The Principle of Self-Consistency as a Consequence of the Principle of Minimal Action," October 1995, pp. 557–580, and "Time Machines and the Principle of Self-Consistency as a Consequence of the Principle of Stationary Action (II): The Cauchy Problem for a Self-Interacting Relativistic Particle," October 1996, pp. 445–479, both in *International Journal of Modern Physics D*.

[72]I. D. Novikov, "An Analysis of the Operation of a Time Machine," *Soviet Physics JETP*, March 1989, pp. 439–443.

[73]R. D. Driver, "Can the Future Influence the Present?" *Physical Review D*, February 15, 1979, pp. 1098–1107.

[74]R. F. Young, "The Dandelion Girl," *The Saturday Evening Post*, April 1, 1961. See also note 49 in Chap. 2.

inventor of a time machine and agrees to his request to use it to travel into the future. Once he is in the future, alas, the machine breaks. The man then finds another machine that, though it is too small for him to fit in it, is able to hold a recording that he sends back into the past to himself, to a time *before* he started his forward journey. The message on the recording (which he did not receive the 'first' time) is, of course, *not* to make a deal with the inventor. This advice he follows, and so the principle of self-consistency is violated twice in this story.[75]

An ability to play a role in history is not without some constraints. You can't save Jesus with a rifle (see note 45), or Joan of Arc with a fire extinguisher, or knock-out John Wilkes Booth with a baseball bat outside of Ford's Theatre, or blow-up Hitler with a bomb, and you can't prevent either the Black Death in the London of 1665 or the Great Fire the following year. But it *is* logically possible for a time traveler who has an infected rat sneak into his time machine, or who carelessly discards a match, to be the *cause* of the last two examples. That was the fate, for example, of the time traveling historian from A.D. 2461 who was the cause of the plague in A.D. 562 Rome, as well as of that in England nearly 800 years later.[76]

Michael Moorcock's 1969 novel *Behold the Man* gets the impossibility of changing the past, and the possibility of affecting it, right. When a disturbed man journeys backward in time to ancient Galilee to meet Christ, only to discover that there is no such person, *he* assumes the role and lives out the Biblical accounts up to and including dying on the Cross. He has not changed the past, but he certainly plays an important role in it!

An early science fiction story that got this right, long before the philosophers and physicists thought of it, was the clever tale whose artwork I have reproduced at the end of the Introduction.[77] In that story, a time traveler journeys back from 1943 to 1870 and shoots his then 14-year old grandfather in the head. Leaving his victim lying on the ground with "blood oozing all over the youth's forehead," the would-be killer returns to 1943. Once back, however, he finds himself in a strange place where he learns from two men that the Germans destroyed New York in 1920 with poison gas! Suddenly realizing that the death of his grandfather has apparently changed history (a curious oversight for anyone smart enough to invent a time machine and then to use it to force the 'grandfather paradox'), he decides he'd rather be dead than be cut off for all time from the world he remembers. So, he shoots himself dead. Then we learn that the two men he encountered are actually inmates in an asylum who like to make-up stories for unsuspecting strangers. We also learn that the time traveler's grandfather's photographs always did show him with a "white, furrowed scar on his forehead that might have been caused by a glancing bullet."

---

[75]R. Wilson, "The Message," *Astounding Stories*, March 1942.

[76]G. C. Edmondson, "The Misfit," *Fantasy and Science Fiction*, February 1959.

[77]M. Weisinger, "Thompson's Time Traveling Theory,' *Amazing Stories*, March 1944.

Fig. 4.4 The inventor of a time machine about to commit autoinfanticide in the past (the youngster holding the teddy bear is a younger version of the time traveler). Illustration reproduced by the kind permission of Frank Arntzenius (Professor of Philosophy at Oxford University), from his paper "Time Travel: Double Your Fun," *Philosophy Compass*, November 2006, pp. 599–616

Well, okay, you might say at this point, 'I'm convinced you can't change the past, but let's get back to the autoinfanticide (grandfather) paradox. So *why can't* a time traveler kill his baby-self in the past?' A possible answer, one now generally accepted by philosophers and physicists alike, appeared first in science fiction. In a tale[78] that appeared just the year after Gödel's 1949 discovery of time travel in general relativity, we find a character saying "The answer is quite simple. When the man goes back in time and kills his grandfather, and returns to his own time again, he finds to his surprise that he made a mistake. It was not his grandfather at all! And no matter how many times he goes back and kills his grandfather … he *always* [my emphasis] finds he made a mistake." Or, perhaps, some noise distracts him as his finger tightens on the trigger, or the grenade he tosses at granddad is a dud, or a gust of wind deflects the arrow, or (most ludicrous of all) he simply slips on a discarded banana peel!

Okay, that works for *that* time traveler. But suppose, someone objects, that we arrange to have a *lot* of time travelers go back in time, each with murder in his heart for his grandfather. Then, as one philosopher has observed, "Since [killing one's grandfather in the distant past] is impossible, each assassin fails. Some change their minds, others slip on banana peels, yet others kill the wrong target, and so on. But there is something odd about the idea that such coincidences are guaranteed to happen, again and again!"[79]

Early science fiction avoided invoking banana peels by providing an even more extreme 'explanation' for the failures: the time police, who are charged with foiling would-be grandfather killers. (See, for example, the many stories by Poul Anderson (1926–2001) of the 'Time Patrol.') These time commandos are imagined to roam the corridors of time, disrupting the attempts of all those who would change

---

[78]"Typewriter from the Future": see note 106 in Chap. 1.

[79]Theodore Sider, "Time Travel, Coincidences, and Counterfactuals," *Philosophical Studies*, August 2002, pp. 115–138.

recorded history. Stories of these temporal cops are simply westerns, mysteries, police procedurals, or some other similar type of specialty genre story wearing thin camouflage. This story device, whose main purpose is to allow both time travel and free will,[80] has been correctly called "boring" by at least one philosopher (see note 5 in the Introduction), an evaluation shared by modern philosophers, physicists, and (I think) even most modern science fiction writers.

So, we seem to be back to banana peels to save grandfather—but it *is* difficult to deny that vast hordes of murderous grandsons *do* appear to require an unlimited number of strategically placed banana peels, strewn all about the past, to trip-up every one of those potential assassins. This problem, of repeated, improbable coincidences to thwart murderous descendants from the future, was first commented on by the philosopher Paul Horwich in 1975 (see note 19 in the Introduction), and then given a convincing resolution by another philosopher in 1997.[81]

To explain the argument, I'll first use the philosopher's less deadly example of dated objects. "Suppose," he writes, "that every object has written upon it the date on which it will cease to exist … perhaps a time traveler travelled into the future, observed the demise of objects and then travelled back [to just after he left for the future] and wrote the dates."[82] If now the time traveler tries to destroy an object *before the date written on it*, then he *will* fail. As the philosopher amusingly described his attempts to destroy a pen 'before its time,' "I take it outside to place under the wheels of a passing train, but there is a train strike that day. The telephone rings just as I am about to drop the pen into a vat of acid. I slip on a banana peel on my way to put the pen in the microwave. My dog eats my designs for a pen grinder. And so on, for as long as you please. However many attempts I make, the attempts in no way require the occurrence of the coincidences that foil them." To put it bluntly, 'Stuff happens.' The pen has the observed date of its destruction on it, and that date is still in the future and so it is simply impossible to destroy it now.

Now, here's the point: the date on the pen is there *because* all those attempts to destroy it before that written date fail. But the presence of the date is *not* the reason for any of the weird (?) occurrences that disrupted all the attempts to destroy the pen, but rather it's *because* all those attempts failed that the date is what it is. This same argument applies to the grandfather paradox. The only time travelers available, today, to go back into the past to *try* to kill their grandfathers, are precisely those time travelers whose grandfathers were *not* killed. Or, to paraphrase our philosopher (note 81), to ask 'why do coincidences always foil the time traveller's

---

[80]See, for example, David King, "Time Travel and Self-Consistency: Implications for Determinism and the Human Condition," *Ratio*, September 1999, pp. 271–278.

[81]Nicholas J. J. Smith, "Bananas Enough for Time Travel?" *British Journal for the Philosophy of Science*, September 1997, pp. 363–389.

[82]This does present us with the curious (although non-paradoxical) situation that the time traveler will find, upon his appearing in the future, the date he *will write* (in his personal future) when he returns to just after he left on his time trip.

attempts to kill [grandfather in the distant past], is to get things back to front. It is only because the murder attempts fail that the time traveler is alive in the future to even make the *attempt.*'

In other words, not only is the grandfather 'paradox' not a paradox, it isn't even surprising!

4.4 Causal Loop and Bootstrap Paradoxes

"My dear Collingwood, don't drive yourself crazy trying to resolve the paradoxes of time travel. The [time machines] are gone ... have a drink."[83]

The grandfather paradox might finally have been put to rest, but there are still plenty of other logical minefields left to be negotiated. One of the more puzzling is that of the closed loop in time, a conundrum nicely illustrated by one philosopher[84] as follows, as an explanation of the journey one time traveler makes to 3000 B.C.: "In our time travel story it just may be that the traveler's interest in going back to ancient Egypt is stimulated by recently discovered documents, found near Cairo, containing the diary of a person claiming to be a time traveler, whereupon our hero, realizing it is himself, immediately begins ... construction of a rocket in order to 'fulfill his destiny.'" In other words, (1) he builds a time machine and goes back to the past because of the discovered diary, and (2) the diary is discovered because he goes back to the past. Each of these points by itself has logical clarity, but together they form a closed time loop (a *causal loop*) of enormous mystery.

Science fiction was strewn with causal loops long before the philosophers and physicists began to ponder them, however, with (for example) one early tale on a time traveler who journeys a century into the past because she finds an old, yellowed newspaper story describing her arrival.[85] But this tale wasn't the first to use a causal loop, as we can find one of the first sophisticated treatments of this device in a story that appeared even earlier (in the same publication).[86] A time traveler in 1930, about to start his journey into the future in an airplane/time machine, wonders at the last moment if he should really go—then he sees himself returning and thus *knows* he will successfully make the trip.

As he later tells a friend, "That decided me ... Paradoxical? I should say so! I had seen myself return from my time-trip *before* I started it [just like Marty McFly in the original *Back to the Future* film]; had I *not* seen that return, I would *not* have

---

[83]A science fiction suggestion that in certain situations (particularly causal loops), might actually be good advice! From L. Sprague de Camp's "The Best-Laid Scheme," *Astounding Science Fiction*, February 1941.

[84]L. Dwyer, "Time Travel and Some Alleged Logical Asymmetries Between Past and Future," *Canadian Journal of Philosophy*, March 1978, pp. 15–38.

[85]P. Bolton, "The Time Hoaxes," *Amazing Stories*, August 1931.

[86]F. J. Bridge, "Via the Time Accelerator," *Amazing Stories*, January 1931.

commenced that strange journey, and so could *not* have returned in order to induce me to decide that I *would* make the journey!" And later, when he finds himself in a dangerous situation in the future, he draws hope from that initial experience: "I *would* escape ... It was so decreed. Had I not, with my own eyes, seen myself appear out of the fourth dimension back there in the Twentieth Century, and glide down to my landing-field? Surely, then, I *was* destined to return to my own age safe and sound."

Even more dramatic is the second, internal time loop that ends the story. When the time traveler arrives in a ruined city in the year A.D. 1,001,930 he is greeted, *by name*, by an old man who says he (the old man) is the Last Man alive. He knew the time traveler was coming because an ancient history book had said the Last Man had, in fact, appeared in the year A.D. 502,101 in the very time machine out of which the time traveler has just stepped. The time traveler is so startled by all this (and who could blame him!?) that he decides to mull over what he has been told until the next day. As he wakes up in the morning, he is just in time to watch the Last Man depart for 502,101. Stranded in the future, the time traveler wanders the empty city in despair until he chances upon a museum. And in the museum, sealed in a glass case, is his time machine (!)—it has been there for half a million years, since the end of the Last Man's journey. And so the time traveler is saved; he merely adds some oil to the still-functional engine (if you can accept time travel, I suppose this is no more difficult to believe) and returns to 1930—just as he saw himself do at the beginning of the story.

Since Bridge's astonishing story, the idea of a causal loop in time has been used many times in science fiction. Here's a representative sampling:

(1) Time travelers arrive in the forty-sixth century, only to find that they are expected. Their host tells them why: "I have been awaiting your arrival from the past. I have a written record of your coming. You see, I have a time machine myself ... With my time machine, I recently went a year into the future and read the written account I had made, or will make after you leave. Then I came back, awaiting your arrival."[87]

(2) Armed travelers return to the Triassic age to uncover the secrets of a mysterious artifact that has been recently discovered; at the end we learn it is the remains of their own automatic rifle[88];

(3) A time traveler journeys back 500 years, where he suffers an accident that results in his being "agelessly stuck" in his time-traveling gadget until he is freed—by himself, 500 years later. He then gets into the gadget to journey back 500 years[89];

(4) The world's time suddenly loses 5 min, an astonishing event that comes to be called "the time drop." After 2 weeks of investigation, a reporter traces this

---

[87]E. Binder, "The Time Cheaters," *Thrilling Wonder Stories*, March 1940.

[88]J. Blish, "Weapon Out of Time," *Science Fiction Quarterly*, Spring 1941.

[89]A. B. Chandler, "The Tides of Time," *Fantastic Adventures*, June 1948.

event to a reclusive (but brilliant, of course) scientist who reveals that he has invented a time machine. The reporter decides to test this claim by using the machine to return to just before the start of the time drop, to observe precisely what caused it—it is, in fact, a malfunction of the machine that is at fault and the reporter finds himself caught in a 2-week long causal loop[90];

(5) A physicist knows something odd lies in his future when he is confronted with a 700 year-old museum copy of a book. The puzzle is how to explain a message penned in ancient, faded ink, in modern English *and in his handwriting*, on the back side of one of the recently unglued endpapers! How, too, to explain his own fingerprints all over the same endpaper? How, indeed, to answer these questions is his problem when he is presented with all of this and is asked, "Have you, by any chance, been visiting the thirteenth century?" At the end of the story the time loop is closed when the physicist finds himself writing that same message on a *brand new* copy of the book that has been sent from the past (and that he returns to the thirteenth century via a "time portal")[91];

(6) A time traveler from 1964 is secretly observed by one of the 'locals' when he arrives in 1683. The oddness of the sudden appearance of the time traveler and his machine ("It were a kind of Dazzle") makes the local think it might be that the stranger is the man who stole some items from his home the previous night, the same night he had an "ill Dream." Stealing the time machine after the time traveler has gone exploring, the local travels to 1964 where he learns how valuable antiques are. So back he goes to 1683, to the night before the time machine first appeared, to get some 'antiques' from his house. And thus he realizes who the thief *really* was. Before leaving again for the future, he enters his own bedroom to see himself asleep and then to awaken. And so he also learns the cause of what he called his "ill Dream."[92];

(7) A movie production crew goes into the past to make a film. At the end of the story it becomes clear that their presence in the past was not an insignificant event, as one character realizes after seeing the evidence of how they affected (*not* changed!) the past: "If this is true, then the only reason that the Vikings settled in Vinland is because we decided to make a motion picture showing how the Vikings settled in Vinland"[93];

(8) A private college, endowed decades before by a generous but mysterious benefactor, experiments with a time machine. Suddenly, one of the college's graduates is accidently sent a hundred years into the past—where she becomes the benefactor. The college comes into existence, therefore, because it will exist[94];

---

[90]W. Sheldon, "A Bit of Forever," *Super Science Stories*, July 1950.

[91]M. Leinster, "The Gadget Had a Ghost," *Thrilling Wonder Stories*, June 1952.

[92]D. I. Massor, "A Two-Timer," *New Worlds SF*, February 1966.

[93]H. Harrison, *The Technicolor Time Machine*, Doubleday 1967.

[94]C. Simak, "The Birch Clump Cylinder," *Stellar 1* (J. del Rey, editor), Ballantine 1974.

(9) A man gets the money to support his experiments in time travel by selling a large collection of old, rare comic books he has discovered in his late mother's attic. Later we learn how the comic books came to be there; after his experiments are successful, the inventor travels back into the distant past, buys the *newly* published comic books right off newsstands, and stores them in his mother's attic where, decades later, he knows his younger self will grow up and then find them (and thus get the money to make it all happen)[95];

(10) In the 1980 film *Somewhere in Time* (based on the 1975 novel *Bid Time Return* by Richard Matheson), a man in the present is visited by a mysterious old woman who gives him a watch. Later, he travels back to 1912 where he meets a girl to whom he gives the watch. He then returns to the present, and she lives out her life from 1912 on, until she too reaches the present, where we discover she is the (now old) woman who gives the man the watch.

Once philosophers discovered the bizarre nature of causal loops, they quickly proved themselves to be the equal of science fiction writers in imagining strange doings. Here's one example of that, one which any writer would be proud of: "If James cannot decide whether to marry Alice or Jane, he simply travels to the future and learns that he is to choose Alice; he then chooses her for this reason. One wants to object that the decision to marry Alice was never really made at all! But this is not true; the decision was made—as a result of the knowledge that this was the decision ... It is not the case that the prospective bridegroom could visit the future and compare the results of marrying Alice with those of marrying Jane in order to decide between the alternatives. For if he visits the future, he will learn only that in fact he chose Alice, for better or for worse!"[96]

This same philosopher elaborated on his view of causal loops in a later paper, where he wrote "What if time travel becomes commonplace, so that we must deal with a constant stress of time travelers returning from the future to reveal what they have seen?"[97] His answer is "I think it is clear that the ... causal loop we have been discussing would become very common, and would play a prominent role in human affairs." He denied, however, that such causal loops would mean the loss of free will. As he explained his position, knowledge of a rigged roulette will not prevent you from putting your money on the table *if you want to*, but perhaps that

---

[95]D. Knight, "The Man Who Went Back," *Amazing Stories*, November 1985. This same idea was used earlier in the story "Compounded Interest," (*Magazine of Fantasy and Science Fiction*, August 1956) by Mack Reynolds, in which the inventor of a time machine has the money to build his gadget because he uses it to go back into the past where he deposits a small sum, which then grows (through the 'magic' of compound interest) into the cash he needs to fund his time machine.

[96]G. Fulmer, "Understanding Time Travel," *Southwestern Journal of Philosophy*, Spring 1980, pp. 151–156.

[97]G. Fulmer, "Time Travel, Determinism, and Fatalism," *Philosophical Speculations in Science Fiction and Fantasy*, Spring 1981, pp. 41–48.

knowledge will influence your *freely* made decision making. Whether you learn that the roulette wheel is rigged by traditional means (perhaps you see magnets being installed under the table) or by means of time travel is irrelevant—even with this knowledge, you act freely. Other philosophers have not been so generous. One disliked causal loops so much, for example, that while he believed them to be conceptually possible, he also thought them to have "a queer smell,"[98] so much so that he simply preferred to avoid thinking about them!

One concern that many philosophers and physicists have had with closed loops in time is that they fear that would mean being trapped on an endless cycle of repeating events. For example, one philosopher long ago wrote

> "There is nothing contradictory in imagining causal chains that are closed, though the existence of such chains would lead to rather unfamiliar experiences. For instance, it might then happen that a person would meet his own former self and have a conversation with him, thus closing a causal line by the use of sound waves. When this occurs the first time he would be the younger ego, and when the same occurrence takes place a second time he would be the older ego. Perhaps the older ego would find it difficult to convince the younger one of their identity; but the older ego would recall an identical experience long ago. And when the younger ego has become old and experiences such an encounter a second time, he is on the other side and tries to convince some 'third' ego of their physical identity. Such a situation appears paradoxical to us; but there is nothing illogical in it."[99]

What has been (erroneously) described with that is the beginning of an endless succession of encounters around a closed causal loop. There is, however, just *one* encounter on such a loop in spacetime (but, of course, the mind of the time traveler experiences the encounter twice), subject to the constraint of self-consistency. Some physicists, too, have been so concerned about multiple trips around closed timelike curves (CTCs), because they think such trips would allow the past to be changed, that they have felt it necessary to specifically forbid such a possibility. As one paper put it, "That the principle of self-consistency is not totally tautological becomes clear when one considers the following alternative: The laws of physics might permit CTCs; and when CTCs occur, they might trigger new kinds of local physics which we have not previously met. For example, a quantum-mechanical system, propagating around CTCs, might return to where it started with values for its wave function that are inconsistent with the initial values; and it might then continue propagating and return once again with a third set of values, then a fourth, then a fifth ... The principle of self-consistency by fiat forbids changing the past."[100] This last statement is, of course, in agreement with the position I have taken in this book, a position that has generally been accepted by most philosophers for several decades now, but the proponents of the principle of self-consistency

---

[98]M. MacBeath, "Communication and Time Reversal," *Synthese*, July 1983, pp. 27–46.

[99]H. Reichenbach, *The Direction of Time*, University of California Press 1956, p. 37.

[100]J. L. Friedman et al., "Cauchy Problem in Spacetimes with Closed Timelike Curves," *Physical Review D*, September 15, 1990, pp. 1915–1930. Another physicist, however, has flatly rejected this need for the Principle, calling it redundant: see D. Deutsch, "Quantum Mechanics Near Closed Timelike Lines," *Physical Review D*, November 15, 1991, pp. 3197–3217.

seem to have been driven to it by a fear of the past 'happening again' over and over, as in the 1993 film *Groundhog Day*.

Science fiction writers have stumbled into the error of endless cycling on a closed time loop, too. In one such tale,[101] the inventor of the first time machine travels 500 years into the future where he finds a bronze statue of himself that honors his discovery of time travel. Suddenly injured, fatally, he returns to the present with the statue and then dies. As a memorial, the statue is placed in the very spot where the inventor found (will find) it. As the tale ends, the late inventor's lab assistant wonders to himself what will happen 500 years later: "Suddenly a strange machine will come out of the past and [the inventor] will be here again—although he is dead and has been dead 500 years. [He will take the statue] and go back to the past ... to die. And once again that maddening cycle will begin, to go on and on forever as long as time spins its threads."

That story illustrates yet another puzzle associated with those causal loops that contain a circulating, physical object. That is, *who made the statue?* We can ask the same question about the watch in the time loop of *Somewhere in Time/Bid Time Return* as, at every instant of its existence, the watch is in the possession of either the man or the woman? So, *when was the watch constructed?*

There have been some science fiction writers who specifically recognized this question, long before either the philosophers or the physicists paid attention to it. In one early tale,[102] for example, we read of a time machine that travels from 1935 to 1925. When the question of the origin of the time machine comes up, we read

"One time machine, found in 1935 and brought back to 1925—found in 1935 *because* brought back to 1925. That is all."

"But who made it in the first place?—Oh, skip the 'in the first place.'[103] Just plain: who made it?"

"No one. It was never made ... It is here because it is here."

This same puzzle was addressed in *The Technicolor Time Machine* (note 93), when one character is perplexed over a piece of paper in his wallet with a diagram on it, a piece of paper he got from himself (an *older version* of himself, who traveled into the past to give it to his younger self). In frustration, he asks a friend:

"Then no one ever *drew* this diagram. It just travels around in this wallet and I hand it to myself. Explain that."

His friend replies:

"There is no need to, it explains itself. The piece of paper consists of a self-sufficient loop in time. No one ever drew it. It exists because it is, which is adequate explanation. If you wish to understand it, I will give you an example. You know that all pieces of paper

---

[101] S. Mines, "Find the Sculptor," *Thrilling Wonder Stories*, Spring 1946.

[102] R. M. Farley, "The Man Who Met Himself," *Top-Notch Magazine*, August 1935 (*Top-Notch* was an adventure pulp published between 1910 and 1937).

[103] The reason for this line in the story is that earlier the question of "Where did the time machine come from *originally?*" was raised. The answer: "There was never any 'original.' ... There is no round-and-round circle of events, no repetition. Merely *one* closed cycle." This is, in fact, the modern view of causal loops, expressed in a 1935 (!) science fiction story.

have two sides—but if you give one end of a strip of paper a 180-degree twist, then join the ends together, the paper becomes a Möbius strip that has only one side. It exists.[104] Saying it doesn't cannot alter the fact. The same is true of your diagram; it exists."

"But—where did it come from?"

"If you must have a source, you may say that it came from the same place that the missing side of the Möbius strip has gone."[105]

The undeniable mystery of causal loops is the reason behind the philosopher I cited earlier (note 98) who thought they have a "queer smell" and so viewed them with much suspicion. He wasn't alone in that feeling, and another philosopher said as much whe he wrote "despite [strong] arguments for the consistency of time travel stories [with causal loops], the impression is apt to remain that something is wrong with them. I think this impression is correct."[106] One story that this philosopher could well have had in mind is a classic,[107] a tale that describes a knife brought from a museum in the future back to the present. It arrives in the present with a flawless blade, but soon thereafter gets a nick in the blade. How, wonders the narrator, can the time loop be completed "again"? I do not find this quite the puzzle that either the author (and perhaps the philosopher) do: it is simply a variation of the grandfather paradox (which has been shown not to be a paradox at all). *If* the knife is found flawless in the future, *then* it was not (will not) be nicked in the past. As written, the story is not logically consistent as it involves changing the past but, *if* one removed the detail of a *nicked* blade, then we *would* have a true (paradoxical) causal loop, with the question the story, itself, asks about the knife: "How was this knife created ... when its existence has no beginning or end?"

The nicked knife does illustrate a subtle problem that bedevils *any* causal loop containing a physical object. Consider once again the watch in the film *Somewhere in Time*. Assume the watch received by the man in the present is bright and shiny. He then takes it back into the past and gives it to his love. It remains with her after his return to the present until, decades later, she gives it to him—bright and shiny. Why didn't it tarnish? Is there some peculiar anti-tarnish property to a watch in a causal loop? Well, if so, is that anymore odd than a causal loop itself?[108]

None of that, however, provides a means for rejecting time travel *if* one can argue that it is possible to have time travel *without* causal loops. Indeed, Professor Hanley (see note 105) argues that it is possible, and presents what he claims is an

---

[104]See note 99 in Chap. 1, and the related discussion there.

[105]One philosopher calls this bit of dialog "unhelpful," while ignoring the fact that it appeared in a science fiction pulp magazine and not a scholarly journal, and was clearly meant to dazzle teenage boys (see note 39 and related discussion in "Some First Words") with the concept of a causal loop, rather than to break new ground in metaphysical thought. See Richard Hanley, "No End in Sight: Causal Loops in Philosophy, Physics and Fiction," *Synthese*, July 2004, pp. 123–152.

[106]G. Nerlich, "Can Time Be Finite?" *Pacific Philosophical Quarterly*, July 1981, pp. 227–239.

[107]P. S. Miller, "As Never Was," *Astounding Science Fiction*, January 1944. This is the same Miller who appears in note 15 (and see note 46, too).

[108]It is *not* sufficient to say that perhaps she polished the watch. Polishing would remove material from the watch, which means she gives him a watch different from the one he gives her in the past.

example of how to do it. Alas, another philosopher convincingly showed that the example is flawed and that Hanley's claim that there is no causal loop in his story "is unjustified."[109] From an entertainment point of view, however, eliminating causal loops is going in the wrong direction, as it is the inclusion of causal loops that gives a feeling of mystery to a good science fiction story.

To finish this section, then, we can do no better than to discuss causal loops that are even *more* bizarre than are those with a physical object; that is, loops that involve time traveling *information*. (Since information doesn't 'tarnish,' however, such a loop avoids that particular puzzle associated with a physical object in a causal loop.) A classic example of such a loop is a mathematician who is visited in his youth by a time traveler from the future (perhaps himself), who gives him the proof of a theorem for which the mathematician is (will be) famous in the future. *Where*, then, did the proof actually come from? In what *mind* was it *created*?[110]

The philosopher David Lewis wrote with particular insight on causal loops, especially ones that involve information transfer, such as a time traveler going back in time to tell his younger self how to build a time machine so that once its constructed he can go back in time and tell himself how to do it.[111] (This was item (3), you'll recall, in Jim Nicholson's 1931 letter to *Science Wonder Stories* magazine, quoted at the end of the first section of this chapter.) As Professor Lewis wrote (see note 5 in the "Introduction"), "But where did the information come from in the first place? Why did the whole affair happen? *There is simply no answer* [my emphasis]. The parts of the loop are explicable, but the whole of it is not. Strange! But not impossible, and not too different from inexplicabilities we are already inured to. Almost everyone agrees that God, or the Big Bang, or the entire infinite past of the Universe, or the decay of a tritium atom, is uncaused and inexplicable. Then if these are possible, why not the inexplicable causal loops that arise in time travel?"

A few years later, another philosopher[112] gave a similar response to a paradox involving a causal loop similar to Lewis', a loop involving a time machine containing a book with instructions on how to make the time machine. The book travels into the past on the machine so it can be read—in order to make the machine.

---

[109]Bradley Monton, "Time Travel Without Causal Loops," *The Philosophical Quarterly*, January 2009, pp. 54–67.

[110]Professor Hanley (note 105) says the answer to such questions is "straightforward": the information comes "from itself." I think the issue is rather deeper than that.

[111]See D. Franson, "Package Deal," in *Microcosmic Tales*, Taplinger 1980. The British philosopher J. R. Lucas had a similar scenario in mind when he wrote, in his book *A Treatise on Time and Space* (Methuen 1973, p. 50), "It is very important, not only for reasons of modesty, that I should not be able to use a Time Machine to go into a public library and read my own biography." Robert Heinlein didn't agree with Lucas: in his 1956 novel *The Door Into Summer* the protagonist, an inventor, travels thirty years into the future, where he reads some patent disclosures for inventions that he doesn't remember, even though they are in his name. He then returns to his own time and promptly files the patents!

[112]M. R. Levin, "Swords' Points," *Analysis*, March 1980, pp. 69–70.

In answer to the question "Who wrote the book about building a time machine?" the philosopher says this question is "no different from questions about where *anything* originally came from. We can ask about the origin of the atoms ... their time line is not neatly presented to us. The atoms either go back endlessly, or if the Universe is finite, they just start. In either case the question of ultimate origin is as unanswerable as the question of the book's origin. What makes us think that when such questions are asked about the loop they are different and *ought* to be answerable is that the entire loop is open to inspection." While the instructions in the book don't tarnish, the book itself of course brings us back to our previous antique watch 'problem.' Suppose the book is brand-new at the start of the trip backward in time. Later, when the machine (and the book) have reached the end of the loop, just before beginning the trip back in time, have the pages turned yellow and brittle? If so, how do we account for the brand-new version? And if not, why not?

An analyst who takes strong exception to these two philosophers is Oxford physicist David Deutsch, who wrote (note 100) "the real problem with closed timelike lines under classical physics is that they could be used to generate knowledge in a way that conflicts with the principles of the philosophy of science, specifically with the evolutionary principle." What Deutsch is referring to is the metaphysical claim, attributed to the philosopher Karl Popper (see note 36 in Chap. 3), that *knowledge comes into existence only by evolutionary, rational processes* and that solutions to problems do not spring fully formed into the universe. One might call this the physics version of the work ethic—the creation of knowledge demands hard work!

Deutsch's idea had actually appeared *decades* earlier in a science fiction tale.[113] Time travel, discovered in the year 2007, is found to have a limited temporal reach into the future of 50 years, a limit due a law passed in 2057 banning time travelers from the past. To try to go past 2057 leads to a prompt arrest of the time traveler and a 'deportation' trip back to his own time. The story eventually explains that the law was passed precisely because of Deutsch's concern. As one character in the story explains, "Suppose [that one could travel more than 50 years ahead], then a time traveler from the past could get [new inventions], carry them back to his own time, and give them to scientists—which[would] cancel all the long period of invention which [produced the inventions]. Which [would] violate causal laws."

More recently, a philosopher has offered a quite interesting response to the Deutsch/Popper assertion. He writes (note 81), of information "appearing out of nowhere," that "These cases are puzzling, but they by no means show that the time travel scenarios in question are impossible or incoherent, *or even improbable*. We think it very improbable that ... information should come from nowhere—but only because this does not happen very often. It does happen *sometimes*—for instance, when you say something and I mishear you. I think that you said something very

---

[113]P. Anderson and G. Dickson, "Trespass," *Fantastic Story Quarterly*, Spring 1950.

Fig. 4.5 A curious
paradox. CORNERED
©2005 Mike Baldwin.
Reprinted with permission
of UNIVERSAL UCLICK.
All rights reserved

There it was: the same piece of cake he
ate yesterday. His time-machine really
worked. Think of the possibilities.
He could have his cake and eat it too.

profound—something which neither of us would, in fact, ever have thought of. Where does the idea come from? If this sort of thing were to start occurring regularly [as via causal loops], then we would simply accept it without raising an eyebrow."

In the final chapter I'll discuss a dramatic example (due to two Russian physicists) on how an information-creating time loop might be constructed using a wormhole time machine. Such a time loop wouldn't pass muster with Deutsch, of course, and he would consider such a thing as being as objectionable as is creationism, the anti-evolution claim that purports to 'explain' fossils (with measured ages in the millions of years) by simply *declaring* them as having been made by God just a few thousand years ago.[114] Deutsch's position is considered by nearly all scientists today to be correct *for the specific case of creationism*, but the evolutionary principle may be on shakier ground with respect to declaring causal information time loops to be impossible.

While philosophers have struggled with information in a time loop, and most physicists have carefully stepped around the issue, science fiction has had lots of fun with information in causal loops. Here's a sampling of such tales:

(1) A man receives telephone calls from *two* versions of himself, one ten years in the future saying he absolutely must accept an invitation to fly to the Bahamas

---

[114]*Why* would God do such a thing? Apparently 'just to have some fun with geologists and biologists,' as creationists call such ancient fossils 'sports of nature.'

that he will receive that very day, with the other version calling from tomorrow insisting that the plane will crash. What should he do?[115];

(2) Lovers who are irrevocably separated in time communicate by mail in one tale,[116] while lovers in another story[117] communicate via telephone calls to the ever more distant past (and yet, with the aid of a clever twist at the end, finally meet);

(3) A telephone lineman starts getting telephone calls from himself from 10 days in the future, with the first call telling him how to make the gadget to transmit such calls[118];

(4) A time traveling historian on a visit to A.D. 1528 from A.D. 2211 accidently gives a copy of the predictions of Nostrodamus to the prophet, thus explaining the predictions[119];

(5) A time machine experiment gone wrong allows thirteenth century Roger Bacon to meet twentieth century scientists, an encounter that explains the amazing forecasts in Bacon's *Opus Maius*[120];

Hollywood, too, has had some fun with information causal loops, with the best (in my opinion) example of that being the 1989 movie *Bill & Ted's Excellent Adventure*. In that film (where we learn that even the not very bright can be time travelers), a set of missing keys is necessary for the successful completion of a task. The two time travelers decide that after the task is done, they will go back in time, steal the keys (*that's* why they're missing!), and hide them so they can use them *now*. Where should they hide them? Why, "over there," says one of the boys, pointing at a hiding place—and sure enough, when they go over and look, the keys are there. They agree that once they have finished with the keys, it will be *most* important that they really do put the keys in the hiding place!

All of these examples that I've just given you, however, were *decades* too late to be the first in fiction about information in a time loop; that honor goes to the 1904 novel *The Panchronicon* by the lawyer Harold Steele MacKaye (1866–1928). An Edwardian literary time machine with style, the Panchronicon (literally, a 'machine for all time') swings on a rope tether around a steel post erected at the North Pole. By "cutting the meridians" faster than the sun does, it travels through space and time from 1898 New Hampshire to the London of three centuries earlier.[121] Using

---

[115]G. Klein, "Party Line," *The Best from the Rest of the World* (D. A. Wolheim, editor), Doubleday 1976 (story originally published in France in 1973).

[116]J. Finney, "The Love Letter," *The Saturday Evening Post*, August 1959.

[117]T. N. Scortia, "When You Hear the Tone," *Galaxy Science Fiction*, January 1971. See also L. Padgett, "Line to Tomorrow," *Astounding Science Fiction*, November 1945.

[118]M. Leinster, "Sam, This Is You," *Galaxy Science Fiction*, May 1955. This story was later broadcast as an episode on the "X-Minus One" radio drama program. See also F. A. Reeds, "Forever Is Not So Long," *Astounding Science Fiction*, May 1942.

[119]L. Del Rey, "Fools' Errand," *Science Fiction Quarterly*, November 1951.

[120]N. Schachner, "Lost in the Dimensions," *Astounding Stories*, November 1937.

[121]'Time traveling' by crossing time zones is an idea that one can trace at least as far back as to Edgar Allen Poe's 1841 short story "Three Sundays in a Week."

it, a time traveler fan of Shakespeare journeys from 1898 back to the bard, who is suffering from writer's block. There she whispers the magic lines from a play he is stuck on (lines she has memorized for her literary club meetings) into his receptive ear. Does this make Shakespeare a plagiarist? Of himself!?

4.5 Sexual Paradoxes

"Once time machines exist, no event is low probability if it is needed to make the past consistent."[122]

There are causal loops even stranger than the ones we have already discussed, hard as that may be to believe. These are the sexual paradoxes, first mentioned in 1931 by Nicholson in his letter to Hugo Gernsback. Not only science fiction writers, but philosophers, too, have found these particular paradoxes full of dramatic appeal. For example, as a challenge problem to the readers of a scholarly journal, the British philosopher Jonathan Harrison (1924–2014) posed the following bizarre, indeed astonishing, situation.[123] A young lady, Jocasta Jones, one day finds an ancient deep freezer containing a solidly frozen young man. She thaws him out and learns that his name is Dum, and that he possesses a book that describes how to make both a deep freezer and a time machine. They marry. Soon after they have a baby boy and name him Dee.

Years later, after reading his father's book, Dee makes a time machine. Dee and Dum, taking the book with them, get into the machine and begin a trip into the past. Running out of food during the lengthy journey, Dee kills his father and eats him. Arriving in the past, Dee destroys the time machine, builds a deep freezer (again, using the book), gets into it, and ... wakes up to find that a young lady, one Jocasta Jones, has thawed him out. When asked his name he replies Dum and shows Jocasta his book; they marry, and ....

Harrison concluded this amazing tale with this question for his readers: "Did Jocasta commit a logically possible crime?" That issue is just the *surface* of an ocean of puzzles in this story! Jocasta's crime, of course, is that she has (if unwittingly) committed incest; readers who remember the Greek myth of Oedipus, and who his mother/wife was, will understand why Harrison named his female character as he did. But what of Dee's crime? He has, after all, eaten his father! But perhaps that isn't a crime at all, because Dee and Dum are one in the same, and is it really a crime to eat yourself? According to another philosopher, Murray MacBeath, Harrison's story is "a story so extravagant in its implications that it will be regarded as an effective *reductio ad absurdum* of the one dubious assumption on which the story rests: the possibility of time travel."[124]

---

[122]From Robert Forward's 1992 novel *Timemaster*.

[123]J. Harrison, "Jocasta's Crime," *Analysis*, March 1979, p. 65.

[124]M. MacBeath, "Who Was Dr. Who's Father?" *Synthese*, June 1982, pp. 397–430.

This isn't to say that MacBeath was asserting that time travel is impossible. Indeed, he went on to declare that he did believe in the logical possibility of time travel, and his paper is devoted to discovering what thought to be incorrect in Harrison's story. He did that by retelling the story with what he believed are crucial modifications to make it sufficiently less outrageous that it could be taken at least somewhat seriously. In the new version, our hero, thawed out from a deep freezer, is now named Arthur. Arthur is, unfortunately, suffering from total amnesia (this is MacBeath's way of avoiding the psychological trauma of Dee remembering he ate Dum) and so, when asked his full name, he is himself sufficiently puzzled that he replies "Arthur who?" He is finally called (what else?) Arthur Who. And, as you can no doubt guess, his son (who is a genius and gets a PhD at age 14 on a dissertation dealing with the physics of time travel) becomes Dr. Who!

We are then told of a trip back into the past by the two, of the eating of the father (Arthur Who) by the son, of the entering of the deep freezer by Dr. Who, etc. etc. The whole business is quite entertaining and *at least* as complex as Harrison's original story. Just *how* complex is summed up in MacBeath's last, wonderful line: "The Who who was Dr. Who's father was not Dr. Who—that is, not the Dr. Who whose father he was."

MacBeath wasn't the only one that Harrison's story fascinated, and nearly a dozen replies to it were received in addition to MacBeath's. One, in particular, made the thought-provoking observation (see note 112) that not only has Jocasta committed incest but she has done so with a single act of intercourse. As discussed earlier, the events on a causal loop do not happen endlessly but rather only once; thus, Jocasta thaws Dum (Dee) out just once, she marries him just once, and the two consummate their marriage just once. Ordinarily we think it takes two sexual acts to commit incest, the first resulting in the birth of a child, and the second being a parent's union with that child, but this is not so in a causal loop. Time travel *is* an odd business.

Another philosopher replied to Harrison's story with a quite interesting claim, one that had actually been thought to be true for decades—but which today is recognized to be false. The claim was that, irrespective of physics, Harrison's story was biologically flawed and fatally so. As that philosopher wrote, "The biological problem is the following. Dee is the son of Dum and Jocasta. So Dee obtained half his genes from Dum and half from Jocasta. But Dum is diachronically identical with Dee and is therefore genotypically identical with him (that is, himself). That is, Dee is both genotypically identical and distinct from Dum, which is absurd."[125]

That this isn't true was pointed out by a philosopher many years later. In his paper we read this tale: "Suppose Adam travels [far] back in time ... where he meets his mother Betty, mates with her and has a child which is himself. Is this

---

[125]W. Godfrey-Smith, "Traveling in Time," *Analysis*, March 1980, pp. 72–73. This false claim had already been raised by a physicist (L. S. Schulman, "Tachyon Paradoxes," *American Journal of Physics*, May 1971, pp. 481–484), and even earlier by a science fiction writer (P. Anderson, "Time Patrol," *Magazine of Fantasy and Science Fiction*, May 1955).

possible biologically? Yes ... as follows ... on the grounds that we have total replication of Adam's genome."[126] (The *genome* is the totality of genes taken over all gene sites.) Now, suppose each such site holds two genes and, as Dowe points out, in sexual reproduction the father passes on to his offspring one gene for each gene site, to go with the gene the mother gives to each site. To exactly reproduce himself, then, the time traveling Adam 'simply' has to give his offspring, *at each site*, the gene that he has for that site that did *not* come from Betty. Thus, the offspring—baby Adam—ends up with a genome *precisely identical* to the time traveling Adam. This is, of course, an *extraordinarily unlikely* event, as the human genome has tens of thousands of genes. The probability that each and every site gets the 'right' gene from the time traveling Adam is therefore essentially zero. But it isn't *actually* zero and, as the quotation that opens this section says, a low probability to an event isn't a roadblock to its occurrence if that event is required for consistency.

While certainly instructive, the sexual paradox stories by Harrison and MacBeath are remiss in not indicating that the concepts they are dealing with have long been a staple of science fiction, and that the sexual paradoxes received much critical analysis in that genre long before philosophers (and physicists, too) discovered them. From science fiction, for example, we have a tale of young man who travels backward in time 1250 years, from A.D. 3207 to 1957, to become his own grandfather fifty generations removed.[127] And even that is tame compared to the sexual paradoxes other science fiction writers conjured up before philosophers began to discuss them.

In another story,[128] written decades before Harrison's and MacBeath's papers, we meet a young lady caught up in a mind twisting affair in which the mystery of a causal loop is the least of her troubles. In 1957 a girl is born, and after 20 years of intense competition with her mother (who has an uncanny ability to predict the future), she travels back from 1977 to a few months before her own birth. She becomes pregnant (by a man who she later discovers is her father) and gives birth to a girl. The new mother has, of course, knowledge of all that will happen during the next 20 years, including the fact that she will have an intense competition with her rebellious daughter ....

While writers of stories like these in the early 1950s were there as trailblazers, it is a tale that appeared as the 1950s ended that is today generally acknowledged as the best sexual paradox story ever written.[129] We are given only a hint of what is to come when a character listens to a song called "I'm My Own Grandpaw!" In 1945,

---

[126]Phil Dowe, "The Coincidences of Time Travel," *Philosophy of Science*, July 2003, pp. 574–589. See also J. Berkovich, "On Chance in Causal Loops," *Mind*, January 2001, pp. 1–23, and P. Dowe, "Causal Loops and Independence of Causal Facts," *Philosophy of Science*, September 2001, pp. 89–97.

[127]R. Dee (this is *not* the 'Dee' of Harrison's story!), "The Poundstone Paradox," *Magazine of Fantasy and Science Fiction*, May 1954.

[128]C. L. Harness, "Child By Chronos," *Magazine of Fantasy and Science Fiction*, June 1953.

[129]R. Heinlein, "All You Zombies—," *Magazine of Fantasy and Science Fiction*, March 1959.

a newborn girl, Jane, is found on the steps of an orphanage. At age 18, in 1963, she has a one-night affair with a mysterious stranger that leaves her pregnant. Some months later, during the birth of a daughter, it is discovered that Jane actually has a double set of sexual organs, and because the female set has been ruined by the pregnancy, doctors restore her as a man. Soon after, the baby girl mysteriously disappears from the hospital ward. Years later, in 1970, Jane (now a man, of course) meets another stranger who uses a time machine to transport both of them back to April 3, 1963. By April 24 male-Jane meets female-Jane and impregnates her (and so now we know who the mysterious stranger was during the one-night affair!). Meanwhile, the stranger with the time machine travels forward to March 10, 1964, a little after female-Jane has given birth, kidnaps the baby from the hospital (thus clearing-up another mystery!), takes her back to September 20, 1945, and leaves her on the steps of the orphanage. And so we see that Jane is her own mother *and father*, thus out-doing all previous tales about self-parenting.

This is pretty impressive stuff, but Heinlein still has one more twist for us. After leaving baby-Jane in 1945, the time machine stranger returns to April 24, 1963, retrieves male-Jane (who has just kissed female-Jane goodnight after fathering her-himself in herself), and takes him to 1985 where he recruits him into the Temporal Service—and finally, the stranger jumps forward to 1999, his 'real time.' At the end we at last learn that the stranger is, in fact, an even older version of male-Jane—*all* the central characters in the entire story are the *same* individual at various points along a single, highly twisted world line. The lone character in Heinlein's tale is truly a self-made man/woman in every sense of the phrase! This ultimate act of *creatio ex nihilo* has, correctly I think, been called "smaller than the minimal loop."[130]

Jane, in all her/his versions, is the only character in the story that appears to have purpose. In terrifying words that describe a causal loop, Heinlein ends the tale with an explanation of the story's title: "The Snake That Eats Its Own Tail, Forever and Ever. I know where I came from—but where did all you zombies come from? ... You aren't really there at all. There isn't anybody but me—Jane—here alone in the dark. I miss you dreadfully!" In a December 1958 letter to his literary agent, Heinlein wrote of this amazing tale, "I *hope* that I have written in that story the Farthest South in time paradoxes." In my opinion, he did.

The sexual paradox has continued to fascinate science fiction writers up to the present day. In the novel *Timemaster* (note 122), for example, the hero at one point spends a night with his wife—and with two versions of himself from the future. He will, of course, experience that night two more times! Later, he becomes upset when his wife runs off with one of the older versions, but he quickly calms down when he considers that eventually *he* will be the older version. Consider, too, a story[131] that

---

[130]S. Lem, "The Time-Travel Story and Related Matters of SF Structuring," *Science Fiction Studies*, Spring 1974, pp. 143–154.

[131]G. Benford, "Down the River Road," *After the King: Stories in Honor of J. R. Tolkien* (C. Tolkien and M. Greenberg, editors), Tor 1991.

tells of a young man hunting the father who, years before, had abandoned him in a burning house. The death of the young man's mother in the flames has sent him on a 10 year quest for revenge up and down what is literally a river of time, a river on which to travel in one direction ("up time") is to move into the past, whereas moving "down time" leads to the future. Eventually he corners the father and, despite the man's pleading, kills him. It is only later, after examining papers he finds in his father's pocket, that the young man realizes he has killed his *future* self (Benford, a physicist, knows the pitfalls of time travel, and you'll notice that there is no autoinfanticide paradox here).

4.6 Splitting Universes and Time Travel

"In all time travel stories where someone enters the past the past is necessarily altered. The only way the logical contradictions created by such a premise can be resolved is by positing a Universe that splits into separate branches the instant the past is entered."[132]

One early science fiction technique for allowing backward time travel and a changeable past, while still avoiding paradoxes, is that of alternate universes. According to this idea, if a time traveler journeys into the past and introduces a change (indeed, his very journey may be the change) then, as the above quote states, reality splits into two versions, with one fork representing the result of the change and the other fork being the original reality before the change. (To a fifth-dimensional observer, of course, all conceivable forks, all possible four-dimensional spacetimes, have always existed.) Indeed, according to this view the entire universe is splitting, at every microinstant, along every alternative decision path for every particle in the cosmos! This is often called the theory of *alternate realities with parallel time tracks*.

Such a seemingly fantastic view seems to actually have some scientific plausibility because of the so-called *many-worlds interpretation* (MWI) of quantum mechanics, pioneered in physics by Hugh Everett III (1930–1982), in a 1957 Princeton doctoral dissertation. Everett's theory is the antithesis of what is commonly called the *collapse of the wave function*, the idea that all potential possibilities have a non-zero possibility until a consciousness actually decides or observes which one will actually *be*. That quantum mechanical concept gets its name from the probabilistic wave equation formulated in 1926 by the German physicist Erwin Schrödinger (1887–1961). Before the observation, all possible futures have various values of probability; after the observation (which 'collapses' the wave function) exactly *one* of those futures (*the* future) has probability 1 and all the others have probability 0.

The MWI idea can be seen in Hale's story "Hands Off," discussed earlier, and in art *40 years* before that! With almost certainly a theological twist, consider the

---

[132]M. Gardner, "Mathematical Games," *Scientific American*, March 1979.

Fig. 4.6 Grandville's
Infinity Juggler of many-
worlds

beautiful illustration in the 1844 book *Un Autre Monde* (*Another World*), reproduced in Fig. 4.6. Known either as "The Infinity Juggler" or "The Juggler of Worlds," it is the work of the French artist Jean-Ignace Isidore Gérard (1803–1847), who published under the name 'Grandville.' The juggler—Grandville's version of Hale's mentor—appears as a court jester who is clearly having fun manipulating his multitude of worlds, while the man (humanity?) in the foreground watches. The man appears to be simultaneously fearful and fascinated, involved yet clearly impotent. Is Earth one of the worlds among which the Jester stands, or is it one of those flying through space? Or is Earth, perhaps, simply the unfortunate world ingloriously stuffed down the front of the Jester's pants? (That would surely explain a lot!) If born a hundred years later, Grandville would surely have found work as an artist in the imaginative world of the science fiction pulps.

Early science fiction stories that treat the collapsing wave function concept can be traced back to the late 1930s and early 1940s.[133] A particularly interesting example is the story of an inventor who, while trying to build a radio with which

---

[133] See, for example, Jack Williamson's 1938 novel *The Legion of Time*, and C. L. Moore's, "Tryst in Time," *Astounding Stories*, December 1936. L. Sprague de Camp (1907–2000), too, was an early pioneer in the exploration of the MWI idea in science fiction long *before* Everett. In his 1941 novel *Lest Darkness Fall*, for example, he uses the analogy of a tree (the "main time line") that is always sprouting new branches.

to signal Mars, accidently stumbles on the "temporal-aberrant carrier wave" and thus establishes contact with a universe that forked off of ours in 1863 when Robert E. Lee *won* the Battle of Gettysburg![134] In Everett's MWI, however, the wave function of the universe does *not* collapse. Indeed, it *couldn't*, because there is no observer external to the entire universe (we are talking science now, not of theology and God); instead, the wave function 'splits' at every decision point in spacetime. Although this leads to a multitude of realities *far* beyond comprehension, cosmologists still tend to like the MWI because it avoids the puzzle of having to produce an observer 'outside the universe.'

It's important to understand that the MWI is different from yet another idea popular in science fiction, that of *parallel universes* (see again the third discussion question at the end of Chap. 3). In parallel universes *all* possibilities *always* exist, independent and parallel in time. In the MWI, on the other hand, ever more universes are continually coming into existence. Unlike the MWI, which can at least claim a scientific basis (quantum mechanics), there is no analogous theory for parallel universes. But, of course, even though lacking a theory, nonetheless science fiction writers have been quite inventive with the idea because parallel universes offer a way to avoid (at least some) causal loops.

One clever, early pulp story[135] illustrates how that works. To improve the performance of his time machine, an inventor needs batteries with tremendous energy density, a density far in advance of the batteries in the present. Unable to travel far into the future—if he *could* obtain them there, then of course a causal loop (the very entity we wish to avoid) would be created upon his bringing them back to the present—his assistant first travels back to 1851. There he leaves a note on desk of a well-known experimenter, with a plea for him to devote his life to battery research; a copy of the 1937 *Electrical Handbook* is left with the note as proof that there really has been a visit from the future! Before returning to the present, the assistant takes a sheet of (new) 1847 five-cent stamps from the experimenter's desk.

Returning to the present, which is now different (a new time track, in accordance with the splitting-universe idea), the powerful batteries *are* readily available *because* the experimenter believed the note. Buying several of them, using money obtained by selling the pristine 1847 stamps to a collector, the assistant returns to a slightly earlier 1851 than before (to *before* the fork in time!), watches himself appear[136] and leave the note and the handbook, and *then, unobserved, the assistant removes both*: Thus, upon returning once more to the present, he finds all is as before—except now he and the inventor have the powerful batteries. As

---

[134]N. Bond, "Parallel in Time," *Thrilling Wonder Stories*, June 1940. See also S. N. Faber, "Trans Dimensional Imports," *Isaac Asimov's Science Fiction Magazine*, August 1980.

[135]W. Sell, "Other Tracks," *Astounding Science Fiction*, October 1938.

[136]Just like Marty McFly does at the end of the 1985 film *Back to the Future*. The movie is fun, but pulp science fiction did it first.

before, one might ask where the batteries came from, but unlike the previous mystery of information-creating causal loops, the answer is clear and non-mysterious. *They came from the hard work of the experimenter on a different time track.* Such shuttling back-and-forth between time tracks is the signature of what is called a *cross-time* story, a device to avoid paradoxes while still allowing for changing the past. The first example of this time travel sub-genre had actually appeared 4 years earlier.[137]

In another cross-time tale from modern times, we read of the horrible fate suffered by a man when an experiment in a Princeton physics lab goes wrong.[138] It is discovered, too late, that parallel time tracks are not simply grooves into which you drop, like a ball, after leaving the time track of our world. Each version of a person in each world is not like a ball rolling down a groove from past to future. Rather, each world's time track is just a line on a smooth surface; as the man is told, during a temporary stay in a world still close to his (our) original world, "We gave you a push sideways, and you moved off your original line—but instead of dropping into the next groove, you've just kept on rolling across the surface, from one line to the next, at an angle. There are no grooves, nothing to stop you from sliding on across the different lines forever. You have the same futureward vector as you started with, but you've added a small cross-time vector, as well." And so the man drifts cross-time, and gradually the worlds he experiences grow ever more alien.[139]

The science fiction is undeniably fun, but for this book the underlying *scientific* theory of time travel is classical (that is, non-quantum) general relativity, and that theory has nothing to say about alternative time tracks in multiple worlds. For most time travel theoreticians there is *one* time track, and the past of our world is unique and inviolate. I agree with the great quantum physicist J. S. Bell (1928–1990), who wrote of Everett's theory that "if such a theory were taken seriously it would hardly be possible to take anything else seriously."[140] As Bell further observed, in the MWI "there is no association of the particular present with any particular past," a quite strange idea that had already appeared in science fiction years earlier.[141]

While most early time travel analysts did base their work just on classical general relativity, there *are* now many more who think quantum mechanics itself, independent of its interpretation, has much to contribute as well. Perhaps, in fact, it

---

[137]M. Leinster, "Sidewise in Time," *Astounding Stories*, June 1934. Splitting universes with multiple time tracks and time loops became quite popular after Leinster's and Sell's stories; you can find the basic idea repeated yet again in Alfred Bester's "The Probable Man," *Astounding Science Fiction*, July 1941, for example, in which each new journey into the past causes the future to fan out into an infinity of new time tracks.

[138]L. Watt-Evans, "The Drifter," *Amazing Stories*, October 1991.

[139]In Jack Haldeman's 1990 novel *The Hemingway Hoax* we read that "there is not just one [parallel] universe, but actually uncountable zillions of them."

[140]In "Quantum Mechanics for Cosmologists," *Speakable and Unspeakable in Quantum Mechanics*, Cambridge University Press 1987.

[141]J. R. Pierce, "Mr. Kinkaid's Pasts," *Magazine of Fantasy & Science Fiction*, August 1953.

may make an absolutely crucial contribution to the theoretical basis of time travel. One analyst who believes this (along with an even stronger belief in the MWI) is the British physicist David Deustch (note 100), who holds that general relativity is not the proper theory with which to study the physical effects of CTLs. He believes that the traditional mathematical machinery of general relativity actually obscures, rather than clarifies, the difficult task of separating the merely counter-intuitive from the unphysical.

Indeed, Deutsch calls the conventional spacetime methods, based on general relativity and differential geometry, *perverse*. He also does not like the conceptual problems raised by general relativity's wormholes and singularities. Any non-quantum mechanical discussion, he says, of the "pathologies" of backward time travel is simply not adequate. Deutsch divides these pathologies into two fundamental classes: (1) paradoxical constraints, such as the free-will issue seemingly raised by the grandfather paradox, and (2) causal information loops. Deutsch claims that his quantum mechanical analyses show that the first class of pathologies simply does not occur, because the past that the time traveler enters is the past of a world different from the one he has left. Further, his results also show (to him) that the pathologies of the second class may be "avoidable." These are not the views among the majority of time travel students, however (that does *not* mean Deutsch is wrong!), and general relativity *is* the standard tool used by the majority of time travel theoreticians. When quantum mechanics does enter the calculations of most analysts, it is generally on an *ad hoc* basis.

More concerning for the MWI view is a result reported in 2004, that a macroscopic object (a human time traveler, for example) attempting to traverse a wormhole time machine (to be discussed in some detail in Chap. 6) "must necessarily undergo violent interactions with the time machine," interactions *so* violent that they must "cause the object to disintegrate." The different pieces of the now certainly dead time traveler would emerge from the wormhole in different worlds—this is definitely not a result likely to encourage volunteers for the first time machine trip![142]

So, many physicists and philosophers, not sharing Deutsch's position,[143] tend to agree with Bell, including the late John Wheeler (Everett's thesis advisor!), who wrote of the MWI "I once subscribed to it. In retrospect, however, it looks like the wrong track. ... Its infinitely many unobservable worlds make a heavy load of metaphysical baggage."[144] Agreeing with Wheeler was a philosopher who called the MWI "highly controversial" and declared that "few working physicists take it

---

[142]A. Everett, "Time Travel Paradoxes, Path Integrals, and the Many Worlds Interpretation of Quantum Mechanics," *Physical Review D*, June 25, 2004, pp. 124023–1:124023–14.

[143]For a modern philosophical argument specifically rebutting Deutsch's enthusiasm for the MWI, see Theodore Sider, "A New Grandfather Paradox?" *Philosophy and Phenomenological Research*, March 1997, pp. 139–144.

[144]See note 34 in Chap. 3.

seriously."[145] Perhaps even more damning was a physicist's statement that "the idea of 10^{100}+ slightly imperfect copies of (the universe) all constantly splitting into further copies ... is not easy to reconcile with commonsense. Here is schizophrenia with a vengeance."[146] Or, as one science fiction writer bluntly put it, in a tale of the inventor of the "chronomotive impulse belt" (which allows moving between the *two* parallel worlds that are all that exist), the MWI is the "Doctrine of Infinite Redundancy—which is, of course, utter nonsense."[147]

Deutsch's position does raise the obvious question of what motivates a quantum theoretician to study CTLs at all, given that they originate in general relativity and *not* in quantum mechanics. Deutsch's response is that although CTLs did indeed originate in classical Einsteinian general relativity, the still incomplete theory of quantum gravity *does* predict CTLs, too. And that is an exciting observation for time travel enthusiasts because, as Deutsch writes, the results of his quantum studies of CTLs show that "contrary to what has usually been assumed, there is no reason in what we know of fundamental physics why closed timelike lines should not exist." That view was later endorsed by other physicists who wrote, after a quantum mechanical study of how a particle could transit a time machine spacetime in a physically consistent manner, "there is no contradiction between the postulates of quantum mechanics and the possible existence of causality violation in general relativity."[148]

Long before these scientific endorsements, science fiction had enthusiastically embraced the many-worlds idea and its connection with time travel. The first such tale[149] appeared when Everett was just 3 years old; it put forth the insightful observation that although alternate time tracks may allow changing the past for the better (something that can't be done, for better *or* for worse, with a single time track), in the end any such change may still be futile. As Daniels' time traveler puts it, "I did have an idea to ... go back to make past ages more livable. Terrible things have happened in history, you know. But it isn't any use. Think, for instance, of the martyrs and the things they suffered. I could go back and save them those wrongs. And yet all the time ... they would still have known their unhappiness and their agony, because in this world-line those things happened. At the end, it's all

[145]R. A. Healy, "How Many Worlds?" *Nous*, November 1984, pp. 591–616.

[146]B. S. DeWitt, "Quantum Mechanics and Reality,' in *The Many-Worlds Interpretation of Quantum Mechanics* (B. S. DeWitt and N. Graham, editors), Princeton 1973.

[147]B. Shaw, "What Time Do You Call This?" *Amazing Science Fiction*, September 1971. When a bank robber in one world tries to make his escape into the other world, he literally runs into 'himself' trying to escape after robbing the 'same' bank in the parallel world!

[148]D. S. Goldwirth *et al.*, "Quantum Propagator for a Nonrelativistic Particle in the Vicinity of a Time Machine," *Physical Review D*, April 15, 1994, pp. 3951–3957. See, too, the earlier D. S. Goldwirth *et al.*, "The Breakdown of Quantum Mechanics in the Presence of Time Machines," *General Relativity and Gravitation*, January 1993, pp. 7–13.

[149]D. R. Daniels, "The Branches of Time," *Wonder Stories*, August 1935.

unchangeable; it merely unrolls before us."[150] Many years later, in a critique of the many-worlds idea, a philosopher/physicist echoed Daniels' words: [In the world] that I (subjectively) experience I may blunder, but [in another world], with equal actuality, I triumph gloriously. The Everett interpretation can be used this way to mitigate sorrows, but this use is two-edged, for it equally well implies the speciousness of happiness."[151]

The editorial introduction to Daniels' pioneering tale is quite interesting: the opening line is "To say that this short story contains some revolutionary time-travel theories would be putting it exceedingly mild." That editor then went on to tell his readers, with great enthusiasm, that "when the author ... submitted this story to us, his accompanying letter stated that in it he had settled the time-travel question once and for all. We must admit that a broad, unbelieving grin spread over our countenances when the author dared make this assertion. BUT—the smile soon left our faces ... [T]o our chagrin, Mr. Daniels had really propounded so many brand new ideas about time and time-travel, and such logical ones—*that he has not left one loophole in his argument!*"

John W. Campbell (1910–1971), the first (and only) editor of *Astounding Science Fiction* (today's *Analog*), called alternate time track stories "mutant" because they represented the first new innovation (or mutation) in the time travel concept since H. G. Wells. Campbell incorrectly claimed *The Legion of Time* (note 133) was the first such tale (see Campbell's editorial in the May 1938 issue of *Astounding*) , and that "Other Tracks" (note 135) was the second, but in fact it was Daniels who was first with splitting time tracks in science fiction. After Daniels the concept quickly became part of standard science fiction lore and could be used by other writers with little explanation. For example, just a little more than a decade later one author did not have to say much about his "First Law of Chronistics," which determines the development of "the branches of Fan-Shaped time." It was sufficient for his readers to learn that should a time traveler to the past change anything, a parallel branch of time would be created on which the time traveler would be trapped: "The man who interfered with the space-time matrix, displacing even a comma in the great scroll of time, would be cut-off from his origin forever."[152]

Still, if there is one thing we can say about science fiction, it's that no 'rule' is immune to challenge. Decades after Daniels' tale put forth the MWI, we find the well-known author James Blish (note 100 in Chap. 1) *rejecting* it. In a story about the reception of radio signals from the future, we read of one character telling another "I *was* going to do all those things. There were no alternatives, no fanciful 'branches in time,' no decision-points that might be altered to make the future

---

[150]These sad, resigned words were written when the author, David R. Daniels (1915–1936), was just twenty years old. A year later he committed suicide.

[151]A. Shimony, "Events and Processes in the Quantum World," *Quantum Concepts in Space and Time* (R. Penrose and C. J. Isham, editors), Oxford University Press 1986.

[152]J. MacCreigh, "A Hitch in Time," *Thrilling Wonder Stories*, June 1947.

change. My future, like yours . . . and everybody else's, was fixed. It didn't matter a snap whether or not I had a decent motive for what I was going to do; I was going to do it anyhow. Cause and effect . . . just don't exist. One event follows another because events are just as indestructible in space-time as matter and energy are."[153]

This denial of the MWI is simply an author's choice, of course, for whatever story effect is desired, and others may make different choices. Isaac Asimov, for example, used the MWI idea in in the story of a time traveler who journeys back to 1871 London, to retrieve a lost Gilbert and Sullivan operetta (*Thespis*). When he returns to the present he finds that his wife Mary (who was alive when he left) has been dead for a year on the new time track that his actions in the past have created. As the story ends, the devastated time traveler thinks "I had changed history. I could never go back. I had gained *Thespis*. I had lost Mary."[154] This sad fate is repeated in another story of a time traveler lost in an infinitude of time tracks with no hope of ever finding his way home: "In all of time, how many, many worlds there must be. How to find a single twig in such a forest?"[155]

Splitting universes have been used in literary works outside the genre of science fiction, as well. Examples include "The Garden of Forking Paths" by the Argentine writer J. L. Borges, the first play J. B. Priestly wrote (the 1932 *Dangerous Corner*), John Updike's 1997 novel *Toward the End of Time*, and Gore Vidal's 1998 novel *The Smithsonian Institution*. Typical of these fictional fantasies about splitting universes is a tale (anticipating Asimov's) by Lord Dunsany (1876–1957)—the Irish writer Edward Plunkett—the story of a man who goes back in time to correct "two or three mistakes he had made in his life."[156] This he successfully does, but the result is a new, subtly different subsequent history. The differences are not infinitely subtle, however; after the changes, he finds that his home, his wife, and all the delicate details of his life have vanished. As he relates to a visitor at the lunatic asylum he is now confined to, as the result of his despair, "I tell you I'm lost. Can't you realize that I'm lost in time? I tell you that you can find your way traveling the length of Orion, sooner than you shall find it among the years . . . Don't go back down the years trying to alter anything . . . Don't even wish to . . . [T]he whole length of the Milky Way is more easily traveled than time, amongst whose terrible ages I am lost."

In writing for a mass audience, rather than just for the more limited science fiction and fantasy one, perhaps the best known literary work of alternate history is the classic 1953 novel *Bring the Jubilee* by Ward Moore (1903–1978). In that work Lee wins the Battle of Gettysburg, and the South wins the Civil War. Using a time machine, a historian travels from 1952 (of the world in which the South wins) into the past of 1863 to study the battle, where he inadvertently disrupts events to the point that the North wins; that is, reality splits and the newly created fork represents

[153]J. Blish, "Beep," *Galaxy Science Fiction*, February 1954.

[154]I. Asimov, "Fair Exchange?" *Asimov's Science Fiction Adventure Magazine*, Fall 1978.

[155]M. F. Flynn, "The Forest of Time," *Analog Science Fiction*, June 1987.

[156]Lord Dunsany, "Lost," *The Fourth Book of Jorkens*, Arkham House 1948.

the time track of *our* world. The historian is trapped on this new fork, cut off forever from his original time track. The entire novel is in the form of a discovered manuscript, written in 1873 and found in 1953, and the pathos of the ultimate isolation endows the novel with great emotional impact.

A 1992 novel on the same theme, Harry Turtledove's *The Guns of the South*, begins with a fascinating premise but then misses the crucial distinction between a single versus multiple time tracks. In that work racists from the future (2014) arrive by time machine at Lee's 1864 winter camp. They bring with them AK-47 automatic assault rifles and offer to supply Lee's army with all it can use. Lee accepts and the South wins the Civil War. The future, of course, changes—or does it? The time travelers have brought back books from the future showing that the South lost the war, so the implication is that history must have forked. So far, so good. But all through the novel, the time travelers move back and forth between the nineteenth and the twenty-first centuries, apparently finding their own time unchanged. And if that is so, then the whole point of the story vanishes. Why all the effort to change history when it is clear that nothing has changed? The novel is entertaining reading (Turtledove is a trained historian), but I believe Moore's novel to be the superior work of science fiction.

I'll end this discussion on splitting universes with a startling theological issue raised by a philosopher.[157] Arguing that God cannot branch into multiple time tracks because God is unique, the conclusion seems inescapable that God therefore exists on exactly one of how ever many different time tracks there may be. What if that chosen time track isn't ours? Then, concludes the philosopher, Nietzsche's nineteenth-century metaphorical claim that "God is dead" (for us) might literally be true! He admits that this is "fanciful," but still . . . .

4.7 For Further Discussion

In an afterword to his story "Dead City" (*Thrilling Wonder Stories*, Summer 1946), Murray Leinster muses "You've heard the old argument that a man can't travel backward in time because he might kill his grandfather. I've wondered why nobody has argued that a man can't travel forward in time because he might be killed by his grandson." One possible answer to Leinster is that *if*, at the moment the forward-bound time traveler departs, he has not yet sired a child, *then* there simply wouldn't be a murderous grandson waiting for him in the future. Perhaps, however, Leinster had this somewhat more

(continued)

---

[157]Q. Smith, "A New Topology of Temporal and Atemporal Permanence," *Nous*, June 1989, pp. 307–330.

complicated scenario in mind: After the time traveler arrives in the future he is attacked by a mysterious stranger but survives, and later returns to the present. He then sires a child who will be the parent of that mysterious stranger. (As far as I know, this plot line has not appeared in a science fiction story.) Contrary to Leinster's view, explain why the *possibility* of being killed by a grandson is not a reason for forbidding the possibility of a trip in time (in either direction).

In his causal loop paper (note 105) the University of Delaware philosopher Richard Hanley correctly writes (on p. 146) "physicists have tried to avoid free will problems by ignoring causal loops involving intentional agency," and partly illustrates this claim with the autoinfanticide paradox, writing of the attempt of a time traveler to kill his younger self as inevitably failing because "the past is apparently brought about willy-nilly." (Hanley unfortunately then uses the story "Thompson's Time Traveling Theory" as an example of this—see note 77, and the end of the "Introduction"—when it is that time traveler's *grandfather* who is the intended target.) Discuss the merits of Hanley's claim, keeping in mind the end of Sect. 4.3 (in particular, note 81). If you are interested in genetics and astronomy as well as in time travel, then for extra credit comment on Hanley's claim (p. 137) that "one can extract information about my DNA from … my astrological chart."

The fictional killing of Hitler was imagined in print even before World War II, in Geoffrey Household's intense 1939 novel *Rogue Male* (made into the 1941 film *Man Hunt*). And so it's not surprising that one of the popular change-the-past themes in science fiction is that of a time traveler killing the Führer. (This idea, somewhat oddly, appeared in the debates leading up to the 2016 American Presidential election, when one of the candidates, to show the toughness of his character—even though he opposed abortion—declared "Hell, yes, I'd kill baby Hitler! You gotta step up, man." This candidate did elaborate a bit, stating there might be some risk involved with tampering with the past.) Stories in this sub-genre include E. Norden's "The Primal Solution" (*Magazine of Fantasy & Science Fiction*, July 1977), W. R. Thompson's "The Plot to Save Hitler" (*Analog*, September 1993), L. del Rey's "My Name Is Legion" (*Astounding Science Fiction*, June 1942), and R. M. Farley's "I

(continued)

Killed Hitler" (*Weird Tales*, July 1941). Read some of these tales and compare the various repercussions envisioned by the authors following an assassination of Hitler by a time traveler.

After our discussion of Heinlein's time travel masterpiece "All You Zombies—," you might think it impossible to write a new story that exceeds it in complexity. That might well be true, but a modern masterpiece by Ted Chiang certainly gives it a good run for the money. "The Merchant and the Alchemist's Gate" (*Magazine of Fantasy & Science Fiction*, September 2007) uses a 'wormhole' that connects the present to the future 20 years hence, and it is stuffed with intertwined causal loops and information bootstraps. Read it and keep track of all such occurrences. How many did you find? The last line of the story clearly expresses the view that the past cannot be changed: "Nothing erases the past. There is repentance, there is atonement, and there is forgiveness. That is all, but that is enough." Is the story always faithful to this view of time travel?

In his paper (note 126) on the coincidences of time travel, the University of Queensland philosopher Phil Dowe writes "It's true that remote time travel [into the very distant past] does not allow for causal loops ..." *Is* this true? Consider, as you think about this, the story "Time's Arrow" (*Science-Fantasy*, Summer 1950) by Arthur C. Clarke. In that tale geologists have just discovered, in a remote desert, the fossilized tracks of a monstrous creature, from fifty million years ago, tracks that indicate that the beast was in hot pursuit of fleeing prey. Before the geologists can unearth the entire set of tracks, to see if the pursuit was successfully completed, they are visited by a physicist who just happens to be conducting near-by experiments in time travel. (This proximity is explained by noting what better place to conduct time travel experiments, powered by atomic energy, than in a remote desert?) At one point during the visit, after being told of the ancient pursuit frozen in rock, the physicist muses "It would save you a lot of trouble, wouldn't it, if you could actually *see* what took place in the past, without having to infer it by these laborious and uncertain [geological] methods." This comment results in the Chief geologist paying a visit to the physicist's lab. After driving over in a car equipped with tires having "an odd zigzag pattern" in the tread,

(continued)

an accident suddenly sends the entire lab into the past. Soon after, the other geologists unearth the rest of the fossilized tracks, and learn what the creature's prey had been when they see a zigzag pattern in the rocks, tracks that show "the great reptile was about to make the final leap upon its desperately fleeing prey." Can you see how to modify this story so as to have a causal loop involving the very distant past? (For perhaps even more inspiration on thinking about causal loops, watch the 1980 movie *The Final Countdown*. In it the designer of a modern naval warship that temporarily travels back through time to the Pearl Harbor of December 6, 1941, turns out to be a crew member who was accidently left behind in the past. In the past he *will be* able to design the ship because he already knows how it *was* designed—by himself!)

Comment, at length, on the cartoon shown in Fig. 4.5. (Does it make logical sense?)

In the story "Salvation" by Jerry Oltion (*Analog*, December 2007) a physicist approaches the Universal Church of the Divine Revelation for money to build a time machine. He is blunt in making his case: "You could go back in time and meet Jesus. Assuming he existed." That statement causes (it should come as no surprise) not just a bit of pandemonium but, nonetheless, an influential Church leader decides to provide the funding. Why? Because later, while sitting in his office as he talks with the physicist, a sheet of paper suddenly appears in the air above the leader's desk and then flutters down to land on the telephone. Picking the paper up, the leader sees it is a sheet of his own letterhead, with writing in his own angular, precise handwriting, saying "It works. Give him the money. You almost named the dog Solomon." This convinces the leader because, as we are told, "Paper appearing out of nowhere was a good trick, but it might easily be just that: a trick. Duplicating his letterhead and his handwriting wouldn't be all that difficult either. [On the other hand] knowing the name [the leader] had considered but rejected for his German Shepard 15 years ago was a different level of feat entirely." The physicist seems to be startled by the appearance of the paper, too, and asks "May I see that?" His reaction convinces the leader it wasn't a staged event: "Well, I'll be damned," the physicist replies. Once the time machine is under construction, the two men realize they have to send the enigmatic message back in time to complete the loop. As they prepare to do so, the leader asks a curious question. After retrieving the mysterious sheet of paper from his desk,

(continued)

he says to the physicist "Should I send the original [the one he is holding in his hand], or should I write another?" The physicist replies with "Write a new one. If we send the original, we put it in a closed loop and [we'll] never get it back. We don't want to lose the first object to travel in time. We'll want that for the Smithsonian someday." Does this make sense? Also, comment on whether or not the dog's name is a bootstrap paradox.

A perplexing little time travel paradox, one that I don't think science fiction has yet treated (and I'm pretty sure physicists haven't had anything to say about it either), was cooked-up by the English philosopher Robin Le Poidevin in his 2003 book *Travels in Four Dimensions: the enigmas of Space and Time* (Oxford, pp. 180–181). There he writes "Peter and Jane, both 20 years old, are out for a walk one day in 1999 when suddenly a time machine appears in front of them. Out steps a strangely familiar character who tells Jane that he has an important mission for her. She must step into the machine and travel to the year 2019, talking with her a diary the stranger hands to her. In that diary she must make a record of her trip. Obligingly, she does as she is asked and, on arrival, meets Peter, now aged 40. She tells Peter to travel back to 1999, taking with him the diary she now hands him, and recording his trip in it. On arrival in 1999, he meets two 20-year-olds called Peter and Jane, out for a walk, and he tells Jane that he has an important mission for her." Le Poidevin then writes that "the really tricky question is: how many entries are there in the diary when Jane first steps into the machine? We imagine it blank. But this is the very same diary as the one Jane hands to the 40-year-old Peter, which then contains her entry. And by the time Peter arrives back in 1999, it will contain his entry, too. But then, if the diary already contained two entries when Jane was handed the diary, then it would contain three entries when she handed it to Peter, who would then add another one, so the diary would have contained four entries when it was first handed to Jane, and so on. If the problem is not immediately apparent, this is because we imagine an indefinite number of trips, but in fact there are just two: Jane's trip to 2019 and Peter's trip to 1999. So there ought to be a consistent answer to the question, how many entries are there in the diary? Yet, as we have seen, there does not appear to be a consistent answer." Another philosopher soon claimed he *did* have the answer: namely, 2. Read his paper (Erik Carlson, "A New Time Travel Paradox Resolved," *Philosophia*, December 2005, pp. 263–273), and either explain why you agree with Carlson's reasoning or enthusiastically rebut it.

As in "The Time Eliminator" (Fig. 4.1), other stories have imagined gadgets that simply view the past, rather than visit it as would a time machine. This is done in an attempt to avoid paradoxes—but does it? Two stories that illustrate how just viewing the past risks affecting the past as much as time travel would, are Horace Gold's "The Biography Project" (*Galaxy Science Fiction*, September 1951) and Donald Franson's "One Time in Alexandria" (*Analog*, June 1980). In the first tale the Biotime Camera, operated by the Biofilm Institute, allows teams of biographers to film (alas, no sound!) and study the lives of past notable personages. Of particular interest are the lives of those who developed neurotic psychoses, such as Isaac Newton. And, indeed, the Biotime Camera does capture Newton's image as he begins to display increasingly disturbed behavior. We see Newton, for example, as he begins to peer into dark corners, looking for those who have come to spy on him. On his death bed, the biography team assigned to him reads his lips and discovers that his final words are "My guardian angel. You watched over me all my life. I am content to meet you now." It is then that the Biofilm Institute realizes what it has done. Newton *was* in fact being spied upon—by the Biotime Camera, which has not changed the past but has certainly affected it. In the second tale an archeologist uses a time viewer to read the lost manuscripts in the ancient library at Alexandria before it was completely destroyed in an inferno. The viewer uses an infrared beam—and it is the heat from that beam from the future that proves to be the origin of the fire in the past. Again, the past has been affected, but not changed, by time viewing. Is it true to claim, however, that such viewing gadgets could not be the source of other paradoxes, such as causal loops or information bootstraps? If you think that isn't a valid claim, give a counter-example.

In a story by Francis Flagg and Weaver Wright (a pseudonym used by Forrest J. Ackerman), "Time Twister" (*Thrilling Wonder Stories*, October 1947), we read the following exchange between the inventor of a time machine and his none-too-bright helper:

"You mean to say," he questioned incredulously, "that I could go back a hundred years?"

"If you had the proper machine in which to travel, yes."

"But that'd take me back to before I was born."

The Professor smiled tolerantly.

"Look at this diagram, Hank. This line is the time continuum. It incorporates space, too. [The authors didn't actually print a diagram with the story, but

(continued)

surely the Professor is using a Minkowski spacetime diagram]. This dot is you. It doesn't matter when you were born, or when you will die. You exist right now, that's the fact. Traveling into the past or future wouldn't make you grow any younger or older. Such a thought is naïve. Let me demonstrate the mechanics of it for you. If . . . we calculate with non-Euclidean mathematics . . ."

"It don't sound reasonable," the farmhand objected. "If I went back—"

"I know," interjected the Professor, "if you went back you might meet your own father as a young man and you'd be older than he, or maybe he and your mother would be kids going to school."

"Haw, haw! That'd be funny, that would."

What famous movie, made nearly 40 years later, does the end of the conversation remind you of? Hint: "flux capacitor."

As mentioned in the text, a famous science fiction example of *affecting* (but not changing) the past is "Behold the Man" by Michael Moorcock, the tale of a time traveler who arrives in ancient times during the very years of the ministry of Jesus as reported in the Bible. When he finds there is actually no such person, the time traveler takes the role himself and lives out the events as reported in the Gospels, including the Crucifixion. This is a powerful story, but it had already been done more than 15 years earlier, by Philip K. Dick, in his short story "The Skull" (*If*, September 1952). In his tale, Dick tells of a man from the twenty-second century who is sent by government authority back to the mid-twentieth century to kill the Founder of a religious movement, a movement that 'now,' 200 years later, threatens those same government authorities. History records that the Founder gave a powerful speech just before being arrested and executed, a speech that started the religious movement, and so the time traveling assassin is told to kill the Founder *before* he can give that speech. (The parallel between Jesus and the Founder should be obvious.) Read these two stories and compare and contrast how Moorcock and Dick handled time travel paradoxes. Comment, in particular, on the relationship between the assassin and the Founder. Moorcock's tale should be easy to find, and Dick's is available as a free pdf download (it is in the anthology *The Best of Philip K. Dick*, Halcyon Classics 2010, as well).

A subtle change-the-past sequence appears in the original *Back to the Future* film that is easy to miss. When the hero, Marty McFly, returns to 1955 in the time car, he leaves from the parking lot of the Twin Pines Mall, so named because of the two pine trees that stand nearby. Arriving in the past with literally a bang, the time car inadvertently destroys one of the (then) young pines. Near the end of the movie, when Marty returns to the future (1985), he finds that the mall is now called the Lone Pine Mall. This is charming and fun, indeed clever, but modern scholars of time travel reject it, and other claims of changing the past, as not being logical. (Shakespeare understood this point, when he has Lady Macbeth declare, concerning the murder of Banquo, "What's done cannot be undone: to bed, to bed, to bed.") What Marty's trip *would* explain is why the mall would *always* have had the name of the Lone Pine Mall. Watch the movie and see how many other 'change-the-past' episodes you can find.

Chapter 5
Communication with the Past

"[As for travel to or for signaling the past] you'd have to exceed light speed which immediately entails the use of more than an infinite number of horsepowers."[1]

5.1 Reversed Time

"I have not discovered Mr. Wells' Time Machine."[2]

One way to communicate with the past is to 'simply' live backwards in time. Philosophers and other writers of speculative fiction were the first to wonder what things might be like in a world where the time asymmetry is reversed—that is, in a world where time 'runs backward.' Indeed, fascination with the idea of time reversal actually dates back thousands of years, long before science fiction, as it can be found in Plato's dialogue *Statesman*, written (most probably) 15 years before Plato's death in 347 B.C.

At one point, Plato offers an extended description of the world suddenly running backward in time in the ancient past. After one character is told that at that remote time "all mortal beings halted on their way to assuming the looks of old age, and each one began to grow backward," he asks "But how did living creatures come into being, Sir? How did they produce their offspring?" The answer is shocking: "Clearly ... it was not of the order of nature in that era to beget children by intercourse ... It is only to be expected that along with the reversal of the old men's course of life and their return to childhood, a new race of men should arise, too—a new race formed from men dead and long laid in Earth ... Such resurrection

---

[1] An observation by Haskel van Manderpootz, professor of the "newer physics," in S. G. Weinbaum's "The Worlds of If," *Wonder Stories*, August 1935. Compared to the 'modest' Van Manderpootz, all other physicists in the world are a mere "pack of jackels, eating the crumbs of ideas that drop from [his] feast of thoughts."

[2] W. R. Inge (1860–1954), in his November 1920 Presidential Address to the Aristotelian Society at the University of London Club, in a sympathetic treatment of the possibility of a time-reversed world.

© Springer International Publishing AG 2017
P.J. Nahin, *Time Machine Tales*, Science and Fiction,
DOI 10.1007/978-3-319-48864-6_5

of the dead was in keeping with the cosmic change, all creation being now turned in the reverse direction."

The reversed-time world is an important philosophical concept. Before the turn of the century, for example, Francis Bradley (one of the early proponents of the block universe, you will recall from Chap. 2), thought about reversed-time worlds and concluded that they would be quite odd: "Let us suppose ... that there are beings whose lives run opposite to our own ... *If* in any way *I* could experience *their* world, I should fail to understand it. Death would come before birth, the blow would follow the wound, and all must seem to be irrational."[3] A half-century later the South African philosopher J. N. Findlay (1903–1987) took Bradley's position of supporting a skeptical attitude towards the possibility of time-reversed worlds. Writing in a book review, Findlay declared "The reversed world in question wouldn't merely strike us as queer, but definitely crazy: it would be a world where what is wildly and intrinsically *improbable* was always occurring. It would, in fact, be much more startling than the original asymmetry that led us to think of it."[4] (Findlay was almost certainly thinking of things like a tea cup, shattered due to a fall, spontaneously reassembling itself.)

The question of backward-running time so fascinated Findlay that, some years later, he posed it as a problem for the readership of a scholarly journal (*Analysis*). This led to a number of responses, and subsequently he presented both his own negative view of time-reversed worlds and that of the best reader response he had received to his posed problem (which came from McGechie).[5] While Findlay showed admirable open-mindedness by awarding the title of *best* to an argument that refuted his own position, he remained unconvinced about the concept of reversed-time worlds, stating that "I continue to feel that a total reversal of my experiences is a terrifying possibility."

The terror aspect of living backward in time had been nicely captured in a science fiction story years before Findlay wrote. A scientist who is involved in an accident with radioactive materials has his sense of time flow reversed, and the story carefully and logically analyzes what his life would be like in such a situation. For example, the scientist can talk (backwards for others, of course), so he is understandable only if his words are recorded and then played in reverse. He cannot eat, because for him that would involve the regurgitation of food. He cannot answer questions because "if he should answer any questions put to him, it would mean he was giving the answer before he heard the question, on his time scale." And finally, he can't pick anything up because the normally stable position and velocity error-correction mechanism between eye, hand, and brain, which is a negative feedback system in normal time, has become an unstable positive feedback system in reversed time. The horror of his existence is contained in the only words the man

---

[3] F. H. Bradley, *Appearance and Reality* (2nd edition), Oxford University Press 1897, p. 190.

[4] J. N. Findlay, *Philosophy* (25) 1950, pp. 346–347.

[5] J. N. Findlay and J. E. McGechie, "Does It Make Sense to Suppose That All Events, Including Personal Experiences, Could Occur in Reverse?" *Analysis*, June 1956, pp. 121–123.

utters (deciphered after reversed playback): "Where am I? What's happened? Why are things so different? Why? Why?"[6]

Despite Findlay's "terror," however, the prevailing view today is that to the inhabitants of a time-reversed (or what is sometimes called a *counterclock*) world, *nothing would look odd*! This is a fairly new idea,[7] and not so long ago the philosophical literature displayed a misunderstanding of how a time-reversed world would appear to its occupants.[8] More recent analyses than Smart's (note 7) advocating the 'normality' of a time-reversed world are more compelling.[9] One concern about a time reversed world however is not easily dismissed: matter with a time reversed time sense is thought by many to be antimatter in our world, and any interactions between the two worlds would be spectacular, indeed![10]

But let's ignore that possible difficulty. The philosopher J. R. Lucas argued that even if beings from two such time-reversed worlds could meet, they still could not communicate: "If two beings are to regard each other as communicators, they must both have the same direction of time. It is a logical as well as a causal prerequisite."[11] Now, this matter is well worth some effort to understand, because it is intimately tied to time travel. At first blush, Lucas' words seem almost self-evident, and after a little thought they might seem to be absolutely irrefutable. The philosopher Murray MacBeath, however, took exception.

MacBeath opened his analysis[12] with a story to demonstrate that the persuasive power of Lucas' position is only superficial. In that story of Jim and Midge, Jim is one of us, whereas Midge is a 'Faustian time'[13] being. In his analyses MacBeath uses capitalized words and symbols for the time-reversed Midge, and lower case for Jim, as shown in Fig. 5.1. As MacBeath explains, "While Jim and Midge are together a face-to-face conversation is hardly likely to get off the ground. To make this clear let us say that they are together from t_0 until t_{10} on Jim's time-scale, and from T_0 until T_{10} on Midge's TIME-scale; t_0 is then the same temporal instant as T_{10} and, in general, $t_n = T_{10-n}$. If Jim at [his time] t_2 asks Midge a

---

[6]M. C. Pease," *Astounding Science Fiction*, "Reversion," December 1949. See also R. A. Banks, "This Side Up," *Galaxy Science Fiction*, July 1954, for a tale about the confusion caused by projecting a film the wrong way in time.

[7]The modern view that a time-reversed world would appear normal to someone living in it can be traced back at least as far as to J. J. C. Smart, "The Temporal Asymmetry of the World," *Analysis*, March 1954, pp. 79–83, an analysis, alas, that may not convince everyone.

[8]M. Dummett, "Bringing About the Past," *Philosophical Review*, July 1964, pp. 338–359.

[9]See, for example, D. L. Schumacher, "The Direction of Time and the Equivalence of 'Expanding' and 'Contracting' World-Models," *Proceedings of the Cambridge Philosophical Society* 1964, pp. 575–579; J. V. Narlikar, "The Direction of Time," *British Journal for the Philosophy of Science*, February 1965, pp. 281–285; F. R. Stannard, "Symmetry of the Time Axis," *Nature*, August 13, 1966, pp. 693–695.

[10]This issue is raised, several times, in Robert Silverberg's 1968 novel *The Masks of Time*.

[11]J. R. Lucas, *A Treatise on Time and Space*, Methuen 1973, pp. 43–47.

[12]M. MacBeath, "Communication and Time Reversal," *Synthese*, July 1983, pp. 27–46.

[13]In Goethe's play *Faust*, the normal flow of time is routinely upset.

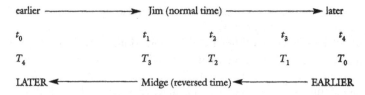

Fig. 5.1 Opposite time flows in counterclock worlds

question, and Midge hears the question at T_8, she will answer at T_9, and Jim will hear his question answered at t_1, *before* he asked it! What is more, if Jim is inexpert at interpreting backward sounds, and at t_4 asks Midge to repeat her answer, Midge will hear that request at T_6, BEFORE she has heard the original question; and her puzzled reply at T_7 will be heard by Jim at t_3 before he has uttered the request." Certainly this is a mess in time, and Lucas seems to be on safe ground with his denial of the possibility of communication between Jim and Midge. MacBeath, however, shows how to refute all of Lucas' arguments if Jim and Midge are allowed to be clever about how they send their messages back and forth—that is, if we give up some of our usual ideas of what a conversation is like. MacBeath's analyses are far too lengthy and detailed to present here, but the simplified diagram of Fig. 5.1 should enable one to follow the logic of his approach.

We imagine that Jim and Midge will not actually talk, and so will not have to decipher backward-spoken language. Rather, they will exchange messages via computer-generated text displayed on monitor screens, screens that are separated by a window that is proof against all penetration but the light emitted by those screens.[14] The nature of this window is not a trivial matter: if we accept the anti-matter nature of Midge's world then it is *essential* to keep her and Jim apart! To that end, MacBeath imagined that the window is double paned, with a perfect vacuum in-between. The exchange of photons between the two worlds should present no problems because photons are their own anti-particles.[15]

Now, imagine that at t_0 Jim brings a computer to the window. He programs it to wait for 4 days, until t_4, and then to display the following message on its screen: "This message is from Jim, who experiences time in the sense opposite to yours. Please study the following questions and display your answers on a computer screen three days from now." Jim's messages ends with the list of questions.

Because all that took place at t_4, Midge sees Jim's message and questions at what we will now call T_0. As requested, she brings her computer to the window, enters the answers to Jim's questions, and programs the machine to display them (after a

---

[14]Communication between beings in counterclock worlds, using written messages displayed through a window, appeared in science fiction *years* before MacBeath wrote: see I. Watson's 1978 novelette "The Very Slow Time Machine."

[15]There is no difference in the time sense of photons in either world because the flow of proper time for a photon—traveling at the speed of light, by definition—is zero (recall the discussion in Sect. 3.6).

3 day delay) on its screen. Thus, at T_3, which is Jim's t_1, Jim sees Midge's computer screen light up with "Hi, Jim. This is Midge. The answers to your questions are at the end of this message. Now, I've got some questions for *you*. Please display the answers two days from now." Midge's message ends with answers to Jim's questions and her list of questions.

Jim sees Midge's message at t_1, enters the answers to her questions and sets the machine to answer after a 2-day delay. At t_3, which is Midge's T_1—and by now you see how the process goes. It's cumbersome, sure, but it works. Or at least it does if everybody follows the rules. What if they don't? MacBeath provides other, increasingly complicated analyses to treat some of the more subtle problems that can be imagined in this method of exchanging messages. I will mention just two of them, which have direct analogs with what we normally think of as 'time travel.'

For the first problem, consider Jim's initial message, created at t_0 to be sent at t_4. He receives Midge's answer as described above at t_1, *before* his message is displayed through the window. So, what happens if at t_2 Jim cancels the message and it is *not* displayed? He has already gotten Midge's reply, but how can that happen if he does not send his message? This is, of course, a bilking paradox, with an explanation that we discussed in the previous chapter.

A second problem is the apparent possibility of creating a causal message loop. For example, let's say that at t_0 Jim suddenly decides to send a message through the window. (His reason for this sudden urge will be explained in the next few lines.) He thinks all night about what to send and, at t_1, finally settles on the following: "Greetings to the people on the other side of the window. This message comes from Jim, who hopes you will reply." Midge immediately sees Jim's message through the window (at her time T_3) and so is suddenly caught up with the desire to respond. She thinks all night about what to send and, hoping to be witty, she finally decides (at time T_4) on the following echo to Jim's message. "Greetings to the people on the other side of the window. This message comes from Midge, who hopes you will reply." Jim immediately sees this through the window (at what is his time t_0) and so now we know why he decides to send his original message! And Jim *will* send his original message—he has to because Midge replied to it.

In a review of a book on the direction of time, the philosopher Hilary Putnam restated the problems of a time-reversed world in the form of a provocative question: "How do you know that one man's future isn't another man's past?"[16] He began by making the interesting observation that for us to be able just to observe a backward-running universe, we would have to provide our own normal radiation source, because the counterclock stars in such a universe *absorb* radiation rather than emitting it. This point was elaborated on by another philosopher some years later, who wrote "We have uncritically imagined someone looking in on ... two worlds having opposite time directions ... Part of the story we tell, of the process of seeing, involves the emission of photons from objects [for example, the computer screens of Jim and Midge] and the subsequent impinging of these photons on our

---

[16]H. Putnam, *The Journal of Philosophy*, April 1962, pp. 213–216.

retinas. But this process is obviously directed in time. In a world where time ran opposite to ours, we could not see objects at all: objects would be photon-sinks, not photon-emitters."[17]

Putnam concluded his comments about reversed time with a cautious warning: "It is difficult to talk about such extremely weird situations without deviating from ordinary idiomatic usage of English. But this difficulty should not be mistaken for a proof that these situations could not arise." That challenge is no doubt why so many writers of science fiction and fantasy have tackled the question of what it would be like if time ran backward. We can find such a tale long before the science fiction pulps, in fact, in a tale that appeared when Einstein was just 7 years old. In that story[18] the narrator (a professor of astronomy and higher mathematics) suddenly finds himself on Mars. There he encounters beings who know the future up to their deaths, and whose memories of the past are "scarcely more than a rudimentary faculty." The entire tale is in the form of a conversation between the professor and one such being, who argues (quite persuasively) for the virtues of his 'backward' existence compared to that of earthlings (the Martian name for Earth is the story's title). Bellamy realized that his story implies a fatalistic block universe: "No one could have foresight ... without realizing that the future is as incapable of being changed as the past," he wrote.

Other writers, too, were fascinated by the implications of reverse time. When Merlyn the magician makes his first appearance in T. H. White's 1939 masterpiece *The Once and Future King*, for example, he explains how he knows the futures of others: "Ordinary people are born forward in Time, if you understand what I mean, and nearly everything in the world goes forward, too ... But I unfortunately was born at the wrong end of time, and I have to live backwards from its front, while surrounded by a lot of people living forwards from behind. Some people call it having second sight."

What may have put the idea in White's mind for his time-reversed magician is only speculation today, but perhaps it was something he might have read a decade before, in a fellow Englishman's writing, the 1929 book *The Nature of the Physical World* by Sir Arthur Eddington, where one finds the following passage: "In "The Plattner Story" H. G. Wells relates how a man strayed into the fourth dimension and returned with left and right interchanged ... In itself the change is so trivial that even Mr. Wells cannot weave a romance out of it [but see one of the *For Further Discussion* questions at the end of Chap. 2]. But if the man had come back with past and future interchanged, then indeed the situation would have been lively."

Whether or not those words influenced English fantasy, they certainly had some effect on American science fiction. In his 1979 memoir *The Way the Future Was*, pulp editor Frederik Pohl wrote that Eddington's book (which Pohl incorrectly attributed to Sir James Jeans) had given him the idea for a story using the reversed-time twist. But before Pohl could publish it, an even better (claimed Pohl) tale

---

[17]N. Swartz, "Is There an Ozma-Problem for Time?" *Analysis*, January 1972, pp. 77–82.

[18]E. Bellamy, "The Blindman's World," *The Atlantic Monthly*, November 1886.

arrived from Malcolm Jameson. Jameson, too, had read Eddington's book, and the result was the novella-length "Quicksands of Youthwardness," which Pohl published in *Astonishing Stories* as three-part serial during 1940–1941. Unfortunately, Jameson's tale is both pretty awful *and* devoid of *any* connection with Eddington's suggestion. One can only wonder about what might have been in the story Pohl says *he* discarded—did *it* have the future remembered? Alas, Pohl wrote that he couldn't remember!

Remembering the future does occur in one tale where everybody knows what will happen (as they live backward) by reading "prediction books." What distinguishes that story from many others on the same theme is an interesting, ironic conversation a student in a reverse-time world has with a philosophy professor about how things would be if time went the 'other way,' as in *our* world:

> "How can we tell? The reverse sequence of causation may be just as valid as the one we are experiencing. Cause and effect are arbitrary, after all."
>
> "But it sounds pretty far-fetched."
>
> "It's hard for us to imagine, just because we're not used to it. It's only a matter of viewpoint. Water would run downhill and so on. Energy would flow the other way—from total concentration to total dispersion. Why not?"[19]

The student is unconvinced, however, and when he tries to visualize such a peculiar world (*our* world, don't forget!) it gives him a "half-pleasant shudder." Imagine, he thinks in wonder, never knowing the date of your own death.

A few years later, in Wilson Tucker's 1955 novel *Time Bomb*, we find the intriguing idea of political assassination by time bomb, with the bombs actually *time traveling* to their targets. A policeman begins to suspect what is happening when it becomes evident that one of the explosions was actually an implosion: "The time bomb ... had been going in and had carried the force of the blast with it. Inward. Into the past. He frowned at that. A backward explosion? An explosion which ran counter to the normal flow of time, to the normal method of living? ... How would an explosion appear to a man if the blast happened in the opposite manner? If it began exploding now, in this moment, but continued backward instead of forward? Would it be an implosion?"

There is little doubt that the definitive treatment of a reversed-time world is that of Philip K. Dick's 1967 novel *Counter-Clock World*. Dick's world was once our world, but then, as in Plato's tale that began this chapter, time suddenly begins to run backward. People still alive reverse their direction of aging (but still think, walk, and talk in forward time), and dead, buried people come alive again and emerge from graveyards as the "Sacrament of Miraculous Rebirth" is intoned by a priest; all live their way back to the womb, just as in Plato's tale of 1600 years earlier. Such imagery is powerful stuff, but the physicist John Wheeler (of black hole fame) would have none of it. As he wrote, "Most of us would probably agree

---

[19]D. Knight, "This Way to the Regress," *Galaxy Science Fiction*, August 1956.

that the universe has not contained and will not contain any backward-looking observers. We do not expect to see caskets with corpses in them coming to life, nor do we expect to find bank vaults in which a gram of radium will integrate rather than disintegrate."[20]

5.2   Multi-dimensional Time

"If there are extra time dimensions we get violations of causality, because one could sneak to yesterday through the extra dimensions, and ... if you had sneaked to yesterday, you would have disappeared from today."[21]

There are those, however, who haven't been quite so sure as Wheeler about the impossibility of a reversed-time world. One physicist, for example, showed how (under certain initial conditions at the Big Bang) there is a possible solution to the gravitational field equations that gives an oscillating universe that temporally runs backward during the contraction phase.[22] And a philosopher has argued that the *direction* of time is local, not global (just as special relativity showed is the *rate* of time) and that the arrow of time can point in opposite directions at different locations.[23] As odd as such ideas may seem, a generalization of reversed-time— *multi-dimensional* time—makes it seem small potatoes. This is the idea that there might be *many* possible directions to the arrow of time, not just two. At first this may seem an absurd idea, something akin to a man jumping onto his horse and riding off in all directions at once. But philosophers (and perhaps just a few physicists[24]) have started to take at least a semi-serious look at the concept; as

---

[20]Wheeler's comments can be found in the General Discussion at the end of *The Nature of Time* (T. Gold, editor), Cornell University Press 1966.

[21]An intriguing (if somewhat mysterious) thought from F. J. Yndurain, "Disappearance of Matter Due to Causality and Probability Violations in Theories with Extra Timelike Dimensions," *Physics Letters B*, February 28, 1991, pp. 15–16.

[22]H. Schmidt, "Model of an Oscillating Cosmos Which Rejuvenates During Contraction," *Journal of Mathematical Physics*, March 1966, pp. 494–509. An elaboration of Schmidt's ideas is in A. Walstad, "Time's Arrow in an Oscillating Universe," *Foundations of Physics*, October 1980, pp. 743–749.

[23]G. Matthews, "Time's Arrow and the Structure of Spacetime," *Philosophy of Science*, March 1979, pp. 82–97.

[24]More than half-a-century ago one writer asserted that two-dimensional *complex* time was old hat in the theories of spinning particles—see M. Bunge, "On Multi-dimensional Time," *British Journal for the Philosophy of Science*, May 1958, p. 39. For a summary of many of the objections to multi-dimensional time see J. K. Kowalczynski, "Critical Comments on the Discussion About Tachyonic Causal Paradoxes and the Concept of Superluminal Reference Frames," *International Journal of Theoretical Physics*, January 1984, pp. 27–60 (and the reply by E. Recami, September 1987, pp. 913–919).

with so many other of the radical concepts associated with time travel, though, science fiction writers were dealing with multi-dimensional time long before it became a respectable topic in learned philosophical and physics journals.

In one pulp story, for example, we find a professor asking his redundantly named class in speculative metaphysics "Why shouldn't time be a fifth, as well as a fourth, dimension?"[25] In response to a generally skeptical reception to that, the professor goes on to say "I believe in the existence of a two-dimensional time scheme ... Ordinarily, most people think of time as a track they run on from their births to their deaths ... Think of this time track we follow over the *surface of time* as a winding road [it is the imagery of a surface that gives the professor *two* time dimensions] ... Once in a while another road crosses at right angles. Neither its past nor its future has any connection whatsoever with the world we know."[26]

The year before, the same pulp had published another tale[27] that went well beyond a mere two time dimensions. We are told in that story of two countries on an alien planet at war in the distant future. The war is a stalemate until one side begins to fire a gun at its foe from just two miles from its target, in the heart of enemy territory—*from the middle of next week*! The gun's shells are true 'time bombs.' This is not mere 'ordinary' time travel along one time track, however, but a multidimensional effect. Using a photograph of the gun in actual operation to support his astonishing discovery, an agent for the side being shelled reports to his superior that "the gun and its crew are existing along another time axis at right angles to the direction of our 'normal time,' so that from our point of view they are existing perpetually in the same instant."

That explains why the gun crew can (will?) operate without interference in next week's future, as they are in their adversary's time only for the instant that the two time tracks intersect. Indeed, the spy used the same trick to obtain his undetected photograph: "I secured the photograph by orienting myself along still another time axis at right angles to that of the gun, and approached it as an instantaneous, invisible entity." By the story's end both sides are using and counter-using this technique, evading each other "to and fro along an ever increasing complexity of mutually perpendicular time axes." In fact, the final count exceeds 75 time axes, making Heinlein's two-dimensional time look rather skimpy by comparison.

Well, of course, 75 time directions *is* science fiction (I think), and physicists are not so enamored of multidimensional time as are science fiction writers. For example, Eddington wrote that he found the idea of any region of spacetime involving two-dimensional time to "defy imagination."[28] Another physicist showed that the extremal property of timelike geodesics (look back at Fig. 3.15 and its

---

[25]R. Heinlein, "Elsewhen," *Astounding Science Fiction*, September 1941.

[26]This story has an amusing scene in which one of the professor's students accidently 'jumps time tracks' and so enters a new track with his arrow of time pointing backwards.

[27]N. L. Knight, "Bombardment in Reverse," *Astounding Science Fiction*, February 1940.

[28]A. S. Eddington, *The Mathematical Theory of Relativity* (2nd edition), Cambridge University Press 1924, p. 25.

discussion) would fail for multidimensional time, which he then associated with the stability of matter and a failure of causality.[29] Yet another physicist, however, was just a bit more willing to consider multidimensional time, and suggested that a viable theory of quantum gravity *might* support the idea of multiple time dimensions.[30] Of just *what* more than one time dimension might actually *mean*, however, this same physicist echoed Eddington by writing "Physics in a spacetime of . . . two timelike dimensions would be very weird indeed." Agreeing with this physicist was a philosopher who called the idea that there could be more than one dimension to time a "rather wild possibility" and a "fairy-tale."[31] These are probably fair statements of how most physicists presently think of multidimensional time.

But not all philosophers are of that persuasion, and many are in fact as fascinated by the possibilities of multidimensional time as are science fiction writers. So, *why* this interest in something so different from anything we actually experience? Where does the motivation come from? Of what *use* is multidimensional time? I think the answers to those questions all derive from how multidimensional time offers a theoretical model for giving meaning to the view that the past can be changed (take a look back at note 20 in the Introduction).

In a certain trivial sense, of course, the past is always changing. For each of us the past is the set of all events that have happened, arranged in a before/after temporal order, and this set is continually increasing (and so changing). That is not, however, what most people mean by a changeable past. What *is* meant is that there may be some kind of change in the temporal ordering of events, or that an event that once was (or wasn't) a member of the set of past events no longer is (or is now) a member. Two-dimensional time offers a way to make sense of such possibilities, which one-dimensional time simply cannot do. To see how that works, I'll follow the presentation in a paper that forcefully argues that it does make sense to talk about altering the past.[32]

Meiland was aware that some might find his model *ad hoc*, even "incredibly weird" (in his own words), but he justified his efforts by taking a refreshingly enlightened, non-Humean view of what he thought would be the proper response to meeting purported time travelers: "If strange machines containing people in futuristic garments and speaking strange tongues (or perhaps using ESP instead of speech) were to appear and were to claim to be from the future, we might very well begin to search for a theory of time that allows their claim to be true." In Fig. 5.2 you can see how Meiland tried to do just that.

---

[29]J. Dorling, "The Dimensionality of Time," *American Journal of Physics*, April 1970, pp. 539–540.

[30]C. Isham, "Quantum Gravity," in *The New Physics* (P. Davies, editor), Cambridge University Press 1989.

[31]D. Zeilicovici, "Temporal Becoming Minus the Moving-Now," *Nous*, September 1989, pp. 505–524.

[32]J. W. Meiland, "A Two-Dimensional Passage Model of Time for Time Travel," *Philosophical Studies*, November 1974, pp. 153–173. Jack Meiland (1934–1998) was a professor of philosophy at the University of Michigan.

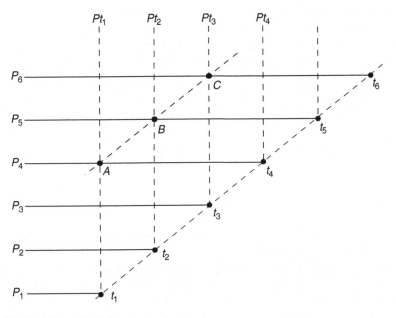

Fig. 5.2 Two-dimensional time

The dashed diagonal line, marked with the points $t_1, t_2, \ldots$ represents our usual one-dimensional image of time. The horizontal lines $P_1t_1, P_2t_2 \ldots$ (which we can simply call $P_1, P_2, \ldots$, for short) are the pasts for the present instants $t_1, t_2, \ldots$. That is, P_1 is the past with respect to the present t_1, P_2 is the past with respect to the present t_2, and so on. The dashed vertical lines allow us to locate any moment in the past. For example, the intersection point A of P_4 with Pt_1 is the location of t_1 in the past with respect to the present t_4.

With this model, Meiland then analyzed in detail several interesting special cases. Suppose that t_1 and t_2 are 1 year apart and that there is a similar time separation between all adjacent, marked present moments on the diagonal. Let us further suppose that a time traveler at t_4 journeys backward 3 years to t_1, to arrive at point A. Assume he stays in the past 2 years; then his temporal locations lie along the dashed diagonal line ABC; that is, at B he is 3 years in the past of t_5, and at C he is 3 years in the past of t_6. From Fig. 5.2, then, we can imagine the time traveler saying, as he climbs into his time machine at t_4, "One year from now I'll be two years from now." That rather astonishing statement makes sense when we take both uses of now to be t_4 and observe that B (1 year from A) is 2 years in the past with respect to t_4.[33]

---

[33] A critic of time travel (see note 119 of Chap. 2) used what he claimed to be the absurdity of such a statement to support his ejection of time travel. One of Meiland's reasons for developing his two-dimensional model of time was, in fact, to be able to reply to that critic (Donald Williams).

Meiland's two-dimensional time model is undeniably fascinating, but it simply has no theoretical justification (as far as I know[34]). It is not necessary to assume two-dimensional time to explain Meiland's "strange machines containing people in futuristic garments"; it is possible to do that with one-dimensional time in a four-dimensional spacetime. The classic paradoxes, too, are understandable without two-dimensional time, as discussed in the previous chapter.

5.3 Maxwell's Equations and Sending Messages to the Past

"Communication with a world exactly, to minutest detail, a duplicate of our own [but] twenty thousand years ahead of us might ruin the human race as effectively as if we had fallen into the Sun."[35]

Every physicist and electrical engineer knows that the mathematical description of the electromagnetic field is given by Maxwell's equations. In particular, radio engineers know that the waves of energy their antennas launch into space follow the predictions of those equations with astonishing accuracy. Indeed, when Einstein's relativity theory was completed, it was found that Maxwell's equations automatically satisfy relativity because magnetic effects *are* relativistic effects; in other words, relativity is built into Maxwell's equations. Whereas Newton's laws of dynamics had to be patched up, Maxwell's equations were untouched by the discovery of relativity.

Thus, it was a puzzle when physicists discovered that careful study of the seemingly perfect Maxwell equations, when applied to antennas, apparently results in the prediction of causality violation. It is found, in fact, that the equations have *two* solutions. One, as expected, contains the feature of time delay; that is, creating an electromagnetic disturbance at the antenna *now* causes a detectable effect at a distant point in space *later*. This is the so-called *time-retarded-solution*, and its common-sense physical interpretation is that of energy waves traveling away from the antenna as they also travel into the future. The shock was that Maxwell's equations also accept an *advanced solution*; energy waves *arriving* at the antenna from infinite space.

The physicist Paul Renno Heyl (1872–1961) wrote the perhaps first scientific work discussing advanced electromagnetic effects, in his 1889 University of Pennsylvania doctoral dissertation with the provocative title "The Theory of Light on the

---

[34]Many of the arguments against multi-dimensional time can be found in M. MacBeath, "Time's Square," in *The Philosophy of Time* (R. Le Poidevin and M. MacBeath, editors), Oxford University Press 1993. MacBeath concludes, however, with "I would not want to rule out the possibility ... that time is three-dimensional. Or worse." See also Alasdair Richmond, "Plattner's Arrow: Science and Multi-dimensional time," *Ratio*, September 2000, pp. 256–274.

[35]The Victorian writer Samuel Butler (1835–1902), in the "Imaginary Worlds" entry of *The Notebooks of Samuel Butler* (published posthumously in 1912), commenting on the chaos that communication across time might cause.

Hypothesis of a Fourth Dimension." Heyl cited the scientific guru of the fourth dimension, C. H. Hinton (look back at the last "For Further Discussion" assignment in Chap. 2) as his inspiration. The situation described by Heyl is something like that of a child standing at the edge of a pond. She throws a rock into the middle of the pond and watches ripples spread out and away from the splash. Suddenly, she sees ripples appear all around the edge of the pond and then travel inward toward the center, where they all converge at once. A spout of water then erupts from the surface of the pond at the simultaneous meeting of the inward-traveling ripples, and she watches as a rock is ejected from the spout to land back in her hand. She is, of course, open-mouthed with astonishment! How absurd, you think, as you read this, and who could blame you? This amazing imagery of advanced effects we owe to the philosopher Karl Popper, and it has come to be called "the fable of the Popperian pond."[36]

Pursuing the mathematics of wave motion in the fourth dimension, Heyl wrote "We are led to the curious conclusion that, in Hinton's aether,[37] the nature of the central disturbance *after* a given instant can influence the form of the aether *before* that instant. In other words, the aether seems to be endowed with an uncanny faculty of foreknowledge." We can avoid such a counter-intuitive implication of advanced effects, but only at the price of something many physicists and philosophers consider equally unacceptable: information traveling from the future into the past. We can still think of the advanced solution as representing electromagnetic waves of energy traveling *away* from the transmitting antenna—that is, as being *broadcast*, just like the retarded solution, rather than being received from infinity—*if* we also think of the waves traveling backward in time. Thus, the advanced solution to Maxwell's equations holds out the possibility of sending messages to the past, a sort of poor man's time travel. It may seem that we have simply traded one problem for another, however, because just sending information into the past can cause many of the same paradoxical, causality-busting situations that physical time traveling is claimed to cause.

The cosmologists Fred Hoyle (1915–2001) and J. V. Narlikar commented on the potential problems posed by communication backward in time, in their 1974 book *Action at a Distance in Physics and Cosmology*: "The [Maxwell] equations supply us with both advanced and retarded solutions (and, because of the linearity, with any linear combination of them) ... With so many solutions theoretically possible, why does nature always select the retarded one? That this question cannot be answered within the framework of Maxwell's theory must be regarded as one of its intrinsic weaknesses."

---

[36]K. Popper, "The Arrow of Time," *Nature*, March 17, 1956, p. 538. See, too, note 110 (and its discussion) in Chap. 2.

[37]It was thought, in nineteenth century physics, that electromagnetic waves need a *medium* through which to propagate (like ocean waves need water, and sound waves need air), a mysterious substance called the *aether* (or *ether*) that exists even in a vacuum. The 1871 Michelson-Morley experiment, however, implied that the aether simply does *not* exist.

That branding of Maxwell's theory as having an "intrinsic weakness" because of its prediction of an advanced solution was, I think, unwarranted. Indeed, the advanced solution can be given a perfectly reasonable physical interpretation.[38] Imagine a transmitting antenna sending electromagnetic waves to an identical receiving antenna. At any point in space between the two antennas there are electric and magnetic fields. Maxwell's equations allow us to calculate the fields produced by alternating currents in the two antennas. When we do an analysis of the relationship between the transmitting antenna's current and the fields, we use the retarded solution because the current is the cause and the fields are the effect. But in the analysis of the relationship between the fields and the receiving antenna's current, the situation is reversed, and the fields are the cause and the current is the effect. That is, the advanced solution is simply the mathematics relating the current in the receiving antenna *now* to the fields *in the past*.

An acceptance of both solutions has, in fact, been in the physics literature for nearly a century. As the Yale physicist Leigh Page (1884–1952) wrote decades before Hoyle and Narlikar, "While the advanced potentials, as well as the retarded potentials, satisfy the electromagnetic equations, the former has generally been discarded for the reason that it has been more in accord with the trend of scientific intuition to consider that the present is determined by the past course of events than by the future. However, if it is once admitted that the present state is uniquely determined by any past state, it follows that the future is also so determined, and hence the employment of a future state as well as a past state in specifying the present marks no inherent departure from our accustomed methods of description . . ."[39]

It may still be tempting, however, to just dismiss the advanced solution as a mere anomaly of the mathematics and to discard it on physical grounds. This is the traditional approach taken by physicists when confronted with non-causal solutions in any physical theory and, indeed, that was what Swiss physicist Walter Ritz (1878–1909) did with the advanced solutions to Maxwell's equations (an approach that involved Ritz during the last year of his life in a dispute with Einstein). For Ritz, the reversal of cause and effect simply did too much violence to his intuition to be taken seriously, and so he thought one must *impose* causality on Maxwell's equations (a condition they do not inherently contain) by *a priori* rejecting the advanced solution.[40]

Still, electrical engineers make similar kinds of judgements all the time, as when the solution of a quadratic equation for a passive (energy-dissipating) resistor gives

---

[38]See, for example, S. L. Schwebel, "Advanced and Retarded Solutions in Field Theory," *International Journal of Theoretical Physics*, October 1970, pp. 347–353, and L. M. Stephenson, "Clarification of an Apparent Asymmetry in Electromagnetic Theory," *Foundations of Physics*, December 1978, pp. 921–926.

[39]L. Page, "Advanced Potentials and Their Application to Atomic Models," *Physical Review*, September 1924, pp. 296–305.

[40]For more on this, see O. Costa de Beauregard, "No Paradox in the Theory of Time Anisotropy," *Stadium Generale* 1971, pp. 10–18.

both a positive value (which 'makes sense') and a negative value (which doesn't 'make sense' and so is simply ignored). There *were*, however, those who encouraged caution on this issue. Seventy-five years ago, for example, the eminent MIT electrical engineer Julius Adams Stratton (1901–1994) echoed Leigh Page's warning when he wrote of the disturbing advanced solution, "The familiar chain of cause-and-effect is thus reversed and this alternative solution might be discarded as logically inconceivable. However, the application of 'logical' causality principles offers very insecure footing in matters such as these and we shall do better to restrict the [Maxwell] theory to retarded action solely on the grounds that this solution alone conforms to the *present* [my emphasis] data."[41]

And in a famous paper I'll discuss in the next section, Wheeler and Feynman declared that "We conclude advanced and retarded interactions give a description of nature logically as acceptable and physically as completely deterministic as the Newtonian scheme of mechanics. In both forms of dynamics the distinction between cause and effect is pointless. With deterministic equations to describe the event, one can say: the stone hits the ground because it was dropped from a height; equally well, the stone fell from a height because it was going to hit the ground."[42] For Wheeler and Feynman, the reversal of cause and effect inherent to backward causation and time travel to the past offered no conceptual difficulties.

The elimination of an appeal to causality, or to the 'weirdness' of advanced waves, in arguing for the naturalness of the retarded solution to Maxwell's equations was first done in 1976.[43] The only auxiliary condition applied to the equations was simply the natural one of requiring the initial field energy to be finite.[44] And yet, today, there is still no experimental evidence for the physical reality of the advanced solution. Now and then one does run across speculations that the advanced waves of *something* traveling backward to us from the future might explain the so-called ESP 'talent' of precognition, but that is all it is, speculation.[45]

Advanced waves appeared in the pulp science fiction of the late 1930s, a decade before Wheeler and Feynman. For example, one story actually specifically invoked the advanced solution to Maxwell's equations, with a gadget (making use of what the author called the "anticipated potentials") displaying the near future on a television-like screen.[46] The author's by-line proudly gave his academic credentials as including a master's degree and, in fact, John Pierce was a graduate student in electrical engineering at Caltech. He received his doctorate just months after this

---

[41]J. A. Stratton, *Electromagnetic Theory*, McGraw-Hill 1941, p. 428.

[42]J. A. Wheeler and R. P. Feynman, "Classical Electrodynamics in Terms of Direct Interparticle Action," *Reviews of Modern Physics*, July 1949, pp. 425–433.

[43]P. C. Aichelburg and R. Beig, "Radiation Damping As An Initial Value Problem," *Annals of Physics*, May 1976, pp. 264–283.

[44]J. L. Anderson, "Why We Use Retarded Potentials," *American Journal of Physics*, May 1992, pp. 465–467.

[45]See M. B. Hesse, *Forces and Fields*, Philosophical Library 1961.

[46]J. R. Pierce, "Pre-Vision," *Astounding Stories*, March 1936.

story was published, and then went on to a highly distinguished career at Bell Telephone Laboratories and then later at the Jet Propulsion Laboratory operated by Caltech for NASA. Pierce knew all about Maxwell's equations, of course, and he actually opened his tale with a quote from Page's 1924 article (note 39) on the advanced solution. How many fictional pieces include quotes from the *Physical Review*?

And it would probably take something like advanced waves to explain the funny doings 2 years late in a story of a man caught in a time machine accident. Nearly all of his body ends up 4 years in the future—but only *nearly* all, because his eyes remain the present! As one of the puzzled observers of this odd business wonders, "Strange, that his eyes, now, can convey a message to his brain, four years hence, and his brain tells the eye muscles to move the eyeballs which are four years behind them —."[47]

Some of the most intriguing paradoxes of time travel involve no traveler—only information. Of course, any information flow at all, independent of time travel, involves the flow of energy and, as Einstein showed, energy and mass are different aspects of the same thing. Accordingly, information time travel involves the transfer of mass/energy. Thus, a man in the twenty-fifth century who sends a backward-in-time 'temporal radio' message to a twentieth-century woman stating that he loves (will love?) her is sending much more than mere emotion. Just *how* to send a message backward in time is, of course, the puzzle.

Indeed, all forms of present-day communication are transmissions only to the future. If you speak to someone, or if you send a radio message, there are always delays depending on the distance of separation and the speed of transmission of sound and light, respectively. If you want to send a message to the one hundred and twenty-fifth century, you can; just write a letter and seal it in a pressurized bottle of helium. This basic idea is dramatically presented in a novel about a scientist who is accidently transported from 2162 back to the late-Cretaceous, 80 million years into the past. There he leaves a written record on seven sandstone slabs of his brutal, lonely life among the dinosaurs—letters across time, if you will—found by twenty-second century geologists some years after his disappearance.[48] The transmission of such letters, *forward* in time, while dramatic, is not a puzzle. But what could be more astonishing than the message received *from the future* by the young inventor of the first time machine: after his initial experiment of sending his pilotless machine into the future, it returns with an envelope inside. Eagerly tearing it open, he finds the note is from the National Academy of Sciences: "We know from old records and museum models that this is the Cullen Foster experimental machine. Fifty years looks down on you and says 'Good work'."[49]

Heady stuff, that, but lots of other possible messages are capable of competing with young Foster's when it comes to generating excitement. For example, suppose

---

[47]M. Schere, "Anachronistic Optics," *Astounding Stories*, February 1938.

[48]G. G. Simpson, *The Dechronization of Sam Magruder*, St. Martin's Press 1996.

[49]D. Stapleton, "How Much to Thursday?" *Thrilling Wonder Stories*, December 1942.

you had a gadget that is superficially similar to a telephone but that calls telephones in the *distant* future. You can hear the person (in the future) on the other end, but they can't hear you (in their past). That is, information can flow only from future to past. It is then easy to imagine situations in their use of this device that at least seem paradoxical. For example, suppose you call your own private number 1 month ahead. You hear your future-self first answer the phone, and then recite the winning lottery for the 'previous' day, which is a month in your present self's future. (Your future self does this somewhat odd recital because a month from now you will remember, when your private phone rings, just *who* is calling!) So, now in the present you know you'll make a fortune by winning the lottery a month later.

This example *is* admittedly somewhat mysterious since, for the gadget to call far ahead in time, some sort of signal (as yet unspecified) must travel into the distant future because *something will* make the future phone ring. For the present discussion I am ignoring this crucial issue for the sake of the dramatic impact of the example. Soon, however, we'll get a little way into describing how one might, in principle, actually build this gadget, which in the physics literature is called an *antitelephone*. (Such a device is an *anti*telephone because the person who is the receiver is in the sender's past, the opposite of the situation for an ordinary telephone.) An interesting fictional illustration of such a gadget, despite being told as a hard-boiled detective murder mystery, appeared in science fiction some years ago (alas, while called a "time telephone," no theory for its operation was given, but instead was 'explained' as a "straightforward application of an impressive, but limited, technology").[50] So far, there is nothing paradoxical (or even illegal) in all of this, but what if, when the phone rings in the future the day after you won the lottery, you perversely decide *not* to recite the winning number? This apparent paradox has, in fact, already been treated with the aid of the block universe view of spacetime; that is, if the future-you spoke the lottery number when you originally called, then the present—you (now in the future) *must*, inevitably recite it.[51]

Let's now make things a bit more involved. Suppose that instead of calling your future self, you call the weather service and listen to the recorded message telling you the weather, 30 days hence. You do this day after day, and after a while you get a reputation for being able to predict, *perfectly*, the weather for every day to come, up to a month into the future. Your reputation spreads far and wide, and after a while more the weather service hears about you. Meteorologists check and find you are *never* wrong. Their computer models are only 80 % accurate out to 3 days, and for a week's prediction and beyond, the general public might as well flip a coin on whether it will rain or not on any particular day. But you are 100 % correct out to ten times their range. And so they hire you—and as a secondary job, you also make the

[50]S. Schmidt, "Worthsayer," in *More Whatdunits* (M. Resnik, editor), DAW 1993. The author, Stanley Schmidt, has a Ph.D. in physics and is a former editor of *Analog Science Fiction* magazine.

[51]For a fictional illustration of this (a so-called *bilking paradox*), see W. Tevis, "The Other End of the Line," *Magazine of Fantasy and Science Fiction*, November 1961.

daily weather recordings. (The voice on the other end of the gadget *has* sounded sort of familiar!) Here, then, is the puzzle we encountered earlier in causal loops carrying information: from *where* is the information in the flawless weather predictions coming from?

One easy answer is that the question is meaningless because such a future-to-the-past information flow must be impossible. Indeed, if I am to avoid telling a 'philosopher's fairy tale,' like those I criticized earlier in the book, I must admit that one consistent, non-paradoxical answer is found in recognizing that I have *assumed* that those 30-day weather reports are correct. Maybe, however, they are no better than anybody else's predictions. And so you *don't* become famous, and you *don't* get hired—and so there *isn't* any paradox. Is that the way to avoid paradoxes involving information flowing backward in time?

Perhaps not. As long as 1917 it was realized that special relativity does not preclude such an apparent backward flow. That is, if information could be transmitted faster than light, *then* messages could travel backward in time. That was the year Richard Tolman (1881–1948), a professor of physical chemistry at the University of Illinois and later at Caltech, wrote "The question naturally arises whether velocities which are greater than that of light could ever possibly be obtained."[52] He then answered that question, with his general conclusion being that if such velocities are possible, then a faster-than-light (FTL) observer could see the time order of two causally related events reverse. And thus the observer would see an affect *before* its cause. Alternatively, a *sub*luminal (slower-than-light) observer could see the two events, which are connected via an FTL interaction, reversed in time order from what a stationary observer would see.

Either situation has come to be called Tolman's paradox, but Tolman himself was careful with his words: "Such a condition of affairs might not be a logical impossibility; nevertheless its extraordinary nature might incline us to believe that no causal impulse can travel with a velocity greater than that of light." That was an astonishing statement, given that Einstein himself had specifically stated in his original 1905 paper on special relativity that such a thing simply could not occur. There is nothing, it would seem, to be "inclined" about.[53]

This rather technical connection between FTL speeds and backward time travel made the transition from theoretical physics to popular culture very quickly. It was in the British humor weekly *Punch*, for example, that the famous (but nearly always misquoted) limerick by A. H. R. Buller (1874–1944) first appeared:

There was a young lady named Bright
Whose speed was far faster than light;
She set out one day

---

[52]R. C. Tolman, *The Theory of the Relativity of Motion*, University of California Press 1917.

[53]Take a look back at Sect. 3.5, where we showed that the time order of two events can appear reversed for a subluminal observer if the two events are not causally related. Introducing FTL motion results in extending reversal to causally connected events; that is, FTL motion, reversed causation, and time travel to the past, go hand-in-hand-in-hand.

In a relative way
And returned on the previous night.[54]

Where *Punch* dared to go, Hollywood could not be far behind. Indeed, in this case it was actually there first, with the 1922 one-reel silent comedy movie *The Sky Splitter*. This was just a short film (feature pictures generally had at least four reels), so it is not clear how widely distributed and viewed it may have been. The story is that of a scientist testing a new spaceship: when it exceeds the speed of light, he begins to relive his life.

The linkage between time travel to the past and FTL motion is a central one in science fiction, and its fascination was nicely illustrated by one writer who has a time machine experimenter in the twenty-seventh century wonder "Was the speed of light the core of the mystery? At the speed of light did the past and the future become a shining, merging road down which men could walk—in their ears the thunder of time passing …?"[55] Not everybody was excited with the idea of FTL motion and travel backwards in time, however, with one eminent scientist declaring that "the limit to the velocity of signals is our bulwark against the topsy-turvydom of past and future."[56]

The obvious question at this point, of course, is whether it is even conceptually possible to build a gadget to send FTL messages backward in time? Einstein himself thought not, saying "We cannot send wire messages into the past."[57] But was he right? One hint at the possibility of achieving FTL speeds is in Dirac's 1938 paper (note 52 in Chap. 2). There, in his remarks about pre-acceleration, Dirac wrote "Suppose we have a pulse sent out from place A and a receiving apparatus for electromagnetic waves at a place B, and suppose there is an electron on the straight line joining A to B. Then the electron will be radiating appreciably [because accelerated charges radiate] before the pulse has reached its centre and this emitted radiation will be detectable at B at a time … earlier than when the pulse, which travels from A to B with the velocity of light, arrives. *In this way a signal can be sent from A to B faster than light* [my emphasis]."

This exciting conclusion goes a step beyond the usual examples of 'things that go faster than light.'[58] Dirac had an equally exciting reaction (and here the emphasis is his): "This is a fundamental departure from the ordinary ideas of relativity and is to be interpreted by saying that *it is possible for a signal to be transmitted faster than light through the interior of an electron. The finite size of the electron now reappears in a new sense, the interior of the electron being a region of*

---

[54]On page 591 of the issue of December 19, 1923.

[55]F. B. Long, "Throwback in Time," *Science Fiction Plus*, April 1953.

[56]A. S. Eddington, *The Nature of the Physical World*, Macmillan 1929.

[57]A. Einstein, "La Théorie de la Relativité," *Bulletin de la Société Francaise de Philosophie* 1922, pp. 91–113.

[58]Such as, for example, the intersection *point* of two very long, closing scissor blades. The explanation for how this can be is that the *point* is massless and does not participate in a causal chain (and so carries no information). Thus, special relativity is not violated.

failure, not of the field equations of electromagnetic theory, but of some elementary properties of space-time." This last line sounds very much like the things people say today about the singularity inside a black hole event horizon. And yet, Dirac was careful to point out that as weird as FTL speed may appear, special relativity is not violated because "in spite of this departure from ordinary relativistic ideas, our whole theory is Lorentz invariant." That is, even though 'faster-than-light' means 'backward in time,' which means 'causality failure,' special relativity still holds true and nothing awful happens to physics, only to our intuitions. The reason for this is that causality is *not* a premise or starting point for the special relativity.[59]

Of course, like any scientific theory, Dirac's theory is not necessarily the last word, and we have to admit the possibility that at least some of its implications (in particular, the possibility of FTL speeds) just aren't so. In all electronic communication systems that we use, information is transmitted by *modulating* a so-called *carrier wave*, and there is some reason to believe that such modulated waves cannot be sent at FTL speeds.[60] We must admit that it is one thing to talk of 'advanced wave radios'—often called *Dirac radios* in science fiction—and quite another to see how physics might actually enable one to talk to the past. FTL communication (without the time travel aspect) appeared in pulp science fiction before 1940, as in one story published the year after Dirac's paper.[61] In it we learn of a man on Pluto who has invented a way to send messages to Earth at twice the speed of light. He uses this gadget to warn of a would-be dictator who is on his way to Earth in a 'mere' light-speed rocket ship, and only an FTL message can warn Earth in time.

5.4   Wheeler and Feynman and Their Bilking Paradox

> "We find it difficult if not impossible to imagine waves that go into the future and *return to the present* [my emphasis] bearing information about where (and when) they have been."[62]

---

[59]See, for example, G. Nerlich, "Special Relativity Is Not Based On Causality," *British Journal for the Philosophy of Science*, December 1982, pp. 361–388. This same point was made nearly two decades earlier, in a study of the possibility of superluminal sound in superdense matter, by D. A Kirzhnitz and V. L. Polyachenko, "On the Possibility of Macroscopic Manifestations of Violation of Microscopic Causality," *Soviet Physics JETP*, August 1964, pp. 514–519.

[60]See G. Diener, "Superluminal Group Velocities and Information Transfer," *Physics Letters A*, December 16, 1996, pp. 327–331. For more on the modulation of a light-speed carrier wave in everyday AM radio, and in a more sophisticated single-sideband transmitter, see my book *The Science of Radio*, Springer 1999.

[61]N. Bond, "Lightship, Ho!," *Astounding Science Fiction*, July 1939. The author provides an interesting, detailed description of the gadget, and I think it would make a good question on a Ph. D. qualifying exam in physics or electrical engineering to explain the flaw in it.

[62]Bob Brier, *Precognition and the Philosophy of Science: An Essay on Backward Causation*, Humanities Press 1974. Brier is an Egyptologist (!) at Long Island University—with a Ph.D. in philosophy—who specialized at one time in parapsychology.

In 1941, at a meeting of the American Physical Society, Princeton University physicist John Wheeler and his student Richard Feynman discussed a seemingly outrageous idea that provided a possible clue to how a Dirac radio might function. The idea was that the advanced wave solutions to Maxwell's equations are not mere mathematical curiosities, but rather have profound physical significance. At the time, their talk received only a small abstract notice in the *Physical Review*, but after World War II they wrote it all up in a beautiful paper.[63]

Their primary goal was to explain the origin of the force of radiative reaction discussed by Lorentz earlier in the century. This reaction force is the cause of the energy loss suffered by an accelerated, charged particle. Lorentz, who thought of charged particles as having a finite size, attributed this reaction force to the retarded (by the time required for light to cross the width of the charged particle) coulomb repulsion force between one side of the particle's charge to the charge on the opposite side. This view, however, leads to various conceptual and mathematical problems, including an arbitrary assumption on how the charge is distributed over/ through the finite volume of the particle, as well as the problems of infinite self-interactions and the issue of what keeps the charge from blowing itself apart by internal coulomb repulsion.

Wheeler and Feynman's theory, on the other hand, avoided those problems by postulating point charges, because a *point* charge cannot repel itself. But then whence the reaction force, if there is no repulsion? Their revolutionary explanation was first to imagine the accelerated point charge as emitting retarded radiation outward in space, eventually to be absorbed by distant matter. This distant matter, which itself consists of point charges that are accelerated by the retarded radiation, then radiates *backward* in time, back toward the original charge that started the chain of events. This backward-in-time, or *advanced*, radiation arrives in the past of the original charge, and *it* is the cause of the observed reaction force. Indeed, Wheeler and Feynman proposed that an accelerated charge will not radiate unless there is to be absorption at some other distant place and future time. That is, the *future* behavior of a distant absorber determines the *past* event of radiation; there is simply no such thing as just radiating into empty space. The entire universe, spatially *and* temporally, is a very 'connected' place!

Astonishingly, this non-causal view of spacetime had been around in physics for at least 20 years before Wheeler and Feynman's talk. They had independently developed their ideas but, after their 1941 talk, Einstein (who perhaps recalled his 1909 debate about advanced effects with Ritz) brought a 1922 paper by the Dutch physicist Hugo Tetrode (1895–1931) to their attention. In his paper Tetrode had written that "the Sun would not radiate if it were alone in space and no other bodies could absorb its radiation ... If for example I observe through my telescope yesterday evening that star which let us say is 100 light years away, then not only did I know that the light which it allowed to reach my eye was emitted 100 years

---

[63]J. A. Wheeler and R. P. Feynman, "Interaction with the Absorber as the Mechanism of Radiation," *Reviews of Modern Physics*, April–July 1945, pp. 157–181.

ago, but also the star or individual atoms of it knew already 100 years ago that I, who then did not even exist, would view it yesterday evening at such and such a time."

Tetrode's vivid imagery had been, curiously, itself captured even decades earlier in words from the nineteenth century English poet Francis Thompson (1859–1907), in his "The Mistress of Vision":

All things ... near and far,
Hiddenly to each other linked are,
That thou canst not stir a flower
Without troubling of a star.

None of this, of course, is obvious! As a tutorial paper appearing just 2 years after Wheeler and Feynman's 1945 paper expressed it, "Any physical theory which seriously proposes that events in the future may be the efficient cause of events in the past may be regarded—at least at first glance—as rather revolutionary doctrine."[64] Indeed!

It is interesting to note that Einstein apparently said nothing to Wheeler and Feynman about a paper that pre-dated Tetrode's by 3 years. In 1919 the Finnish physicist Gunnar Nordström (1881–1923) had suggested that the advanced solution might offer an explanation for a perplexing problem in atomic theory. Maxwell's theory says that an accelerated electric charge radiates energy, which implies that the orbital electrons in the classical model of the atom should quickly spiral in toward the nucleus, that is, all matter should collapse. This cataclysmic event (of course!) has not happened, and Nordström's idea was that if one took into account not only the usual retarded solution but the advanced one as well, then perhaps things could be understood. Indeed, Nordström was able to show that such an analysis does give zero for the *average* energy radiated by an orbiting electron. Later, however, Page (note 39) showed that the instantaneous radiated energy is not zero, and that this would lead to observable effects that in fact are *not* observed.

Now, to be sure that the 'doctrine' discussed in note 64 is clear, let me restate what Tetrode, and later Wheeler and Feynman, had in mind. Imagine we have an electric charge (the *source*) that we mechanically shake, that is, accelerate. This allows us to assign a definite *cause* to the charge's acceleration which, of course, radiates energy. This radiation travels outward into space as observed retarded fields until they are eventually absorbed by distant matter. The charges in that distant matter are thus accelerated, and they in turn therefore radiate energy. This induced radiation again consists, according to Wheeler and Feynman, of both retarded and advanced fields. The advanced fields radiate outward but *backward* in time toward the original charge, collapsing upon it at the precise instant we first shook it, thereby producing the radiative reaction force. At any instant of time, at any point in space, the observed field is the sum of the retarded field traveling away

---

[64]C. W. Berenda, "The Determination of Past by Future Events: A Discussion of the Wheeler-Feynman Absorption-Radiation Theory," *Philosophy of Science*, 1947, pp. 13–19.

from the source into the future and the advanced field traveling toward the source in the past.

But, argued Wheeler and Feynman, there is one last point that has been left out of this picture—there is also an advanced field (traveling away from the source and backward in time) because of the original, mechanical shaking of the source charge. Equivalently, a field traveling *forward* in time will *converge* onto the source because we *will* shake it. Wheeler and Feynman showed that before the mechanical shaking that starts this whole process, the advanced radiation field of the source and the advanced radiation fields of the absorbers exactly cancel each other at every point in space and every instant of time (if there is total absorption in the future), which accounts for the experimental fact that we observe a zero total field before the mechanical shaking occurs.

Wheeler and Feynman showed that if we accept these (strange) ideas, then everything we actually observe is predictable: radiative reaction, the direction of the electromagnetic arrow of time from past to future (retarded-only effects), and the absence of infinite self-interactions. The claim by Wheeler and Feynman to have avoided self-interaction problems via the use of the advanced solution was, however, soon challenged. Indeed, the self-interaction of the electron is *needed* to explain the 1947 experiment by Willis Lamb (1913–2008) that measured the deviation (the *Lamb shift*) of the spectrum of hydrogen from what Dirac's theory of the electron predicts. Ironically, it was that experiment that helped motivate the renormalization of quantum electrodynamics (to get rid of the infinities then plaguing it) which led to Feynman's share of the 1965 Nobel Prize. In fact, just 4 years after their 1945 paper, Feynman expressed a revised view that self-interactions could not be avoided.[65]

In any case, we gain the rewards originally claimed by Wheeler and Feynman *only if* we accept backward time travel, a step too big for many in 1945 (and for nearly as many today) because of the resulting time travel paradoxes that seem to be unavoidable. For the same reason, Tetrode's earlier work, published in a German journal, also went virtually unnoticed during the two decades before Wheeler and Feynman's work. In fact, Tetrode wasn't the only anticipator of Wheeler and Feynman, as they had been anticipated, too, in America. In 1926 the chemist G. N. Lewis (1875–1946) had written "I'm going to make the … assumption that an atom never emits light except to another atom, and to claim that it is absurd to think of light emitted by one atom regardless of the existence of a receiving atom as it would be to think of an atom absorbing light without the existence of light to be absorbed."[66] Wheeler and Feynman were aware of Lewis by 1945. Certainly Wheeler and Feynman must have been intrigued by Lewis' paradox: "I shall not

[65]R. P. Feynman, "Space-Time Approach to Quantum Electrodynamics," *Physical Review*, September 15, 1949, pp. 769–789. See also C. Teitelboim, "Splitting the Maxwell Tensor: Radiation Reaction Without Advanced Fields," *Physical Review D*, March 15, 1970, pp. 1572–1582.

[66]G. N. Lewis, "The Nature of Light," *Proceedings of the National Academy of Sciences*, January 15, 1926, pp. 22–29.

attempt to conceal the conflict between these views and common sense. The light coming from a distant star is absorbed, let us say, by a molecule of chlorophyll which has recently been produced in a living plant. We say that the light from the star was on its way toward us a thousand years ago. What rapport can there be between the emitting source and this newly made molecule of chlorophyll?"

The paradox in that, of course, arises from the issue of what happens if, at some intermediate time and place, the star's light is blocked, thus preventing its absorption by the chlorophyll? Could refusing to look at a star *now* affect the emission of the star's light in the *past*? Lewis was obviously making a clear statement of backward causation when posing this bilking paradox. His very next words show that he understood the probable reaction of his readers: "Such an idea is repugnant to all our notions of causality and temporal sequence." Like Tetrode's work, Lewis' ideas were ahead of the times but, actually, their ideas were *not* repugnant to everyone.

In fact, similar puzzles were an inspiration to Wheeler and Feynman and almost certainly motivated them to create their own famous bilking paradox, which they presented in their 1949 paper (note 42). They opened the presentation of their paradox as follows: "If the present motion of *a* is affected by the future motion of *b*, then the observation of *a* attributes a certain inevitability to the motion of *b*. Is not this conclusion in direct conflict with our recognized ability to influence the future motion of *b*?" This question clearly states the conflict between free will and determinism, and to sidestep this *human* concern Wheeler and Feynman constructed a "paradox machine," a machine that operates totally automatically and which has come to be called the "logically pernicious self-inhibitor"![67]

In their description of the paradox machine, Wheeler and Feynman ask us to imagine two charged particles, *a* and *b*, positioned five light-hours apart. As shown in Fig. 5.3, *a* is attached to the arm of a pivoted shutter, toward which a pellet is moving from initially a great distance away. Now, normally we would think of what happens next in terms of just retarded fields. That is, the pellet hits the arm, knocking it downward and thereby accelerating charge *a*; this acceleration of charge *a* creates a retarded radiation field that arrives at charge *b* 5 h later, resulting in the acceleration of charge *b*; the acceleration of charge *b* creates a retarded radiation field that arrives back at charge *a* 5 h later (10 h after the pellet hit the arm).

The Wheeler and Feynman view, however, claims that this description leaves out half the story—the *advanced* fields. Specifically, suppose the pellet will hit the arm and so accelerate *a* at 6 p.m. Then, *b* will be affected not only 5 h later at 11 p. m., but also *earlier* at 1 p.m. This advance acceleration of *b*, in turn, sends out an advanced field that arrives at *a* at 8 a.m. The paradox is now easy to see. As Wheeler

[67]P. Fitzgerald, "Tachyons, Backwards Causation, and Freedom," *Boston Studies in the Philosophy of Science* (volume 8), 1970, pp. 415–436. Even more extreme examples of such paradox machines are described in Tim Maudlin, "Time Travel and Topology," *PSA 1990*, Philosophy of Science Association, volume 1, pp. 303–315.

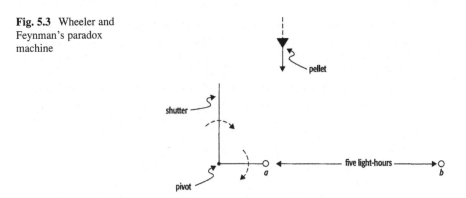

Fig. 5.3 Wheeler and Feynman's paradox machine

and Feynman described events, we see *a* exhibit a premonitory movement at 8 a.m. Seeing this motion in the morning, we conclude that the pellet will hit the arm in the evening. We could then return to the scene a few seconds before 6 p.m. and block the pellet from acting on *a*, a task automatically accomplished by the shutter in Wheeler and Feynman's paradox machine. But then we are faced with the puzzle of explaining just *why a* moved in the morning!

Wheeler and Feynman claimed they had resolved their bilking paradox by observing that *discontinuous* forces (more generally, *signals*) are never seen in nature. They concluded that the shutter does not completely block the pellet, but rather the shutter suffers a "glancing blow." That is, a very weak advanced signal is received by charge *a*, which moves the shutter just enough to induce the "glancing blow," and it is this partial interaction that results in the weakened signal in the "first place." This is, in fact, the very same explanation that was rediscovered decades later in answer to similar bilking paradoxes that involve self-interacting billiard balls transiting wormhole time machines (a topic we'll take-up in the next chapter).

5.5 Absorber Theory and Signaling the Past

"If advanced waves [could be used to signal the past] then our grip on reality would become more tenuous. The past could never be considered over and done with, because anyone with the proper hardware could send messages back in time and alter what had already happened."
    —John Cramer, a University of Washington physicist, taking the minority position on the possibility of changing the past[68]

---

[68]Cramer has written provocatively on advanced waves. See, for example, "The Arrow of Electromagnetic Time and the Generalized Absorber Theory," *Foundations of Physics*, September 1983, pp. 887–902, and "Generalized Absorber Theory and the Einstein-Podolsky-Rosen Paradox," *Physical Review D*, July 15, 1980, pp. 362–376.

Wheeler and Feynman's argument *is* logically and physically sensible; it is, after all, simply an early statement of the principle of self-consistency in physics. Wheeler summed matters up nicely, years later, when he wrote "Interconnections run forward and backward in time in such numbers as to make an unbelievable maze. That weaving together of past and future seems to contradict every normal idea of causality. However, when the number of particles is great enough to absorb completely the signal starting out from any source, then this myriad of couplings adds up to a simple result: the familiar retarded actions of everyday experience, plus the familiar force of radiative reaction with its familiar sign."[69]

Their analysis was based on classical physics but, many years later, Feynman wrote (in his famous autobiographical work *Surely You're Joking, Mr. Feynman*) that at one time he and Wheeler thought it would not be too difficult to work out the quantum version of their theory. But then first Wheeler failed in the task, and then Feynman tried his hand at it and, as he stated, "I never solved it, either—a quantum theory of half-advanced, half-retarded potentials—and I worked on it for years." Their paradox (if indeed it *is* a paradox, since if advanced fields don't actually exist then there is no problem) remains unsolved.

Is Wheeler and Feynman's view of nature correct? Could we use advanced waves to send signals to the past? Or, if that requires some yet-to-be-developed technological breakthroughs in transmitter design, and if receivers are easier to construct, could we at least listen-in to the future (since we are, *now*, the future's past)? And if we could do that, could the future send us the details of the transmitter breakthrough (thus creating a causal information loop in time)?

The first experimental search for advanced waves seems to have been a 1973 effort.[70] The very next year, flaws in that search process prompted two physicists to discuss an experiment designed to detect advanced waves (if they exist). As they wrote, in a grand understatement, the exciting possibility of a positive result "would have such far-reaching consequences on our ideas of the unidirectionality of time and causality that ... the experiment justifies a large amount of effort, even if no conclusive result is obtained for years."[71] Alas, all of the searches for advanced waves have, as I write (2016), given negative results and so the world still awaits the first Dirac radio.

Over the years the Wheeler and Feynman view of nature has been the target of some theoretical concerns. One physicist, for example, complained that Wheeler and Feynman had assumed a static, time-symmetric spacetime for the universe, in which the properties of all past and future absorbers are identical. That is obviously

---

[69]See note 34 in Chap. 3. There Wheeler also wrote "The particles of the absorber are either at rest or in random motion before the acceleration of the source. They are correlated with it in velocity after that acceleration. Thus radiation and radiative reaction are understood in terms, not of pure electrodynamics, but of statistical mechanics."

[70]R. B. Partridge, "Absorber Theory of Radiation and the Future of the Universe," *Nature*, August 3, 1973, pp. 263–265.

[71]M. L. Herron and D. T. Pegg, "A Proposed Experiment in Absorber Theory," *Journal of Physics A*, October 1974, pp. 1965–1969.

not so in an expanding (or contracting) universe and, as he wrote, "No serious modern cosmological theory is framed in [terms of] a static Universe."[72] Another puzzle for that writer was that Wheeler and Feynman took a time-symmetric theory of half-retarded/half-advanced waves in a time-symmetric universe and arrived at a *non*-time-symmetric solution! They performed that trick by supposing not only that the universe is static, but also that it was created with the initial condition of low entropy. Thus, for Wheeler and Feynman, the one-way thermodynamic arrow of time is the primary arrow, with the electromagnetic arrow following as a consequence. (The *how* of a low entropy initial cosmological condition was left unexplained—certainly no mention of the hand of God appears in their work!) This ordering of the primacy of the temporal arrows was, in fact, in agreement with the view adopted by Einstein in his 1909 debate with Ritz, a view taken decades later by Hawking, as well.[73]

Wheeler and Feynman had shown that both the advanced and retarded solutions taken together are self-consistent in a static universe; Hogarth's question was whether the observed retarded solution, *alone*, would be self-consistent in an *expanding* universe (which is the universe we actually observe). His conclusion? It depends on the details of the expansion. Two years after Hogarth, two physicists expanded on his work and claimed to have shown that the retarded solution alone *is* self-consistent *if* the expansion is steady-state via the continuous creation of matter.[74] That would be the case because, if only retarded effects are to occur, then each emitter of radiation needs a large number of absorbers (such as ionized intergalactic gas) in its future light cone to provide for complete absorption. This, in turn, requires that the density of matter not decline "too fast" with the expansion. That is, the future universe must not be "too transparent" and the continual appearance of new matter in the ever-increasing volume of the expanding universe is required to maintain the necessary density.

That conclusion was embraced with particular enthusiasm by Hoyle, a British cosmologist whose name has long been identified with the idea of continuous creation of matter. Since then, however, continuous-creation cosmologies have fallen into disfavor because it was in 1965, just a year after Hoyle and Narlikar wrote, that the cosmic microwave background radiation was detected. That is now taken as very strong evidence for the occurrence of a primordial explosion (or Big Bang) that started the expansion of the universe, and as equally strong support for therefore rejecting a steady-state universe. Not by Fred Hoyle (1915–2001), though,

---

[72]J. E. Hogarth, "Cosmological Considerations of the Absorber Theory of Radiation," *Proceedings of the Royal Society A*, May 22, 1962, pp. 365–383. Hogarth, however, rejected the static universe, asserting instead that the *observed expansion* of the universe provides the required asymmetry, resulting in the cosmological arrow of time as the primary arrow and the electromagnetic arrow as a consequence.

[73]S. W. Hawking, "Arrow of Time in Cosmology," *Physical Review D*, November 15, 1982, pp. 2489–2495.

[74]F. Hoyle and J. V. Narlikar, "Time Symmetric Electrodynamics and the Arrow of Time in Cosmology," *Proceedings of the Royal Society A*, January 1964, pp. 1–23.

who had an almost fanatical devotion to non-Big Bang cosmologies.[75] Real puzzles remain for the Big Bang universe, however. One is that it expands from a dense, opaque past into a less dense, ever-more-transparent future, with each emitter having a large number of absorbers in its *past* light cone. That should result, noted Hoyle and Narlikar (almost certainly with some glee), in an observed advanced solution and thus in a reversed electromagnetic arrow that would allow communication with the past. The fact that we have not (yet?) discovered how to perform such communication might be taken to mean that the idea of an expanding, Big Bang universe is somehow faulty. A related question about absorber theory is that of the puzzle of neutrino absorption. Neutrinos are particles that interact so weakly with matter that a beam of them would have to travel through many hundreds of light-years of lead for there to be a significant attenuation of the beam. How can such 'ghost-like' particles find enough future absorbers to make possible their observed journeys into the future of a Big Bang expanding universe?

For such an exciting idea as communication with the past, it is not surprising that advanced-wave radio has appeared in science fiction. Just 6 years after Wheeler and Feynman's paper, a story by a well-known author hinted at such a gadget based on something called the "ultrawave effect": "While gravitational effects were produced by the presence of matter, ultrawave effects … did not appear unless there was a properly tuned receiver somewhere. They seemed somehow 'aware' of a listener even before they came into existence."[76] It is difficult to believe that such a story idea was conjured-up out of nothing, but rather that the author had read Wheeler and Feynman's paper. Anderson had a 1948 undergraduate honors degree in physics from the University of Minnesota, and so he may *well have* read Wheeler and Feynman's 1945 paper.

The potential bilking paradoxes produced by sending messages backward in time have been treated in at least one novel-length discussion. The puzzles presented are undeniably fascinating, but the story's answer to them is to allow the changing of the past, as argued in this section's opening quote from John Cramer. Indeed, the title comes from the plot device of twice changing the past by sending messages to the past to save the world from terrible disasters. Thus, we read through entire time periods *three* times before finishing the novel. As one character blurts out, "We can monitor the actual consequences of our decisions and actions, and change them until they produce the desired result! My God … it's staggering!"[77] Quite so.

One of the most interesting science fictional uses of backward-in-time signaling is, I think, found in a classic tale by James Blish. There the "Dirac radio" for instantaneous transmissions is described, and we learn that at the beginning of each received message there is always an irritating audio beep (hence the title) that is

---

[75]Besides his scientific work, Hoyle also wrote science fiction. One work, the 1966 novel *October the First is Too Late*, deals with travels in time but fails to say anything about paradoxes.

[76]Poul Anderson, "Earthman, Beware!," *Super Science Stories*, June 1951.

[77]J. P. Hogan, *Thrice Upon a Time*, Ballantine 1980.

seemingly a useless artifact of the mysterious workings of the gadget. Its only obvious characteristic is a continuous spectrum from 30 to well above 18,000 Hz. It is only at the end of the story that the main character learns that this spectrum is the "simultaneous reception of every one of the Dirac messages which [has] ever been sent, or will be sent."[78]

Blish was actually pretty close to the mark with that, as a composite signal with a continuous spectrum (with energy distributed uniformly in frequency), such as one might expect the overlay of many independent signals to be, does indeed have a narrow time structure. If applied to a loudspeaker, such a signal would sound like a sharp pulse or click—or even a *beep*. In the limit of an infinitely wide spectrum, the time signal becomes one of infinite amplitude and zero duration, a singular *impulse* function called, by theoretical physicists and radio engineers alike, the *Dirac delta function*.

There is no mention in the story of advanced waves, but clearly Blish knew that instantaneous (infinite-speed) signals would travel into the past and he does a masterful job of presenting the mystery of listening to the future. At one point characters in the twenty-first century hear the commander of a time-traveling "world-line cruiser" transmit a poignant call for help from 11,000,000 light years away and from sixty-five centuries in the future. Most interesting of all, however, is Blish's statement of a technical issue that I have not seen raised before: if signals arrive at a receiver, simultaneously, from all future times, how can they be separated? Blish resorts to some scientifiction babble-talk to answer that question, but I believe it remains a puzzle.[79]

5.6   Tachyonic Signals and the Bell Quantum Antitelephone

"We cannot fight the laws of nature."

"Nature be damned! Feed more fuel into the tubes. We must break through the speed of light ... Give me a clear road and plenty of fuel and I'll build you up a speed of half a million miles in a second ... What's there to stop it?"[80]

Science fiction writers have often used FTL motion to reverse time, often without much (if any) regard to the fact that, to just *reach* the light barrier, requires (according to special relativity) infinite energy, much less to exceed light speed.

---

[78]J. Blish, "Beep," *Galaxy Science Fiction*, February 1954.

[79]The signal separation problem is also hinted at by physicist/science fiction author Gregory Benford, in a tale that was a precursor to his famous 1980 novel *Timescape* (in which the present attempts to warn the past of a future ecological disaster that threatens life on Earth). See Benford's "Cambridge, 1:58 A.M.," *Epoch*, Berkeley 1975.

[80]Words exchanged by the first officer and the captain of a starship on its way to Alpha Centauri in a story by N. Schachner, "Reverse Universe," *Astounding Stories*, June 1936. The captain, we are told, "had heard, of course, of the limiting velocity of light, but it meant nothing to him."

Therefore, goes the reasoning, because we can't get through the 'light barrier' means that time travel to the past must be impossible, as well. Or, so goes this line of argument. But, could there be a way around this conclusion? After all, while relativity theory indeed precludes the acceleration of a *massive* particle *up to* the speed of light, it does allow a zero rest mass particle (like the photon) to exist *at* the speed of light. Photons are emitted during various physical processes, and they move from the instant of their creation at the speed of light; the only way to slow a photon is to destroy it by absorbing it. Advocates of the possibility of the existence of FTL particles make a similar argument when asking if there might not be particles, emitted during various (as yet unknown) physical processes, that move from the instant of their creation at speed *greater* than that of light?

An affirmative answer would neatly avoid the 'acceleration through the light barrier' problem, but then there are other concerns. For example, such FTL particles would have to have an imaginary rest mass if they were to carry real-valued energy and momentum, and what could *imaginary mass* mean? That question was answered by the proponents of FTL particles, who replied that the rest mass of a superluminal particle would be unobservable because (like the photon) there is no subluminal frame of reference in which the particle could be at rest! That is, there is no frame of reference in which the mysterious imaginary mass could be measured and, anyway, it is only *observable* changes in the real energy and momentum that characterize particle interactions.

The key idea to this line of thought is a *supposed* FTL particle, called a *tachyon*, a name coined by the American physicist Gerald Feinberg (1933–1992) from the Greek word *tachys* for "swift."[81] It is interesting to note that Feinberg admitted[82] that his interest in such a thing was sparked by reading Blish's story "Beep" (see note 78). The idea of the tachyon is actually a very old one that is hinted at in the work of the Greek poet and philosopher Lucretius (who died 20 years before the birth of Christ). In his discussion of visual images, in Book 4 of his giant (well over 7400 lines) science poem *De Rerum Natura*, we find the following words about particles of matter originating deep inside the Sun: "Do you see how much faster and farther they must travel, how they must run through an extent of space many times vaster in the time it takes the light of the Sun to spread throughout the sky?"[83]

The first attempt (later found to be flawed) in the physics literature of a relativistic treatment of FTL particles appeared some years before Feinberg gave

---

[81]G. Feinberg, "Possibility of Faster-Than-Light Particles," *Physical Review*, July 25, 1967, pp. 1089–1105. Feinberg was anticipated in this name by Edward Page Mitchell (the Victorian pioneer in the time travel paradox genre who was discussed back in Sect. 4.2 and its note 37). In his story "The Tachypomp: A Mathematical Demonstration" (*Scribner's Monthly*, March 1874), he describes a gadget for reaching any speed, no matter how great (*tachypomp* is literally "quick sender").

[82]G. Benford, "Time and *Timescape*," *Science-Fiction Studies*, July 1993, pp. 184–190.

[83]A poetic allusion to something traveling faster than light appears, in of all places, Shakespeare's *Romeo and Juliet*. In Juliet's words (Act II, scene 5), "... love's heralds should be thoughts, Which ten times faster glide than the sun's beams, Driving back shadows ..."

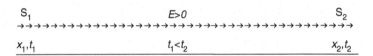

Fig. 5.4 The emission of a positive energy particle, followed by absorption

them a name, where it was observed that special relativity is *not* necessarily violated by FTL motion.[84] But nonetheless, concerns about the physical possibility of FTL particles continued. For example, a serious problem for tachyons, at least for those who dislike the ideas of backward causation and time travel, is the observation that in some frames of reference an FTL particle would appear to have *negative energy*. Feinberg, himself, explained (see note 81) this concern as follows: "By the principle of relativity, any state which is possible for one observer must be possible for all observers, and hence FTL particles can exist in negative-energy states for all observers ... The occurrence of negative energy states for particles has always been objected to on the grounds that no other system could be stable against the emission of these negative-energy particles, an entirely unphysical behavior."

This objection to FTL particles was raised early in the history of tachyons, even before they were named, and it was addressed by three physicists who proposed the so-called *reinterpretation principle* (what I'll refer here to as the RP).[85] To see how the RP works, consider Fig. 5.4, in which a source S_1 at x_1 emits an FTL particle at time t_1. This particle then travels to an absorber S_2 at x_2, arriving there at the later time t_2. S_1 and S_2 are in the same reference frame and, for an observer in the frame, the particle energy E is positive. However, it is always possible to find another observer in a relatively moving frame for whom this process would look as though t_2 is less than (that is, earlier than) t_1 with $E < 0$. In other words, for the moving observer the particle would appear as a negative energy particle moving backward in time. (In the next chapter I'll show you that the particle speed must not be just superluminal, but the even faster *ultraluminal*.)

Note that for the moving observer, the emission by S_1 of negative energy *increases* the energy of S_1, and the absorption of negative energy by S_2 decreases the energy of S_2. S_2's energy decrease (for the moving observer) occurs before the increase in S_1's energy because, as noted before, for the moving observer $t_2 < t_1$. The moving observer naturally interprets this process as the emission of positive energy by S_2, followed by absorption by S_1. This reinterpretation would thus seem to preserve our common-sense idea of causality, as well as avoiding any mention of backward time travel. The RP appears to have slipped around those problems merely by redefining which source is transmitting, and which is receiving, the

[84]S. Tanaka, "Theory of Matter with Super Light Velocity," *Progress of Theoretical Physics*, July 1960, pp. 171–200. See also O. M. Bilaniuk and E. C. G. Sudarshan, "Particles Beyond the Light Barrier," *Physics Today*, May 1969, pp. 43–51 (and the resulting discussion in the December issue, pp. 47–52).

[85]O. M. P. Bilaniuk, V. K. Deshpande, and E. C. G. Sudarshan, "'Meta' Relativity," *American Journal of Physics*, October 1962, pp. 718–723.

tachyon. Indeed, Feinberg claimed (note 81) that the RP avoids the creation of causal loops and their associated paradoxes, a claim repeated nearly 20 years later by a physicist who used the RP to eliminate paradoxes from Gödel's time travel rotating universe.[86]

There is, however, a curious twist to all this. Even if we grant that the RP may avoid causal paradoxes, the fact is that physics isn't fooled as easily as a human observer. That is, the receiver does actually lose energy upon the arrival of the tachyon, which is the opposite of what happens in a radio receiver when it receives a photon. In other words, the receiver must be in an elevated energy state prior to the tachyon's arrival; the receiver must be prepared beforehand to receive a message. If the receiver were instead sitting in its lowest energy state, then it could not accept (or *eject*, according to the RP) the tachyon. So, it's not surprising that, despite the enthusiastic embrace of the RP by some, other physicists took exception, with one arguing that the effectiveness of the RP in avoiding causal loops is "illusory" and "irrelevant,"[87] while others concluded that the causal paradoxes would actually preclude any possibility of tachyons interacting with ordinary matter in the first place (which is just a polite way of saying that tachyons have no more reality than do unicorns!).[88]

The RP's effect of flipping the roles of transmitter and receiver has attracted particular concern. Some analysts have pointed out that if one can modulate a superluminal signal to send a message into the past, then certainly one could *sign* the message. To quote a delightful example, "If Shakespeare types out *Hamlet* on his tachyon transmitter, Bacon receives the transmission at some earlier time. But no amount of reinterpretation will make Bacon the author of *Hamlet*. It is Shakespeare, not Bacon, who exercises control over the content of the message."[89] The last line of this quote is of central importance. The authors emphasize it by immediately observing that a signature is a *relativistic invariant* and that, indeed, it establishes a causal ordering quite independent of any temporal ordering. This example, alone, explains why one analyst said of the RP that it is "laughed to scorn,"[90] while another said of the RP that it "sounds merely like the endorsement of what can only be characterized as a fantastic delusion."[91]

---

[86]A. Italiano, "How to Recover Causality in General Relativity," *Hadronic Journal*, January 1986, pp. 9–12.

[87]R. G. Newton, "Particles That Travel Faster Than Light," *Science*, March 20, 1970, pp. 1569–1574.

[88]W. B. Rolnick, "Implications of Causality for Faster-Than-Light Matter," *Physical Review*, July 25, 1969, pp. 1105–1108, and D. J. Thouless, "Causality and Tachyons," *Nature*, November 1, 1969, p. 506.

[89]G. A. Benford, D. L. Book, and W. A. Newcomb, "The Tachyonic Antitelephone," *Physical Review D*, July 15, 1970, pp. 263–265.

[90]P. Fitzgerald, "On Retrocausality," *Philosophia*, October 1974, pp. 513–551.

[91]W. L. Craig, "Tachyons, Time Travel, and Divine Omniscience, *Journal of Philosophy*, March 1988, pp. 135–150.

Even more sophisticated scenarios than the *Hamlet* one (note 89) were devised to show how problems with FTL signals could arise that the RP could *not* resolve. The American physicist Bryce DeWitt (1923–2004) created such an example (the 'DeWitt Gambit') that involved a sequential, circular chain of tachyon signal transmissions between four observers, all moving in one spatial dimension. DeWitt showed out to arrange the spacetime geometry of the observers so that, at each stage, there is no dispute over who is sending and who is receiving, and so invoking the RP is avoided. Yet, when the signal reaches the first (last) observer, it is *before* he started (will start) the chain! This, of course, sets-up a potential bilking paradox: what if 'now' the first (last) observer decides to *not* send the chain's initiating signal? An even more sophisticated variation on the Gambit, involving four observers moving in *two* spatial dimensions, was soon after put forth by the English physicist Felix Pirani (1928–2015). As he concluded, "It is difficult to see how in the face of this example a classical-particle description of tachyons can be sustained."[92] Confronted by such sharp criticism, from so many, it is understandable why, just before his death, Feinberg co-authored (with two philosophers) a paper in which he seemed to be abandoning his support for tachyons as possible carriers of information backward in time.[93]

Many physicists today reject the possibility of backward-in-time messages, not because of concerns about the RP, but because such messages could create potential bilking paradoxes. To see how this works, the old Wheeler and Feynman idea of explaining (away) bilking paradoxes—that no signal in nature is *really* discontinuous—was examined by one physicist in the context of tachyons.[94] There we are asked to consider the following situation: A human (call him A) has a lamp on a table before him. The lamp is controlled by a tachyon receiver; in other words, the lamp illuminated only when a tachyon signal (a pulse, let us say) is detected. At 3 o'clock A will send a tachyon signal to B (a tachyon echo-transmitter that immediately rebroadcasts everything it receives) if the lamp does not glow at 1 o'clock. Now, the spacetime geometry of A and B is arranged to be such that a signal sent by either A or B to the other travels 1 h backward in time. Thus, if A sends a signal at 3 o'clock, then B will receive it at 2 o'clock (and immediately echo), and the echo will arrive at A at 1 o'clock. The paradox, of course, is that A sends a signal only if the lamp does *not* glow—that is, only if A does *not* send the signal![95]

We are then reminded of the Wheeler and Feynman claim that every pulsed signal is actually continuous; this argument would include the illumination itself of

[92]F. A. E. Pirani, "Noncausal Behavior of Classical Tachyons," *Physical Review D*, June 15, 1970, pp. 3224–3225.

[93]G. Feinberg, D. Albert, and S. Levine, "Knowledge of the Past and Future," *Journal of Philosophy*, December 1992, pp. 607–642.

[94]L. S. Schulman, "Tachyon Paradoxes," *American Journal of Physics*, May 1971, pp. 481–484.

[95]A study of similar situations can be found in L. L. Gatlin, "Time-Reversed Information Transmission," *International Journal of Theoretical Physics*, January 1980, pp. 25–29.

the lamp. Therefore, the lamp is not just on or off, but potentially at any level of illumination between those two extremes. So there sits A, and at 1 o'clock the lamp seems to glow dimly. To that, says Schulman, "A thinks it over, vacillating, finally sending a slightly late signal which isn't full strength." Then the echo isn't full strength either, which accounts for the original dim glow. This conclusion *is* consistent (but what if A's sending device is a toggle switch that *snaps* one way or the other—why then only a partial strength signal?), but it does seem to ask for a lot of supposing. Schulman himself is not so sure about the validity of universal continuity, writing at the end that "it is not clear that the Wheeler-Feynman assumption . . . ought to be made." (For more on the Wheeler-Feynman continuity assumption, in a different, non-time travel context, see the final "For Further Discussion" for this chapter.)

The Wheeler-Feynman continuity idea *is* ingenious, allowing one to find a logically and physically consistent solution in time travel scenarios that, at least at first glance, might seem to have no solution. Consider, for example, the following situation[96]: "We have a camera ready to take a black and white picture of whatever comes out of [a] time machine. The film is then developed and the developed negative is subsequently put in the time machine and set to come out of the time machine at the time the picture is taken. This surely will create a paradox: the negative will have the opposite distribution of black, white, and shades of grey, from the picture that comes out of the time machine. But since the thing that comes out of the time machine is the negative itself we surely have a paradox." But do we?

The answer is no, because "What will happen is that a uniformly grey picture will emerge which produces a negative that has exactly the same uniform shade of grey. No matter what the sensitivity of the film is, as long as the dependence of the brightness of the negative depends in a *continuous* [my emphasis] manner on the brightness of the object being photographed, there will be a shade of grey that produces exactly the same shade of grey on the negative when photographed. This is the essence of Wheeler Feynman's idea."[97] (The conclusion is the same if we move from black-and-white to color photography.)

Nonetheless and despite this apparent success, the supposed ability of a modulated beam of tachyons to send a message into the past still raised concerns among many, particularly about free will and fatalism. Suppose, say those who are concerned about these issues, that you receive a tachyon message from yourself from tomorrow, informing you that a man you plan to kill tonight is still alive (tomorrow). Does that mean it is beyond your power to kill him tonight? According to one analyst (note 67) the answer is no; you *could* kill him—but if you do then the message from tomorrow would not have arrived. And because ignorance is not a

---

[96]Taken from Frank Arntzenius and Time Maudlin, "Time Travel and Modern Physics," in *Time, Reality & Experience* (C. Callender, editor), Cambridge University Press 2002, pp. 169–200.

[97]More on the Wheeler-Feynman continuity idea, *and of its limitations*, can be found in D. Kutach, "Time Travel and Consistency Constraints," December 2003, pp. 1098–1113, and Phil Dowe, "Constraints on Data in Worlds with Closed Timelike Curves," December 2007, pp. 724–735, both in *Philosophy of Science*.

precondition to free will, your newly acquired knowledge does not, by itself, suddenly limit your ability to kill the man. But, this line of arguing went, if you do not attempt to kill the man because you believe the message from your future self, then in fact the message *has* limited you!

Fitzgerald's position was rebutted by Craig (note 91), who argues that it is not your ability to kill that is altered by the message but rather your motivation. Craig points out that such motivational changes can occur without invoking anything as radical as a message from the future. Suppose, he says, that just before you fire the fatal shot into your victim, you learn from him that he is your beloved, long-lost uncle. Clearly, your motivation for killing him is likely to be instantly altered, but equally clearly, your *ability* to kill him is unchanged. The mechanism for obtaining genealogical information, whether via time travel or as a last-minute appeal from your intended victim, is (says Craig) simply irrelevant. Not all buy into that, however, with one unconvinced analyst (note 94) writing that "history is a set of world lines essentially frozen into spacetime. While subjectively we feel strongly that our actions are determined only by our backward light cone, this may not always be the case." That is, Schulman appears open to the possibility that influences originating in the future might indeed have an impact on the present.

With the fading from the physics scene of enthusiasm for tachyons, the romance of communicating with the past using superluminal speeds passed from speedy particles to quantum mechanics via a mathematical result called *Bell's theorem*. John Bell, a physicist mentioned in the last chapter in connection with the MWI of quantum mechanics, published his theorem in a little article in an obscure, now defunct journal. Since then it has become one of the most cited physics papers from the 1980s.[98] The paradox cited in Bell's title refers to a famous 1935 paper in which Einstein (and two of his colleagues at the Institute for Advanced Study) challenged the conventional view of quantum mechanics, the view that there is no objective reality to anything unless it is observed.[99]

In fact, the possibility of quantum mechanics supporting FTL signals had been considered (and rejected) *before* Einstein's paper, by the Italian physicist Enrico Fermi.[100] Fermi concluded that, in a two atom system, the decay from an excited state of one atom (with the emission of a photon) would not influence the other atom before a time lapse of R/c, where R is the distance between the two atoms and c is the speed of light. In 1967, however, the Russian physicist M. I. Shirokov pointed out that Fermi's result was the result of an unjustified mathematical operation (he had replaced an integral from zero to infinity with one from minus infinity to infinity).

[98] J. S. Bell, "On the Einstein-Podolsky-Rosen Paradox," in *Speakable and Unspeakable in Quantum Mechanics*, Cambridge University Press 1987.

[99] A. Einstein, B. Podolsky, and N. Rosen, "Can Quantum-Mechanical Description of Physical Reality Be Considered Complete?" *Physical Review*, May 15, 1935, pp. 777–780. See also N. D. Mermin, "Is the Moon There When Nobody Looks? Reality and Quantum Theory," *Physics Today*, April 1985, pp. 38–47.

[100] E. Fermi, "Quantum Theory of Radiation," *Reviews of Modern Physics*, January 1932, pp. 87–132.

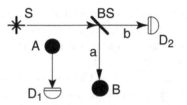

Fig. 5.5 Time travel and the wave function of quantum mechanics. In the above arrangement (reproduced by the kind permission of Francesco Gonella, from his paper "Time Machine, Self-Consistency and the Foundations of Quantum Mechanics," *Foundations of Physics Letters*, April 1994, pp. 161–166), S is a low-intensity source of photons, a source *so* weak that at any instant there is never more than one photon in the system. Each photon begins by traveling toward BS, a half-silvered beam-splitting mirror that, with equal probability, either passes a photon to the right along path b where it is detected by D_2, or downward along path a into mouth B of a wormhole time machine and out of which it emerges from wormhole mouth A *in the past*. Conventional quantum mechanics says that what happens at BS is determined *when the photon reaches BS*—that is, the probability wave function of the photon collapses at BS. But for a photon that is reflected into B, it exits A where it is detected by D_1 *before* the time of its arrival at BS. That means it is known (by detector D_1) what *will* happen at BS *before* the photon arrives at BS—so just *when did* the photon wave function collapse?

Then, nearly 30 years after Shirokov, the German physicist Gerhard Hegerfeldt proved (in the context of conventional quantum mechanics) a very general theorem that establishes a non-zero probability of the second atom responding to the first atom's decay *as soon as* the photon is emitted.[101]

Hegerfeldt's analysis is very general, but it does make one assumption—the non-negativity of energy density (the so-called *weak energy condition*)—which at the time was considered (in Hegerfeldt's words) to be "physically well motivated," but which today is *not* taken to be *a priori* obvious.[102] Einstein's paper, on the other hand, utilized physical assumptions not easily dismissed, and so its conclusions confounded physicists for decades.

The conventional view of quantum mechanics formulates physics in terms of probability wave functions that 'collapse' into specific realities only when measurements (observations) are made of the state of a system (which may be as elementary as a single particle). Until such measurements are made, says this view of quantum mechanics, a system has no specific state; instead, it merely has a probability distribution over a set of possible states (see Fig. 5.5 for a time travel puzzle concerning this claim). Einstein and his co-authors (note 99) strongly rejected this probabilistic interpretation of nature (recall Einstein's famous dictum, "God does not play dice with the cosmos.") Einstein and his colleagues agreed that

---

[101]G. C. Hegerfeldt, "Causality Problems for Fermi's Two-Atom System," *Physical Review Letters*, January 1994, pp. 596–599. This mere *suggestion* of a possible failure of causality so stunned the editor of *Nature* that he felt compelled to quickly write a 'calming' reply: "Time Machines Still Over the Horizon," February 10, 1994, p. 509.

[102]The issue of the sign of energy density is very important in the analyses of wormhole time machines (see note 135 in Chap. 1), and we'll return to it in the next chapter.

quantum mechanics may be valid *as far as it goes*, but they also argued that it leaves out 'something' (as yet unknown) in describing reality. That is, they suggested that quantum mechanics is "incomplete" and that it incorporates "hidden variables."[103] They expressed this idea in the form of a paradox, the famous *EPR paradox*, which they posed as a thought experiment (a *Gedanken-experiment*) in which quantum mechanics declares that the properties of a particular spatially distributed system, when measured at point A, seem to be *forced* to assume specific values at point B (which may be arbitrarily distant from A) *without* there being a measurement at B.

Thus, said Einstein, there are just two possibilities. Either the system properties at B must have been what they are from the very start (even if the measurements at A had not been done) which is the view he held, or there must have been a linkage between the system at A and the system at B such that the wave function collapse at A is instantly transmitted to B to allow the wave function to collapse there as well. Because A and B may be arbitrarily far apart, this second view obviously requires an FTL transmission mechanism,[104] something Einstein called a "spooky action-at-a-distance," a term that eloquently expresses his low opinion of that idea! (Some translations replace *spooky* with *ghostly*, but the negative sentiment remains the same.[105]) For decades the debate between proponents of these two alternatives remained at a metaphysical, non-quantitative level. Then came Bell's paper in 1964.

Bell's theorem mathematically poses the choice between Einstein's hidden variables view and the conventional view of quantum mechanics through the use of an inequality involving certain measurable properties of a system.[106] If these measurements are such that the inequality is violated, then the conventional interpretation of quantum mechanics is vindicated, and Einstein's FTL spooky action-at-a-distance effect simply doesn't exist. Bell's great contribution, then, was to provide the means for removing the debate about quantum mechanics from metaphysics and to place it squarely in the realm of experimental physics.

"All" that needed to be done was to make the required measurements. These technically difficult experimental measurements were eventually performed by the French physicist Alain Aspect and his colleagues at the Institute of Applied Optics of the University of Paris, a decade and a half after Bell showed what had to be

---

[103]For more on hidden variables, see E. P. Wigner, "On Hidden Variables and Quantum Mechanical Probabilities," *American Journal of Physics*, August 1970, pp. 1005–1009.

[104]For an analysis that argues against an FTL mechanism in quantum mechanics, see G. C. Ghirardi, *et al.*, "A General Argument Against Superluminal Transmission Through the Quantum Mechanical Measurement Process," *Lettere Al Nuovo Cimento*, March 8, 1980, pp. 293–298.

[105]You can find a discussion of the possibility of 'explaining' Einstein's "spooky actions" of quantum mechanics by invoking backward causation in R. I. Sutherland, "Bell's Theorem and Backwards-in-Time Causality," *International Journal of Theoretical Physics*, April 1983, pp. 377–384.

[106]The details are not important here, but a lovely exposition (for the lay person) can be found in Bell's essay "Bertlmann's Socks and the Nature of Reality," in *Speakable and Unspeakable in Quantum Mechanics*, Cambridge University Press 1987. See also M. G. Alford, "Ghostly Action at a Distance: A Non-Technical Explanation of the Bell Inequality," *American Journal of Physics*, June 2016, pp. 448–457.

done. The results unequivocally support the conventional view of quantum mechanics.[107] Einstein was simply wrong, there are no hidden variables in quantum mechanics, and his spooky action does seem to exist. Does that mean we have, at last, experimental evidence of the possibility of information transfer at FTL speeds?

Well, maybe, but the issue is still hotly debated. The majority of physicists today, I suspect, are probably more perplexed over what Bell's theorem is saying than were over Einstein's original EPR paradox. In those early days one could agree with Einstein and his colleagues, who argued that quantum mechanics was valid *as far as it went*, but a deeper, more comprehensive theory *would* show the existence of hidden variables. However, because of the work by Bell and Aspect it is definitively known that quantum mechanics *as it stands* leads to correct results, results that can be checked in the laboratory. In other words, there is no need for hidden variables and FTL spooky actions cannot be ruled out.

And so, while tachyons and the tachyon antitelephone may be nothing more than a neat science fiction fantasy, just maybe a quantum mechanical Bell antitelephone can't be so dismissed. Indeed, the possibility of using a quantum mechanical FTL effect was once suggested in a letter written by a senior person at an unspecified California think tank (an organization such as, for example, the RAND Corporation) to the Under Secretary of Defense for Research and Engineering at the Pentagon. Here's what the Under Secretary read when he opened that letter: "If in fact we can control the FTL nonlocal effect, it would be possible . . . to make an untappable and unjammable command-control-communication [C^3] system at very high bit rates for use in the submarine fleet. The important point is that since there is no ordinary electromagnetic signal linking the encoder with the decoder in such a hypothetical system, there is nothing for the enemy to tap or jam. The enemy would have to have actual possession of the 'black box' decoder to intercept the message, whose reliability would not depend on separation from the encoder or on ocean or weather conditions."[108]

One can't help but wonder what might have been the Under Secretary's response to that incredible letter, and what sorts of ultra-mega-super-top-secret experiments it may have prompted. I would be willing to bet, if they did occur, that they failed. As one physicist put it, "Up to now nature has covered her tracks pretty well, blocking all possibilities for using the EPR effect for FTL communication."[109] Of course, the think tank letter, as 'farout' as it may initially appear, actually represents a *failure* of imagination, because the backward causation effect of EPR's spooky FTL effect is certainly a 'quantum jump' beyond a mere unjammable submarine C^3 system.

---

[107]A. Aspect, "Experimental Tests of Realistic Local Theories via Bell's Theorem," August 17, 1981, pp. 460–467, and A. Aspect, *et al.*, "Experimental Realization of Einstein-Podolsky-Rosen-Bohm *Gedanken-experiment*: A New Violation of Bell's Inequalities," July 12, 1982, pp. 91–94, and "Experimental Test of Bell's Inequalities using Time-Varying Analyzers," December 20, 1982, pp. 1804–1807, all in *Physical Review Letters*.

[108]See N. D. Mermin in note 99. For more on the enigmatic letter on the FTL submarine C^3 system, see Jack Sarfatti's letter to *Physics Today*, September 1987, pp. 118 and 120.

[109]J. Cramer, "Paradoxes and FTL Communication," *Analog Science Fiction*, September 1988.

5.7 For Further Discussion

Mark Twain, in his last, posthumously published novel *No. 44, The Mysterious Stranger*, incorporated reversed time as the work of a supernatural being. After the being reverses the world's time direction, the narrator tells the reader that "everywhere weary people were re-chattering previous conversations backward ... where there was war, yesterday's battles were being refought, wrong-end first; the previously killed were getting killed again ... we saw Henry I gathering together his split skull ..." Read the novel, and comment on how well (or not) Mark Twain handled reversed time.

In Philip K. Dick's 1956 novel *The World Jones Made*, we read of a prophet who can see a year into the future. As he says, "To me *this is the past*," and then later we are told "He was a man with his eyes in the present [the world's future] and his body in the past [the world's present]." Read the novel, and then argue either for or against the suggestion that Dick was aware of the advanced wave solution in Maxwell's theory.

The physicist Robert Forward (see the sixth *For Future Discussion* in Chap. 3) argued that one way to send messages into the past is to compress a 15-billion-ton asteroid down to the volume of an atomic nucleus, spin it up, and then aim gamma ray bursts through the resulting near-by region of "unhinged time" (see "How To Build a Time Machine," *Omni*, May 1980). This is, of course, 'simply' an artificially constructed Kerr black hole telegraph transmitter (look back at note 114 and related discussion in Chap. 1). Forward, an optimist of the first rank, thought humans would be able to do this before the end of the twenty-first century. It would seem, then, that what should be done *now* is to build gamma ray frequency *receivers* (well within present-day technology) and listen for such messages from the future. The technical details of such receivers wouldn't matter, as long as their design is widely published. That way, the scientists of the future can learn those details by simply reading of them in old, musty library books and journals, and thus will be able to build their transmitters to be perfectly compatible with our old, musty (to them) receivers! Comment on the likelihood of National Science Foundation funding becoming available to build such receivers.

Suppose a time traveler goes into the future and, while there, discovers that there is an older version of himself living in the home that has been in his family for generations. Explain why this implies that the time traveler will eventually return to the present. Suppose, instead, that after the time traveler arrives in the future he decides to remain in the future, and *not* to return. Explain why this implies there will *not* be an older version of himself living in the family home. In both cases, assume the MWI does *not* apply, that is, assume that there is just a single time line.

One well-known quantum physicist, David Bohm (1917–1992), wrote the following passage in his book *The Special Theory of Relativity* (W. A. Benjamin 1965), concerning the possibility of sending messages into the past: "In effect, S could communicate with his own past [self, M] . . . and tell his past self [M] what his future is going to be. But on learning this M could decide to change his actions, so that his future . . . would be different from what his later self [S] said it was going to be. For example, the past self could do something that would make it impossible for the future one to send the signal. Thus, there would arise a logical self-contradiction." Do you think most physicists, writing today more than 50 years after Bohm, would repeat his words?

The role played by quantum mechanics in time travel studies is broad, deep, profound . . . and mysterious. What I mean by this is nicely illustrated by the final paragraph in a paper by Stephen Hawking ("Quantum Coherence and Closed Timelike Curves," *Physical Review D*, November 15, 1995, pp. 5681–5686). There he wrote "Personally, I do not believe that closed timelike curves will occur, at least on a macroscopic scale. I think that the chronology protection conjecture will hold and that divergences in the energy-momentum tensor will create singularities before closed timelike curves appear. However, if quantum gravitational effects somehow cut off these divergences, I am quite sure that quantum field theory on such a background will show loss of quantum coherence. *So even if people come back from the future, we will not be able to predict what they will do* [my emphasis]." What do you think Hawking meant by his final line? If "people come back from the future" and tell us what they did while in the

(continued)

future, then what's wrong with their memories that causes us to fail to be able to predict what they *will* do? Or, is there perhaps nothing wrong with their memories and Hawking is instead arguing that the future experienced by the returned time travelers will not be the future when *we* 'get there'? If that's the case, then what are the returned time travelers 'remembering'?

If quantum mechanics is actually slightly non-linear physics (as are many other normally linear physical phenomena at sufficiently high energy levels)—physics is *linear* when *superposition* holds, which means the result of two inputs is the sum of the individual outputs resulting from application of the individual inputs—and if one accepts the MWI concept, then at least two physicists (**PHYSICISTS1**) claim to have shown that one could communicate not just with our past, but also with the many pasts in the ever-splitting branches of the many worlds. Another physicist (**PHYSICIST2**) wrote a very funny illustration of what that might be like. Yet another physicist (**PHYSICIST3**) suggested that non-linear quantum mechanics might actually allow one to take *photographs* of the many-worlds. Of that, a physicist (**PHYSICIST4**) wrote (without any exaggeration) that such an achievement would be "perhaps the most amazing discovery in the history of science, indeed in the history of mankind." Or, to quote yet another physicist (**PHYSICIST5**), "interworld communication would lead to truly mind boggling possibilities," some of which have been incorporated in at least one science fiction novel (**SFAUTHOR**). Read the physicists' papers, and the novel, and then summarize with your own commentary.

PHYSICISTS1: N. Gisin, "Weinberg's Non-linear Quantum Mechanics and Superluminal Communication," *Physics Letters A*, January 1, 1990, pp. 1–2, and J. Polchinski, "Weinberg's Non-linear Quantum Mechanics and the Einstein-Podolsky-Rosen Paradox," *Physical Review Letters*, January 28, 1991, pp. 397–400.

PHYSICIST2: J. G. Cramer, "Quantum Telephones to Other Universes, to Times Past," *Analog*, October 1991.

PHYSICIST3: D. Albert, "How to Take a Photograph of Another Everett World," *Annals of the New York Academy of Sciences*, December 30, 1986, pp. 498–502.

PHYSICIST4: M. A. B. Whitaker, "On the Observability of 'Many Worlds'," *Journal of Physics A*, July 11, 1985, pp. 1831–1834.

PHYSICIST5: R. Plaga, "On a Possibility to Find Experimental Evidence for the Many-Worlds Interpretation of Quantum Mechanics," *Foundations of Physics*, April 1997, pp. 559–577.

SFAUTHOR: J. P. Hogan, *Paths to Otherwhere*, Baen 1996.

One upon a time, FTL tachyons (and their associated into-the-past transmissions of information) were strictly in the province of science fiction. In recent years, however, such doings have moved into mainstream fiction, as well, with the most recent (as I write) example being the man-of-action novel by Patrick Lee, *Signal*, St. Martin's Press 2015. (Tachyons are also mentioned in the 2015 film *Tomorrowland*.) That novel is set in modern times, not in the future; its hero is a retired soldier who now works as a self-employed house-flipper. His is definitely not a futuristic science fiction world. Until, that is, be becomes involved with a radio-like gadget that receives signals from 10 h, 24 min in the future. As the novel progresses we eventually learn that the gadget is based on German electronics technology that was being tested in a remote lab in northern Algeria, near the end of World War II. When that lab was overrun by a small American force the equipment was destroyed, but not before one of the Americans heard the gadget playing a song titled "She Loves You" along with the word *yeah* repeated numerous times. He didn't know what that meant until, 20 years later, he watched the Beatles' first American appearance on *The Ed Sullivan Show* TV program, and so suddenly realized that in 1944 he had heard a song that hadn't been written yet! That gadget is at the center of a modern-day, renewed Nazi effort to conquer the world, and there is much 'you've seen it all before' chasing, shooting and other 'James Bond' types of action in the novel, but the author *has* been quite inventive in treating time paradoxes. He does talk a lot about 'changing the future,' worries confusingly about 'changing the past,' and mistakes neutrinos for tachyons, but, still, if one is willing to overlook such issues it *is* an entertaining read. In particular, while the gadget's inherent range to the future is limited to ten-plus hours, the author describes a clever way to arbitrarily extend that value. Read *Signal*, and then describe and critique the method outlined in the novel.

The Wheeler-Feynman assumption of continuity already had a distinguished history in mathematics long before they invoked it in their physics resolution of bilking paradoxes. (Both men were very good mathematicians, and certainly knew what I am about to tell you here.) Imagine a man who is about to walk up a hill, starting at A (the base of the hill) and ending at B (the top of the hill). You know nothing of *how* he walks (perhaps at times he stops for a while, other times he walks slowly, sometimes he walks briskly, perhaps at times he even walks back *down* the hill). All you know is that, starting from A at 10 o'clock in the morning, he arrives at B at 11 o'clock. That is, the walk up

(continued)

the hill takes exactly 60 min. He camps overnight at B and then, the next morning, at exactly 10 o'clock, he walks back down the hill along the same path he followed during his ascent the previous day. He arrives at A at 11 o'clock. That is, the return trip takes exactly 60 min. Again, you know nothing of the details of how he makes the descent. Prove that there is at least one spot on the path that he passes at exactly the same time during his descent as he passed it during his ascent the previous day. Hint: No complicated equations are required. Indeed, no math at all is needed. Just sketch the appropriate, general graphs of the man's ascent versus time, and of his descent versus time, and invoke continuity. That is, sketch the distance (as measured along the path) that he is from A versus time for both the ascent and the descent.

Chapter 6
The Physics of Time Travel: II

"It is very sad to see valuable minds writing such a pile of unmitigated bullshit."

—not all physicists think time travel is worthy of study.[1]

6.1 Faster-than-Light into the Past

"Faster-than-light travel remains a coherent, and possible concept, even though it is forbidden by relativity theory."

—a philosopher makes a physics mistake.[2]

So far we have limited our consideration of relativity theory to speeds below the speed of light—that is, to the condition $v < c$, where v is the relative velocity of two reference frames. There was nothing, however, in the derivation of the Lorentz transformation equations discussed in Chap. 3 that actually used that self-imposed constraint. So, just what, in fact, *does* happen for $v > c$? This is not an empty question, because the second half of the above quote that opens this section is simply not true. That isn't to say we can have FTL for free; there *is* a high price to pay, that of causality violation (although, if you are a fan of time travel, it's a price you are probably happy to pay). If a material object goes FTL, then the mathematics seems to say that the object could travel into the past, just as the caped crusader does in the first (1978) *Superman* movie, in order to change the past (to save Lois Lane from dying in an earthquake). In addition, the mathematics also seems to say that if a signal bearing information could achieve FTL, then that information, too, would travel back into the past (see Fig. 6.1).

---

[1]This rather blunt comment (reported in *Physics World*, December 2009, p. 3) was prompted by a suggestion, from two other physicists, that the Higgs boson might ripple backward through time and thereby stop CERN's Large Hadron Collider from creating the long-sought particle in the first place.

[2]G. Robinson, "Hypertravel," *Listener*, December 17, 1964, pp. 976–977, the printed version of a lecture to the British Association for the Advancement of Science.

© Springer International Publishing AG 2017
P.J. Nahin, *Time Machine Tales*, Science and Fiction,
DOI 10.1007/978-3-319-48864-6_6

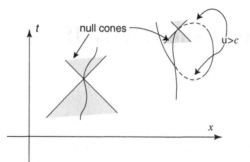

Fig. 6.1 A math-free proof that there can be no closed loops in time (that is, no time travel to the past if $v > c$ is forbidden) in flat Minkowski spacetime, the spacetime of the special theory of relativity. Such a conclusion is far less clear in the curved spacetimes of the general theory of relativity. Figure reproduced by the kind permission of Serguei V. Krasnikov (Polkovo Central Astronomical Observatory in St. Petersburg, Russia), from his 2003 paper "Time Machine (1988–2001)"

The original thinking along these lines visualized such an FTL signal as a modulated beam of tachyons, as mentioned in the previous chapter. In addition to tachyons not having been experimentally observed, even after intense searches for them, there are several theoretical objections (in addition to the bilking paradox problem discussed in the previous chapter) to the likelihood such FTL particles exist.[3] For example, the relativistic expressions for the energy and momentum of a particle with rest mass m_0 moving with speed v are, respectively,

$$E = \frac{m_0 c^2}{\sqrt{1 - (v/c)^2}} \quad \text{and} \quad p = \frac{m_0 v}{\sqrt{1 - (v/c)^2}}.$$

For $v > c$ the radicals in these expressions become imaginary, whereas E and p must always be real-valued (because they can be observed and measured as a result of the interactions the particle has with other matter). The energy and momentum can regain the property of being real-valued if we write $m_0 = \mu\sqrt{-1}$ (that is, $m_0^2 = -\mu^2$) for a tachyon, where μ is the positive, real-valued (but unobservable) meta-mass (that is, $m_0^2 < 0$). This is a radical proposal, of course, (as it is $m_0^2 > 0$ that we are used to; as a Russian mathematician romantically put it, "What binds us to space-time is our [positive] rest mass, which prevents us from flying at the speed of light, when time stops and space loses meaning. In a world of

---

[3]See, for example, "More About Tachyons," *Physics Today*, December 1969, pp. 47–52, a collection of letters received by the journal from its readers. The 'DeWitt Gambit,' mentioned in the previous chapter, was first proposed in one of those letters.

light there are neither points nor moments of time; beings woven from light would live 'nowhere' and 'nowhen'; only poetry and mathematics are capable of speaking meaningfully about such things."[4].

Well, perhaps, but let's continue to pursue the physics of tachyons, at least for a while. It can be shown[5] that the energy and momentum of the tachyon are given by

$$E = \frac{\mu c^2}{\sqrt{(v/c)^2 - 1}} \quad \text{and} \quad p = \frac{\mu v}{\sqrt{(v/c)^2 - 1}}.$$

An interesting consequence of this is that if tachyons *lose* energy, they *speed up*, an observation first made by the German physicist Arnold Sommerfeld (1868–1951) in 1904.[6] This means that if there is a mechanism for continuous energy loss, such as Cerenkov radiation,[7] then tachyons will spontaneously accelerate without limit to *infinite speed*! Curiously, while the above expressions for E and p show that, as $v \to \infty$, tachyons would possess *zero* energy, they would nonetheless have a *non-zero* momentum of μc.[8]

To see how backward time travel and FTL are connected, it is useful to establish a geometrical interpretation of the Lorentz transformation. As stated in Chap. 3, if the x', t' system is moving with speed v in the x (or x') direction relative to the x, t system, then

$$x' = \frac{x - vt}{\sqrt{1 - (v/c)^2}} \quad \text{and} \quad t' = \frac{t - vx/c^2}{\sqrt{1 - (v/c)^2}}.$$

---

[4]Yu I. Manin, *Mathematics and Physics*, Birkhäuser 1981, p. 84.

[5]L. Parker, "Faster-Than-Light Inertial Frames and Tachyons," *Physical Review*, December 25, 1969, pp. 2287–2292.

[6]The consideration of FTL particles already had a long history *before* tachyons were specifically named. You can find late nineteenth century (1888–1889) theoretical analyses of electrically charged FTL particles in the writings, for example, of the English mathematical electrical engineer Oliver Heaviside (1850–1925): see my biography *Oliver Heaviside: the life, work, and times of an electrical genius of the Victorian Age*, The Johns Hopkins University Press 2002, pp. 124–126.

[7]Cerenkov radiation is the energy radiated when a charged particle exceeds the speed of light *in the medium through which it travels*. Since the speed of light *in water* is less than c, it is perfectly okay with special relativity to exceed the speed of light in water, and in fact this commonly occurs for the energetic electrons produced by submerged atomic reactors (swimming pool reactors). The resulting radiation is observed as a blue glow. The radiation is named after the Russian physicist Pavel Cerenkov (1904–1990)—for which he received a share of the 1958 Nobel physics prize—but in fact Heaviside (previous note) had predicted it more than a decade before Cerenkov was born.

[8]This (theoretical) property of tachyons (if they exist) could (perhaps) be used (maybe) to build a revolutionary (to say the least) new rocket propulsion system: see J. Cramer, "The Tachyon Drive: Infinite Exhaust Velocity at Zero Energy Cost," *Analog*, October 1993.

These equations make sense for the case of $v < c$, and we will retain this condition for our two relatively moving *frames of reference*: these frames are the *worlds* of two *human* observers, observers that we'll take to always be *sub*luminal. As one physicist so nicely put it, "The assumption that observers move faster-than-light goes beyond superluminal signaling,"[9] as such observers would have to be thought of as being built not out of flesh-and-blood, but rather out of tachyons.[10] We'll reserve the symbol w to denote the speed of an FTL particle.

Now, recall what we mean by any line parallel to the x-axis; it is a line with a fixed time coordinate. Such a line is a *cosmic moment* line, with the equation $t = $ constant. Similarly, for the moving system we would write the equation of a cosmic moment line as $t' = $ constant which, after the Lorentz transformation is applied, is equivalent to

$$t - \frac{vx}{c^2} = \text{constant.}$$

In particular, the x'-axis ($t' = 0$ cosmic moment line), which passes through the point $x = 0$ at $t = 0$, has the equation

$$t = \frac{vx}{c^2} = vx$$

with the usual convention of $c = 1$.

In a similar way, recall what we mean by any line parallel to the t-axis; it is a line with a fixed space coordinate. Such a line is the world line of a stationary particle in the x, t frame, with the equation $x = $ constant. Similarly, for the moving system we would write $x' = $ constant as the equation of the world line of a particle stationary in that system. From the Lorentz transformation, this is equivalent to

$$x - vt = \text{constant.}$$

In particular, the t'-axis (which is the $x' = 0$ world line of a particle stationary at the origin of the moving system) passes through the $x = 0$, $t = 0$ point, and it has the equation

$$x = vt.$$

Thus, superimposed spacetime coordinate axes for the two frames look like those shown in Fig. 6.2. That is, the relative motion of the two frames results in a

[9]K. Svozil, "Time Paradoxes Reviewed," *Physics Letters A*, April 3, 1995, pp. 323–326.

[10]Science fiction, however, *has* (since the early days of pulp) enthusiastically embraced FTL human travel. In Larry Niven's story "At the Core" (*If*, November 1966), for example, we read of a manned spacecraft that travels 60,000 light years to the center of our own Milky Way Galaxy, and then back, at a speed 420,000 times that of light. As the pilot says (in a grand understatement), "That's goddam fast." Perhaps there is a way to make *some* sense of such an adventure, with the so-called *warp drive* (a'la *Star Trek*), which we'll take-up briefly at the end of this chapter.

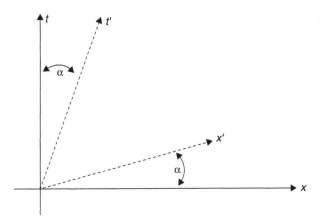

Fig. 6.2 Spacetime coordinate axes rotation by relative motion

rotation of the spacetime axes; but it is a strange sort of rotation, with *opposite* senses for the space and time axes. That is, the *x'* and *t'* axes rotate *towards* each other, as shown in the figure, to make equal angles (α) with the *x* and *t* axes, respectively, where

$$\alpha = tan^{-1}(v).$$

If we limit the moving frame (the frame of a moving observer) to subluminal speeds ($0 \leq v < 1$), then

$$0 \leq \alpha < 45°.$$

At the speed of light ($v = 1$) $\alpha = 45°$ and so the *x'* and *t'* axes coincide—time and space have become indistinguishable.

It is important to realize that observers in either system would measure the same speed for a photon; that is, each would see the world line of a photon as a line with slope 1. This view of the world line of a photon is literally built into the Lorentz transformation because one of Einstein's fundamental postulates for special relativity is the invariance of the speed of light. The truth of this statement for the *x*, *t* system is obvious from Fig. 6.2. It is, perhaps, not so obvious with the *x'*, *t'* system because of the non-perpendicular axes (as shown in the figure) for that system. In Fig. 6.3 the world line of a photon is shown in both systems. In that figure we emit the photon at $x' = 0$, $t' = 0$, and we later measure its coordinates at point A to be $x' = x'_A$ at time $t' = t'_A$. Note carefully how this is done. We draw lines from point A parallel to the *x'* and *t'* axes until they intersect the *t'* and *x'* axes, respectively. This is similar to the way we would get the spacetime coordinates of A in the more familiar *x*, *t* system, where we would draw lines parallel to the *x* and *t* axes.

It should now be obvious that x'_A and t'_A have the same extension, just as they do in the unprimed system, so

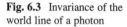

Fig. 6.3 Invariance of the world line of a photon

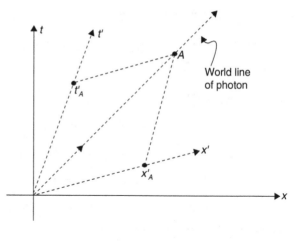

$$\frac{x'_A}{t'_A} \;=\; \text{the speed of light}$$

The speed of light is the only invariant speed under the Lorentz transformation. Indeed, the modern approach to special relativity emphasizes this invariance as the central property of the speed of light, rather than the idea of the speed of light being a limiting speed.

This geometrical interpretation of the Lorentz transformation lets us quickly make another interesting (one, I think, not at all obvious) observation: If a particle is faster than light in the x, t system, then there exists a subluminal x', t' system in which the particle is *infinitely* fast. Figure 6.4 shows the world line of an FTL particle in the x, t system (which is, of course, *below* the world line of a photon; that is, the particle's world line is spacelike). Suppose the FTL particle has speed $w > c$ such that its world line makes angle β with the x-axis. If we now pick v, the speed of the moving x', t' system to be such that $\alpha = \beta$, then the x'-axis will coincide with the world line of the particle, and so the particle will appear to an observer in the x', t' system to be everywhere at once—that is, to be infinitely fast. We have, then,

$$\beta = tan^{-1}(v) = tan^{-1}\left(\frac{1}{w}\right)$$

or $v = 1/w$, which seems to be dimensionally wrong. Recall, however, that with our convention of $c = 1$, the v in this result is a *normalized* speed. To return to the units of everyday use, simply replace v with v/c and w with w/c; this transforms our result to

$$\frac{v}{c} = \frac{c}{w} \quad \text{or} \quad v = \frac{c^2}{w}.$$

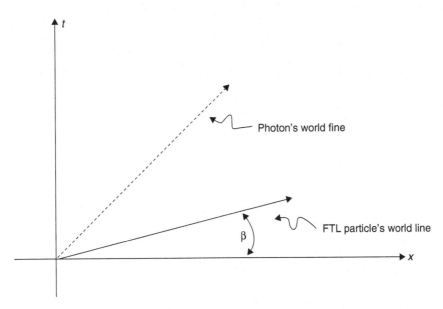

Fig. 6.4 World line of an FTL particle

We can, of course, turn this result around. If an FTL particle moves with speed w in the x, t frame, then to an observer in the x', t' frame moving with subluminal speed v, the particle will appear to be infinitely fast if $w = c^2/v$. A particle with $w > c^2/v$ is said to be not just be superluminal, but *ultra*luminal.

If a particle has infinite speed if $w = c^2/v$, then what physically happens if w is greater than c^2/v? The answer is easy to see from a spacetime diagram, as in Fig. 6.5, where the x' and t' axes have been extended back to negative values. In that figure I have labeled two arbitrary events A and B on the world line of an ultraluminal particle (and so it lies below the x' axis), and have shown the spacetime coordinates of each event in both the x, t and x', t' frames. For the x, t frame we see that A is related to B by the relations $x'_A < x'_B$ and $t'_B < t'_A$; that is, the time order of A and B is reversed for an observer in the x', t' frame. To that observer, the particle appears to be traveling backward in time!

But this isn't quite the end of the story. Following the approach of two pioneering tachyon physicists,[11] we note that if the energy of a particle in the stationary system is E, then the energy as measured in the moving system is given by[12]

---

[11]O. M. Bilaniuk and E. C. G. Sudarshan, "Causality and Space-like Signals," *Nature*, July 26, 1969, pp. 386–387.

[12]The expression for E' is the result of applying the Lorentz transformation to E. You can find all the details of that worked out in A. P. French, *Special Relativity*, W. W. Norton 1968, pp. 208–210.

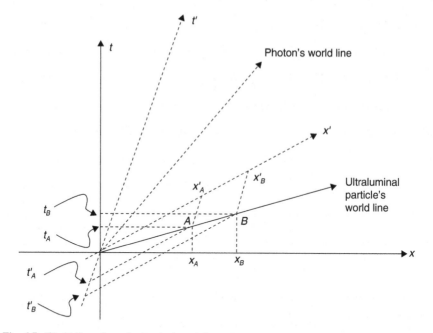

Fig. 6.5 World line of an *ultra*luminal particle

$$E' = \frac{1}{\sqrt{1 - \left(\frac{v}{c}\right)^2}} \frac{\mu(c^2 - wv)}{\sqrt{(w/c)^2 - 1}}$$

Note that sign of E' switches from positive to negative when w (the speed of the particle) exceeds c^2/v, which is precisely the condition for the particle to be ultraluminal and so traveling backward in time (as seen in the primed system). That is, negative energy moving backward in time in one system is positive energy moving forward in time in another. This is, in fact, the original motivation for the RP (reinterpretation principle) discussed in the previous chapter, and the claim was, at one time, that the RP was just what was needed to 'explain' the paradoxical implications of time travel. This is not the majority view today, however, and a number of physicists have been quite inventive in constructing scenarios with causal paradoxes that the RP clearly fails to 'explain.'

For example, the Princeton physicist Shoichi Yoshikawa (1934–2010), in a letter to *Physics Today* (see note 3), was able to create a scenario which *uses* the RP to arrive at a causal paradox. In his construction, the RP allows an observer to transmit an ultraluminal tachyon to a remote observer at time $t = 0$, and to receive a reply from that observer at $t < 0$. This obviously sets-up the possibility of a bilking paradox in which the original observer, upon receiving the $t < 0$ tachyon, then decides *not* to transmit the $t = 0$ tachyon. What makes Yoshikawa's paradox a particularly troublesome paradox is that the RP is the culprit, not the savior.

6.2 Tipler's Rotating Cylinder Time Machine

"... within forty-eight hours we had invented, designed, and assembled a chronomobile. I won't weary you with the details, save to remark that it operated by transposing the seventh and eleventh dimensions in a hole in space, thus creating an inverse ether-vortex and standing the space-time continuum on its head."[13]

                    —this is almost surely *not* the way to build a TM (time machine)

As discussed at the start of this book, the first endorsement of the reasonableness of physicists talking about plausible, scientific time travel, began with Gödel's discovery of closed timelike lines in the mathematics of certain rotating universe models. Such models had been studied as early as 1924 by the Hungarian physicist Cornelius Lanczos (1893–1974), a quarter century before Gödel, but it was Gödel who made explicit the possibility a rotating universe might allow time travel. His realization of time travel as an *inherent* property of a rotating universe is an illustration of a *weak* TM, while what is the central interest concerning time machines (in physics *and* in science fiction) is a *strong* TM. That is, in time machines that can be *intentionally constructed* by manipulating mass-energy in a finite (what physicists call a *compact*) region of spacetime to *create* closed timelike lines where none existed before the manipulation began. Interestingly, the fundamental physical idea behind Gödel's weak TM is the same underlying idea behind the first strong TM, Tipler's rotating cylinder discussed back in Chap. 1 (strictly, of course, only a *compact* TM if the cylinder can be of *finite* length).

The one result from general relativity that we'll use here (without proof) is that the rotation of matter causes a distortion of spacetime that results in the 'tipping over' of light cones, with the future half tilted in the direction of motion. If you imagine a point in the universe about which the rotation takes place, then this tipping effect increases with the radial distance from that point.[14] The fact that rotating masses tip light cones over in the direction of rotation was discovered very early in the history of general relativity (1918), by the Austrian theoreticians Josef Lense (1890–1985) and Hans Thirring (1888–1976). Originally (and naturally) called the *Lense-Thirring effect*, it now generally goes by the name of the *dragging of inertial frames effect*, and it plays a central role in the weak Gödel and the strong (maybe) Tipler time machines. Here's how.

At a certain critical distance from the rotation center (more on this in just a bit), the future half of the light cone at a given point in spacetime will be sufficiently tilted so as to enter the past half of similarly tilted light cones at nearby spacetime points This is illustrated in Fig. 6.6,[15] which shows a circular chain of tilted light

---

[13]L. Sprague de Camp, "Some Curious Effects of Time Travel," in *Analog Readers' Choice*, Dial 1981.

[14]For a picture of this, see S. W. Hawking and G. F. R. Ellis, *The Large Scale Structure of Spacetime*, Cambridge University Press, 1973, p. 169.

[15]Adapted from D. B. Malament, "'Time Travel' in the Gödel Universe," *Proceedings of the Philosophy of Science Association* 1984, pp. 91–100.

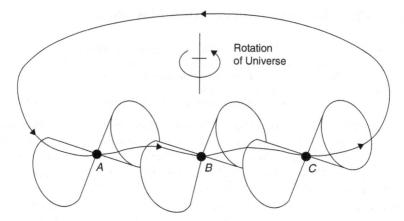

Fig. 6.6 Time traveling through tilted light cones in a rotating universe

cones in a rotating universe. Because light cones are tilted by a rotation-induced twisting of spacetime, a traveler can move around a circular path on a trip that is always directed into his *local* future but nonetheless end-up in his own *global* past, without ever going faster than light. This kind of round trip in spacetime, with a trajectory that winds back into the past without ever becoming spacelike, is a closed timelike curve.

From Fig. 6.6 it should be clear how a time traveler, beginning at A, can weave his way along a circular path (this path needs to have a radius at least as great as the critical value mentioned earlier) to B to C to ... that brings him back into the *past* half of the light cone at A. The traveler's world line is always inside the local light cone; that is, the world line of the time traveler is always timelike, and never FTL. These timelike curves are present in the rotating spacetime from the very beginning of the spacetime, and were not *created* (certainly not by humans) by conscious intent.

You can see how this works mathematically by taking the spacetime metric for Gödel's universe (with, as usual, the convention that the speed of light $c = 1$):

$$(ds)^2 = (dt)^2 - (dr)^2 - (dy)^2 + \sin h^2(r)\left[\sin h^2(r) - 1\right](d\phi)^2$$
$$+ \sqrt{2} \sin h^2(r)(d\phi)(dt)$$

where t, r, y, and ϕ are cylindrical coordinates in four-dimensional spacetime. Now, imagine that our adventurer's world line is the helical curve $r=$constant, $y = 0$, and $t = -\alpha\phi$: if we take the time axis as vertical then the time traveler's world line is a vertical helix in spacetime. For this curve, $dr = dy = 0$ and $dt = -\alpha d\phi$. This last differential means, in particular, that whatever the sign of the constant α, we can choose that one of the two senses of movement in the spatial ϕ dimension that gives $dt < 0$.

Continuing, we have

$$(ds)^2 = \left[\alpha^2 - 2\sqrt{2}\alpha \sinh^2(r) + \sin h^2(r)\left(\sin h^2(r) - 1 \right) \right] (d\phi)^2$$

or, upon letting $u = \sinh(r)$ we have

$$(ds)^2 = \left[\alpha^2 - 2\sqrt{2}\alpha u^2 + u^2\left(u^2 - 1 \right) \right] (d\phi)^2.$$

Now, for $\alpha = 0$ we have

$$(ds)^2 = u^2\left(u^2 - 1 \right)(d\phi)^2$$

which is greater than zero if $u > 1$. This condition holds if $\sinh(r) > 1$, that is, if r is a constant greater than $\ln\left(1 + \sqrt{2} \right)$. In other words, for r sufficiently large (and now we know the critical value for r) we have $(ds)^2 > 0$, the required condition for a timelike spacetime interval. By continuity, then, we will continue to have $(ds)^2 > 0$ even with some small positive or negative value of α different from zero. Because ϕ is a periodic coordinate (we identify $\phi = 0$ with $\phi = 2\pi$), as the traveler moves along the curve she returns repeatedly to the same spatial points, but her time coordinate is increasingly negative. That is, she is traveling into the past. Note, once again, that in Gödel's universe this property holds only for orbits with radii greater than a certain minimum.[16]

Tipler's rotating, infinitely long cylinder is a mechanism for artificially producing the tipped-over light cone effect, thus *creating* closed timelike curves. Figure 6.7, taken from Tipler's Ph.D. dissertation,[17] shows how the cylinder works. The cylinder is represented by the central vertical axis. Far away from the cylinder the light cones in spacetime are upright, but as we move inward they tip over, the future halves opening up into the direction of rotation. (Only the future halves of the light cones are shown.) This direction, which is the direction that far away from the cylinder measures *space*, near the cylinder measures *time* (just as in the Gödelian universe). That is, there has been a dimension reversal! This fantastic possibility has found its way into science fiction, as in Stephen Baxter's 1995 novel *The Time Ships*, wherein the Victorian narrator says "If only one could *twist about* the Four Dimensions of Space and Time — transposing Length with Duration, say — then one could stroll through the corridors of History as easily as taking a cab in the West End!"

To travel back in time, therefore, all the time traveler need do is leave Earth and approach the cylinder until she is near enough to be in the tipped-over region of spacetime. Then she would follow a helical path around the cylinder and could

---

[16]There are other solutions to the Einstein gravitational field equations that have closed timelike lines at *any* radius, no matter how small: see M. J. Reboucas, "A Rotating Universe with Violation of Causality," *Physics Letters A*, March 5, 1979, pp. 161–163.

[17]F. J. Tipler, *Causality Violation in General Relativity*, University of Maryland 1976.

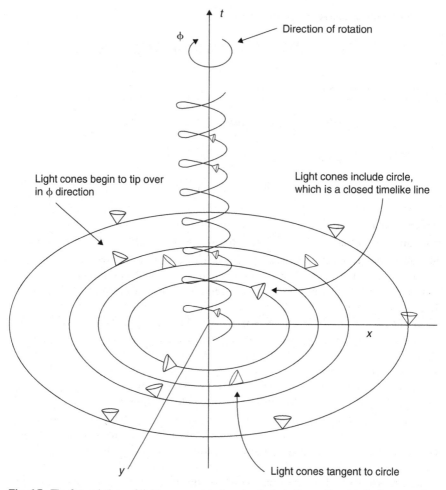

Fig. 6.7 The future halves of light cones point almost entirely in the $+t$ direction far from rotating matter; they begin to tip over as the matter is approached. Note that there is a helical timelike path that moves locally into the *future* in the $-t$ direction, that is, it goes into the past as seen by an observer far from the rotating matter (The world lines of rotating matter are helixes in the $+t$ direction)

spiral along the negative time direction as far back in time as desired (but no further back than to the moment of the cylinder's creation). This motion is such that the time traveler is always moving into her local future, via the tipped-over light cones. Finally, she would withdraw from the cylinder and return to Earth—in the past. The time traveler had better be a good space navigator, of course, because Earth won't be where she left it!

6.3 Thorne's Wormhole Time Machine

"This fact reinforces the authors' feeling that [closed time loops] are not so nasty as people generally have assumed."[18]

The spacetime wormhole is presently the most promising of the approaches that have been advanced for building a time machine. Gödel rotated the entire universe in 1949, and Tipler 'reduced' the problem in 1974 to 'merely' spinning a cylinder of infinite length. In 1988 Kip Thorne scaled things down even more, this time to other extreme. His idea calls for pulling a wormhole on the scale of the Planck length out of the topologically multiply connected quantum foam that spacetime is and then enlarging it (somehow) to human scale, all the while stabilizing it against self-collapse, and then finally using the time dilation effect of special relativity to alter time at one mouth of the wormhole as compared to the other mouth. What a mouthful! What, you almost surely wonder, does all that mean?

First of all, what's 'quantum foam'? The term refers to the idea that the topology of spacetime is *not* a smooth, continuous manifold, but rather (if you look close enough) is a seething 'ocean of fluctuations' that is always changing, changes on the scale of the Planck length (look back at Sect. 1.5). What does *that* mean? Like the ocean surface-in-the-large, large-scale spacetime is simply connected. But just as one sees all sorts of transient structure as one looks at the water more closely (beginning with macroscopic waves and then proceeding downward to the bubbly foam on the waves), spacetime too displays an ever-changing connectivity-in-the-small.[19] That is the 'quantum foam.'

Wormholes have been around in physics for decades, but they have always been thought to be so unstable as to exist only on paper in the mathematics of general relativity. In an analysis[20] published more than half-a-century ago, for example, wormhole instability was shown to be so severe that not only would a human have no chance in getting through one, but also not even a single speedy photon could do so. Even at the speed of light, the photon could not zip through a wormhole before being trapped inside ("pinched off") in a region of infinite spacetime curvature. Wormholes would simply collapse too quickly after formation for even the so-called ultimate speed to save the traveler. Indeed, the presence of mass-energy inside a wormhole actually *accelerates* its collapse. The physics of wormholes, it seemed, made them simply untraversable.

---

[18]The conclusion of a mathematical demonstration that time travel by wormhole does not conflict with the conservation of energy: see J. L. Friedman *et al.*, "Cauchy Problem in Spacetimes with Closed Timelike Curves," *Physical Review*, September 15, 1990, pp. 1915–1930.

[19]Not all accept this view. See, for example, A. Anderson and B DeWitt, "Does the Topology of Space Fluctuate?" *Foundations of Physics*, February 1986, pp. 91–105.

[20]R. W. Fuller and J. A. Wheeler, "Causality and Multiple Connected Spacetime," *Physical Review*, October 15, 1962, pp. 919–929.

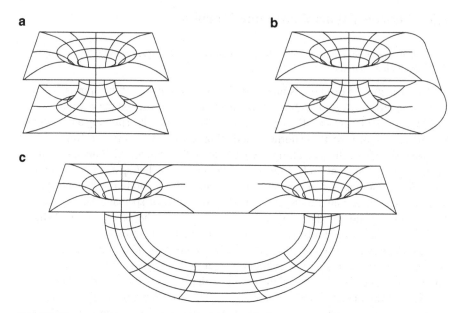

Fig. 6.8 These sketches are unavoidably misleading, being two-dimensional renditions of wormholes that connect two places in a three-dimensional space. Time machine wormholes, on the other hand, connect two places in a four-dimensional spacetime. In particular, the mouths of the wormholes are not 'depressions' into which the time traveler's rocket ship plunges, but rather would appear to be three-dimensional spheres. The wormhole in (**a**) connects two disjoint universes, while those in (**b**) and (**c**) are connections between two places in the same universe. As shown in these last two cases, the wormhole 'handle' can be either long or short, compared to the external distance between the wormhole mouths

And anyway, how would one gain access to a wormhole in the first place? As Thorne and one of his students suggested,[21] one might perhaps imagine someday finding a rotating black hole that mathematically possesses, in its interior, so-called hyperspace tunnels to 'other places'—either in our universe or in other universes (see Fig. 6.8). In the case of a wormhole connecting two places in the same universe, although the external distance in spacetime between the places may be very large (mega-light-years), it is conceivable that the distance *through* the wormhole itself could be very small. The time required to traverse the wormhole, as measured by the traveler's watch might, in fact, be arbitrarily small. This is exciting, but wormholes do not come without some significant problems.

Such problems include the presence of a one-way event horizon, which precludes two-way travel (it seems reasonable to assume that time and space travelers

---

[21]M. S. Morris and K. S. Thorne, "Wormholes in Spacetime and Their Use for Interstellar Travel: A Tool for Teaching General Relativity," *American Journal of Physics*, May 1988, pp. 395–412. This paper was motivated by Thorne's earlier response to a request, from the American astronomer Carl Sagan (1934–1996), for help in making plausible the interstellar travel imagined in his 1985 novel *Contact*.

might wish to eventually return), and enormous gravitational gradients (tidal forces) that dismember anything approaching and/or entering the wormhole. These problems have often been conveniently *ignored* in science fiction. For example, in one tale[22] the time machine is in the form of a wormhole time tunnel that time travelers can simply walk through, and in another characters similarly walk back and forth between two openings in a wall that connect the same place in space but separated in time by 160 years.[23] And in an even older story, we read of a time machine (in the form of a hole in spacetime that is a wormhole in everything but name) through which one can literally step from future to past and back again. That story[24] is particularly notable for having introduced the term *mugwump* for a time traveler who uses a time machine wormhole. To paraphrase the tale's time traveler, as he transits the wormhole his "mug" is in the past and his "wump" is in the future.[25]

Now, before going any further, a note on what physicists are referring to when they write of traversable time travel wormholes. Such wormholes are called *Lorentzian* because they have a spacetime metric with the signature $[+, -, -, -]$ (see Sect. 3.5). Further, the wormhole is taken to be *static*, that is, to have no time-varying behavior. The reason for being specific on the nature of time travel wormholes is that there is another type with what is called the *Euclidean signature*: $[+, +, +, +]$, which is not suitable for time traveling. Motion in a Euclidean signature wormhole involves imaginary momentum or proper time, neither of which is physically plausible for a time traveler. However, *if* you can gain access to a Lorentzian wormhole *then*, as two physicists wrote, "if you manage to acquire even one inter-universe traversable wormhole then it *seems* almost absurdly easy to build a time machine."[26]

In response to that fundamental question of how to gain access to a Lorentzian wormhole, Thorne and his students were bluntly honest—they didn't know. Their best suggestion was that "one can imagine an advanced civilization pulling [a] wormhole out of the quantum foam and enlarging it to classical size."[27] A few years later this dramatic idea found its way into a science fiction novel of a far-future alien civilization able to control the energies of constellations of galaxies: "Spacetime is friable. Wormholes riddle the fabric of spacetime on all scales. At the Planck length and below, wormholes arising from quantum uncertainty effects blur

---

[22]R. C. Wilson, *A Bridge of Years*, Doubleday 1991.

[23]M. Leinster, *Time Tunnel*, Pyramid Books 1964.

[24]H. Kuttner, "Shock," *Astounding Science Fiction*, March 1943.

[25]The term's origin is in 19th century politics, as a description of fence-sitters who try to avoid taking a definite position on some controversial topic under debate. For an illustration of wormhole mugwumping, see K. S. Thorne, *Black Holes & Time Warps*, W. W. Norton 1994, p. 500.

[26]C. Barceló and M. Visser, "Twilight for the Energy Conditions?" *International Journal of Modern Physics D*, December 2002, pp. 1553–1560. See also M. Visser, *Lorentzian Wormholes*, AIP Press 1996.

[27]M. S. Morris, K. S. Thorne, and U. Yurtsever, "Wormholes, Time Machines, and the Weak Energy Condition," *Physical Review Letters*, September 26, 1988, pp. 1446–1449.

the clean Einsteinian lines of spacetime. And some of the wormholes expand to human scale, and beyond — sometimes spontaneously, and sometimes at the instigation of intelligence."[28]

This isn't easy to visualize, but let's plunge ahead and assume we can "pull" a wormhole out of the quantum foam. If so, then once it is (somehow) inflated the wormhole could be stabilized against collapse by threading it with either matter or fields of stupendous negative (outward) tension—and by *stupendous* I mean *STU-PENDOUS*. If b_0 denotes the minimum radius of the wormholes (the size of the so-called *throat* of the wormhole), the tension (radial pressure) at that location must be at least

$$\tau_0 = \frac{3.8}{b_0^2} \times 10^{36} \frac{tons}{in^2}$$

where b_0 is expressed in feet.[29] (This expression shows that for a wormhole with a throat radius of several thousand feet, the value of τ_0 is enormous, of the same magnitude as the pressure at the center of the most massive neutron star.) To stabilize a common sort of everyday wormhole, such as a subway tunnel, we can obtain the required tension/pressure by lining the tunnel with iron plates or concrete. But how, for a hyperspace wormhole, do we obtain 'iron plates' that can achieve the required enormous tension? As Thorne and his students observed (note 27) such stuff could only be called "exotic," a term that had appeared a few years earlier in connection with the observed energy density in the throat of a wormhole.[30]

One possible approach to this problem does not use matter at all. If we make b_0 *very* large, then non-material fields will do the job.[31] Indeed, suppose b_0 equals 1 light year (a large wormhole by anybody's standard!). Then τ_0 is 'only' 4000 tons/ square inch, and that is achievable by threading the wormhole throat with a magnetic field of 'only' 2,700,000 gauss (five million times stronger than the Earth's field).[32] To generate such a field is not impossible, and present-day

---

[28]From S. Baxter's 1993 novel *Timelike Infinity*. That same year the Chinese physicist Liao Liu wrote on how, as a result of a naturally occurring vacuum fluctuation, a wormhole might spontaneously appear: L. Liu, "Wormhole Created from Vacuum Fluctuation," *Physical Review D*, September 15, 1993, R5463–R5464.

[29]The original equation for τ_0 was given in units of dynes/cm^2, with b_0 is expressed in meters (see note 21), but I have converted these units to the more familiar (to the non-physicist) units of tons per square inch, for dramatic purposes, because the units of dynes per square centimeter is so small it's difficult to relate it to anything of everyday significance.

[30]R. Balbinot, "Crossing the Einstein-Rosen Bridge," *Lettere Al Nuovo Cimento*, May 16, 1985, pp. 76–80.

[31]See, for example, Y. Soen and A. Ori, "Improved Time Machine Model," *Physical Review D*, October 15, 1996, pp. 4858–4861, and D. N. Vollick, "How to Produce Exotic Matter Using Classical Fields," *Physical Review D*, October 15, 1997, pp. 4720–4723.

[32]To understand how the calculation of a magnetic field from a pressure requirement is accomplished, note that pressure is dimensionally equivalent to field energy per unit volume, which in turn is given by a well-known result in electromagnetic theory.

experimental, hypervelocity electromagnetic rail guns in development for the US Navy use transient magnetic fields in the mega-gauss range.

As if the stabilization problem wasn't enough of a complication, Thorne and his colleagues (note 27) showed that there is another, even more curious problem. The geometrical requirement that the wormhole interior smoothly connect to the external, asymptotically flat exterior spacetime demands that the wormhole throat flare outward as shown in Fig. 6.8. It turns out that this condition is mathematically equivalent to a requirement that τ_0 exceed the energy density of the throat material[33]; and special relativity, in turn, says that for some timelike observers the energy density will then actually be *negative*. This is a clear violation of the so-called *weak energy condition* (WEC), which says the observed mass-energy density is always non-negative. This is *so* 'obvious' that the WEC was thought, for a long time, to be almost a law of nature. Such a violation is actually not as crazy as it might sound however because, more than half-a-century ago, it was shown[34] that an energy density that is everywhere and everywhen positive is *not* compatible with any quantum field theory that is *local*, as, presumably, will be the yet-to-be-discovered theory of quantum gravity. Over the years since then a variety of other energy conditions have been proposed, such as the *averaged weak energy condition* (AWEC), which says that only the *average* value of the energy density over a complete null geodesic world line has to be non-negative, which leaves open the possibility of *temporary* negativity here, there, then and when.[35]

The traversable, static wormholes studied by Thorne and his colleagues violate even the AWEC (see note 27), however, and in the years since it has become clear to physicists that imposing constraints on the mass-energy density may not be so 'obvious' after all (see note 26). But all may not be lost. Indeed, it has been shown that the violation of the AWEC by traversable wormholes can be made as small as desired, that is, the requirement for the exotic matter required to line a wormhole throat to keep it open can be made as tiny as you want.[36]

Well, tiny the quantity of exotic matter may be but, nonetheless, even a tiny amount of it would be extraordinary weird stuff because a negative energy density can be interpreted as meaning that the exotic material that keeps the wormhole throat open

---

[33]The condition of τ_0 exceeding the energy density in the throat is, in fact, the technical definition of *exotic*—see note 30. In everyday situations, the exotic condition is never even remotely approached. For example, the maximum tension necessary to pull a piece of steel apart—the so-called *tensile strength*, about 100,000 pounds per square inch—is a trillion times less than the mass-energy density of steel.

[34]H. Epstein, V. Glaser, and A. Jaffe, "Nonpositivity of the Energy Density in Quantized Field Theories," *Il Nuovo Cimento*, April 1, 1965, pp. 1016–1022.

[35]T. A. Roman, "Quantum Stress-Energy Tensors and the Weak Energy Condition," *Physical Review D*, June 15, 1986, pp. 3526–3533.

[36]M. Visser, S. Kar, and N. Dadhich, "Traversable Wormholes with Arbitrarily Small Energy Condition Violations," *Physical Review Letters*, May 23, 2003, pp. 201102-1 to 201102-4.

does so by exerting a *repulsive* gravitational force.[37] A repulsive force sounds like a property we'd expect to see associated with *negative mass* and, although such a thing has never been observed (negative matter is *not* anti-matter, which *has* been observed and which does not repel 'normal matter'), it was studied long ago (theoretically, of course) by the English cosmologist Hermann Bondi (1919–2005). Bondi showed[38] that negative mass would indeed have some truly bizarre properties,[39] but there is nothing in general relativity that forbids its possible existence. Wormholes, with negative mass throats, should produce observable effects by which a wormhole *might* be detected. Mathematical analyses of the effect a negative-mass wormhole mouth would have, when crossing the line-of-sight between Earth and a distant star, indicates that there should be an observable *double-spike* in the intensity of the star's light. Astronomical searches for such an optical signature have actually been conducted, with (alas) no success as I write (2017).

There is another interesting implication of a repulsive gravitational force, one that proves to be *essential* to the possibility of a wormhole time machine. Just as Einstein's famous prediction (verified in 1919) from general relativity, that star light passing near the Sun's edge is bent *inward* by the Sun's attractive gravitational field, the repulsive, anti-gravity field of a wormhole will cause any light rays traveling through the wormhole to be bent *outward*. That is, a tight, narrow beam of radiation entering a wormhole will emerge *defocused*. This is crucial because, as you'll soon see, a wormhole time machine would otherwise be destroyed by the light from the dimmest candle.

One might take the failure of astronomical searches for a double-spike light signature to mean that wormholes with negative mass throats (thus violating the WEC) simply don't exist. But not so fast. The first hint that the possibility of a negative energy density might not be such a crazy idea occurred as long ago as 1948, with a theoretical prediction made by the Dutch physicist Hendrick Casimir (1909–2000). As pointed out in Chap. 1, the Heisenberg uncertainty principle allows a temporary violation of conservation of energy to occur, with the magnitude of the allowed violation increasing with decreasing time duration. Even in a vacuum, then, with particle/anti-particle creation and annihilation spontaneously and continuously taking place, the *average* energy density being zero does not

---

[37]The *strong energy condition* says gravity is always attractive—which is clearly not true in a wormhole throat—and so static, traversable wormholes violate both the weak *and* the strong energy conditions.

[38]H. Bondi, "Negative Mass in General Relativity," *Reviews of Modern Physics*, July 1957, pp. 423–428.

[39]For example, general relativity says that a negative mass will *repel* all other masses (positive *and* negative), whereas a positive mass will *attract* all other masses (positive *and* negative). Imagine, then, a negative mass attached to the nose of a positive-mass spaceship. The spaceship tries to move toward the negative mass, while the negative mass tries to move away from the spaceship. So off they both go into the sky, like a cat chasing its tail. This so-called *reactionless anti-gravity drive*, bizarre as it appears, does not violate either of the conservation laws of energy or momentum. See G. Cavalleri and E. Tonni, "Negative Masses, Even if Isolated, Imply Self-Acceleration, Hence a Catastrophic World," *Il Nuovo Cimento B*, July 1997, pp. 897–903.

preclude fluctuations away from zero and so, at times, actually becoming *negative*. What Casimir showed was that if one positioned two perfectly conductive plates parallel to each other, then the normal quantum fluctuations of the energy density in this 'vacuum sandwich' would be altered in such a way as to result in their mutual attraction—and this (tiny) effect was later actually observed.[40]

What does it mean to 'alter the normal quantum fluctuations'? Consider the creation of a photon and its anti-particle, which is another photon. From the wave interpretation of particles, the parallel plates restrict the photons that appear in the vacuum layer to those that have wavelengths that 'fit' because those wavelengths are submultiples of the plate separation (this requirement follows from the fact that a perfectly conducting plate cannot support a non-zero tangential electric field). Photons with longer wavelengths than the plate separation cannot 'fit' and thus do not appear. That is, the parallel plates have created a boundary condition that has quantized the electromagnetic field. The absence of these 'longer wavelength' photons lowers the average energy density between the plates and, because the average without the plates is zero, the altered average energy density must be negative. Indeed, the more the maximum allowed photon wavelength decreases with decreasing plate separation, the more negative the average energy density becomes in the enclosed Casimir vacuum. The negative energy density manifests itself as an inward directed force per unit area (remember, energy density and pressure are dimensionally equivalent).

The experimental detection of the Casimir effect was a remarkable event in physics. As one mathematician put it, "No worker in the field of overlap of quantum theory and general relativity can fail to point this fact out in tones of awe and reverence."[41] Robert Forward, an imaginative physicist who has appeared in this book earlier as an enthusiastic supporter of time travel, has described how the Casimir force might be used to extract energy literally *from a vacuum*. This is an idea as seemingly impossible as is the plan of one science fiction professor to squeeze energy out of *time*. As he asks his assistant, "But tell me, Bob, isn't that a ridiculous thought? To take time, something intangible, invisible, incomprehensible, and contract it — squeeze it together like a sponge?"[42] The story is fun, but Forward's proposal is that as well—*and* good physics.[43] Now, what does all this have to do with wormhole time machines?

---

[40] A complete presentation of Casimir's analysis, with citations to the original literature, can be found in L. E. Ballentine, *Quantum Mechanics*, Prentice-Hall 1990, pp. 399–403. For an historical, tutorial presentation, including Casimir's personal comments on how he was led to make his discovery, see P. W. Milonni and M.-L. Shih, "Casimir Forces," *Contemporary Physics*, September–October 1992, pp. 313–322.

[41] S. A. Fulling, *Aspects of Quantum Field Theory in Curved Space-Time*, London Mathematical Society 1989.

[42] E. Binder, "The Time Contractor," *Astounding Stories*, December 1937.

[43] R. Forward, "Extracting Electrical Energy from the Vacuum by Cohesion of Charged Foliated Conductors," *Physical Review B*, August 15, 1984, pp. 1700–1702.

Thorne and his colleagues (note 27) proposed to use the Casimir effect to achieve the "exotic condition" *without* matter. Their idea was to place identical, conducting, spherical plates that carry equal electric charges at each end of the wormhole (remember, the wormhole mouths are spherically symmetric). The two identical charges repel each other, but the charge size is adjusted so that the gravitational attraction of the plates precisely cancels the repulsion. They then calculated that the Casimir effect results in a negative energy density sufficient to provide the throat tension necessary to prevent wormhole collapse.

There *are* some weird aspects to this (and perhaps that's no surprise). For example, the analysis assumed that the wormhole length is *very* small compared to its radius (10^{-10} cm long and 200 million miles wide!), with the short length required because it represents the separation of the wormhole plates, and the smaller the separation the more negative the average energy density. (The functional dependence is as the *fourth* power of the separation.) Another problem is the balancing of the electrical repulsion and the gravitational attraction of the wormhole mouth plates, as such a balance is clearly an unstable one. Finally, because the two spherical plates completely fill the wormhole mouths, how would a traveler actually get through the wormhole? The 'answer' was to drill a hole through the plates and hope that wouldn't perturb the Casimir vacuum too much.[44]

All the above litany of the difficulties static, traversable wormholes face in simply existing is certainly daunting, but let's now ignore all that and suppose we actually have a wormhole with both mouths in the same universe. (For use as a time machine, it would seem desirable for the time traveler to remain in his/her own universe!) So, how do we turn the wormhole into a time machine? Interestingly, while it is general relativity that gives us the wormhole, it is special relativity that adds the final touch of backward time travel. We begin by imagining that, somehow, one mouth of the wormhole can be moved with respect to the other mouth. One early suggestion, for example, was to use the gravitational attraction of a large asteroid to 'drag' one end of the wormhole, thereby inducing a time dilation effect.[45]

That is, suppose we have two clocks A and B, one in each mouth of the wormhole. These two clocks, and all other clocks in the flat spacetime outside the wormhole, are initially indicating the same time and running at the same rate. Now, recalling the twin paradox from Chap. 3, let each mouth-clock play the role of one of the twins. Imagine that A and B are now separated because the mouth containing B is placed on board a rocket ship. The rocket ship takes a long, high-speed trip out into space along the straight-line path joining A and B in external space, and then returns, just as described in Sect. 3.5. We then unload the space traveling wormhole mouth (with its clock B) and reposition it at its original location. What

---

[44]The very next year it was shown how to construct non-spherically symmetric wormholes to avoid that particular problem: see the two papers by M. Visser, "Traversable Wormholes: Some Simple Examples," *Physical Review D*, May 15, 1989, pp. 3182–3184, and "Traversable Wormholes from Surgically Modified Schwarzchild Spacetimes," *Nuclear Physics B*, December 11, 1989, pp. 203–212.

[45]J. L. Friedman, "Back to the Future," *Nature*, November 24, 1988, pp. 305–306.

is the situation now? We can summarize matters as follows: (1) Clock A, in the non-moving mouth, remains in-step with the local clocks in the space outside the mouth. (2) Clocks A and B, both inside the wormhole, have *not* moved with respect to each other because we are assuming a very short wormhole handle, as in part (b) of Fig. 6.8. We can arrange for the motion of the space traveling mouth (with clock B) to be such that the handle is always short, and so the distance between clocks A and B changes by an arbitrarily small amount. Thus, clocks A and B remain in-step with each other. (3) Clock B, because it has been moving with respect to its external space, arrives back at its starting position reading *behind* (that is, *earlier*) than the clocks outside its wormhole mouth.

For the sake of argument, then, suppose the journey of B is such that there is a two-hour time-slip between clock B and its local, external clocks. Thus, if clock B reads 9 A.M., the clocks outside of mouth B will read 11 A.M. But because clocks A and B are in-step, clock A reads 9 A.M., as do the clocks outside of mouth A. That is, the wormhole connecting mouth A to mouth B is a connection between two parts of the same universe that are two hours apart in time. Now, suppose the journey from mouth A to mouth B can be made through external space in one hour. Then, one could leave mouth A at 10 A.M., rocket to mouth B by 11 A.M., and travel back to mouth A via the wormhole to the starting point—where it is 9 A.M., one hour *before* the trip began! We could, in fact, imagine repeating this process, going back one additional hour for each new loop through the wormhole. One clear restriction, however, is that we could not go back in time to *before* the creation of the wormhole time machine. The wormhole works in the other direction, too. To see this, suppose that the space traveler leaves mouth B at 8 A.M. and rockets through external space to mouth A, arriving at 9 A.M. Entering mouth A, he exits from mouth B (where he started) at 11 A.M., two hours in the future.

Another way to induce a time dilation effect, to convert a wormhole to a time machine, *without moving either mouth*, is to simply place one mouth in an intense gravitational field, that of, say, a neutron star. (Recall, from Sect. 3.3, how gravity influences the time-keeping rate of a clock.) As the physicists who proposed this idea put it, almost *any* interaction with surrounding matter and gravity fields almost inevitably turns a wormhole into a time machine.[46] Others have admitted that the details of the origin of time dilation are probably not issues worth debating, but rather what is called the *back-reaction* is of far more concern.

To understand the back-reaction requires mention of what is called the *Cauchy horizon*, the hyperspace surface in spacetime that separates the region where closed timelike lines can exist, from the region where they cannot exist. The back-reaction is the build-up of unbounded energy levels on the Cauchy horizon, causing its instability and rapid destruction. The name of the horizon comes from the "Cauchy problem"—named after the nineteenth century French mathematician Augustin-Louis Cauchy (1789–1857)—in the theory of partial differential equations. In this

---

[46]V. P. Frolov and I. D. Novikov, "Physical Effects in Wormholes and Time Machines," *Physical Review D*, August 15, 1990, pp. 1057–1065.

theory a Cauchy initial-value problem is said to be well-defined if the initial conditions determine a unique solution, and if a continuous variation in the initial conditions gives a continuous variation in the solution. In that part of spacetime where closed timelike loops are not allowed, backward causation does not occur (by definition) and the laws of physics (all expressed as differential equations) satisfy the Cauchy condition. Outside of this *chronal* region, that is, beyond the Cauchy horizon where physics is *dischronal*, however, the possibility of backward causation raises the possibility of violating the Cauchy condition, and in such a case the Cauchy horizon is also sometimes called the *chronology horizon*.[47]

The instability of the Cauchy horizon is caused by radiation that propagates in closed timelike loops that thread through the wormhole on 'straight lines.' This radiation, as shown a half-century ago,[48] builds-up unbounded energy density levels at the horizon, and thus destroys the horizon. Thorne and his colleagues argued that the defocusing effect of their wormhole time machine's repulsive gravity would be sufficient to counter a disruptive energy build-up on the horizon (note 27). Subsequent analyses have examined other possible ways to avoid unbounded energy density on the Cauchy horizon. For example, in one paper[49] it was imagined that a wormhole time machine has had a circular motion induced for mouth B; that is, mouth B orbits around mouth A. The result is that the Cauchy horizon now does seem to be stable, because now there are no fixed, straight-line timelike loops threading the wormhole from A to B to A to B to . . .. That is, B is a 'moving target' and there is no point on the Cauchy horizon where the energy density becomes unbounded.

Yet another approach for achieving the disruption of destructive, circulating energy loops through a wormhole is by placing a spherical mirror between the two mouths of the wormhole. Proposed by the Chinese physicist Li-Xin Li, a Li mirror would divert all closed null geodesics (represent circulating radiation) that potentially thread through the wormhole.[50] Such potentially fatal geodesics would, instead, be scattered back into space, whereas a purposeful traveler could navigate *around* the mirror and thus use the wormhole as a time machine.

---

[47]A classic work on the mathematics of Cauchy problems is J. Hadamard, *Lectures on Cauchy's Problem in Linear Partial Differential Equations*, Dover 1952. There is a curious bit of irony in this. In a section of his book, Hadamard uses spacetime to illustrate one possible four-dimensional space and, in passing, he casually writes "This conception was beautifully illustrated a good many years ago by the novelist Wells in his *Time Machine*." Hadamard wrote his book in 1923, and he would almost certainly have been astonished to have been informed that less than seventy years later his work would play a central role in the non-fictional theory of time machines.

[48]C. W. Misner and A. H. Taub, "A Singularity-Free Empty Universe," *Soviet Physics JETP*, January 1969, pp. 122–133.

[49]I. D. Novikov, "An Analysis of the Operation of a Time Machine," *Soviet Physics JETP*, March 1989, pp. 439–443.

[50]L.-X. Li, "New Light on Time Machines: Against the Chronology Protection Conjecture," *Physical Review D*, November 1994, pp. R6037–R6040.

Cauchy horizon instability from the back reaction is central to Hawking's Chronology Protection Conjecture, discussed in Chap. 1. His analysis (see note 54 in Chap. 1) led him to conclude that a physical entity—the stress-energy tensor—becomes unphysical on the Cauchy horizon. That is, because of time-traveling quantum field fluctuations of the vacuum, that tensor diverges to infinity at the horizon. This results in a failure of that horizon to form in the first place or, if it does form, in the creation of a singularity that 'seals-off' the horizon to any would-be time travelers attempting to gain access to the closed timelike loops beyond the horizon. Others, however, argued that Hawking was mistaken in claiming that the divergence of the stress-energy tensor on the Cauchy horizon will always forbid time travel.

In a study, for example, of a complex-valued spacetime metric (and such a metric *is* allowed in the so-called 'sum over all possible geometries, path integral' approach to the quantum theory of gravity), that has causal and non-causal spacetime regions separated not by a Cauchy horizon but rather by a region of complex geometry, the stress-energy tensor is always physical and diverges nowhere. The complex geometry region plays the same role as the Cauchy horizon, because such a region would, *classically*, mean that the two regions cannot be reached from one another, but via quantum tunneling an observer *could* travel between the two regions.[51] In fact, studies of stress-energy divergence actually have a long history. For example, the effect of an unphysical (infinite) gravitational and/or electromagnetic energy flux had been analyzed *years before* the wormhole time machine studies began.[52] The authors studied the case of a potential traveler to "new worlds" who tries to cross the Cauchy horizon of an electrically charged, non-rotating black hole. An even earlier computer study had already concluded that, for such a traveler, the attempt to cross the horizon "looks liable to prove a dangerous undertaking."[53]

It isn't at all clear, in fact, if a *theoretical* divergence of the stress-energy *is* the signature of a failure of physics. One doesn't need anything as bizarre as a time machine for the stress-energy to diverge *on paper*. It was shown nearly 40 years ago, for example, that such a theoretical divergence occurs for the electromagnetic field near a perfectly conducting boundary.[54] But it is simply the unphysical nature of a "*perfectly* conducting" boundary condition that causes the divergence, not the fact that the field actually exists near a conducting boundary. Similarly, other real-life considerations (quantum gravity) *may* keep the stress-energy physical everywhere in a time machine spacetime.

[51]L.-X. Li, J.-M. Xu, and L. Liu, "Complex Geometry, Quantum Tunneling, and Time Machines," *Physical Review D*, November 15, 1993, pp. 4735–4737.

[52]S. Chandrasekhar and J. B. Hartle, "On Crossing the Cauchy Horizon of a Reissner-Nordström Black-Hole," *Proceedings of the Royal Society of London A*, December 8, 1982, pp. 301–315.

[53]M. Simpson and R. Penrose, "Internal Instability in a Reissner-Nordström Black Hole," *International Journal of Theoretical Physics*, April 1973, pp. 183–197.

[54]D. Deutsch and P. Candelas, "Boundary Effects in Quantum Field Theory," *Physical Review D*, December 15, 1979, pp. 3063–3080.

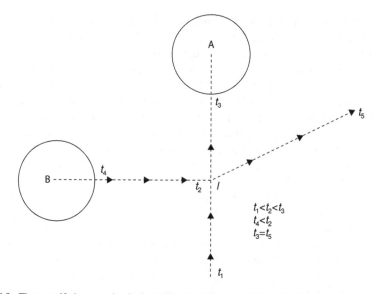

Fig. 6.9 The grandfather paradox in the billiard ball world. A billiard ball approaches mouth A of a time machine wormhole, dead-on center and, just before entering A, it passes without incident through point *I*. The ball then enters A and so exits mouth B in the past, *just in time* to pass through point *I* and *hit its younger self*. This impact knocks the younger ball *away from* A, so we have the familiar paradox of changing the past. That is, the impact did not occur when the ball 'originally' passed through *I* on its way to A and, of course, we also wonder how the ball manages to hit itself after leaving B if it then *doesn't* enter A?

Assuming a wormhole time machine has (somehow) become available, with its Cauchy horizon intact, how do the 'paradoxes' of time travel come into play? In an attempt to study the grandfather paradox, in particular, Thorne and his colleagues studied self-interacting billiard balls traveling backward in time through a wormhole.[55] They used billiard balls—see Figs. 6.9 and 6.10—rather than human time travelers for the same reason Wheeler and Feynman used a pellet and shutter mechanism in their study of advanced electromagnetic waves—to avoid any metaphysical questions about human free will. The central issue for them was the determination of the multiplicity of trajectories for a single, self-interacting time traveling ball, where the Cauchy condition for a well-defined trajectory in spacetime is *unique* self-consistency.

That is, for the trajectory to be well-defined in the Cauchy sense, it was expected there would be exactly *one* consistent trajectory for a self-interacting ball. A multiplicity of zero, of course, would be the physics declaring backward time travel through the wormhole to be nonsense—and that was thought to be a distinct

---

[55]F. Echeverria, G. Klinkhammer, and K. S. Thorne, "Billiard Balls in Wormhole Spacetimes with Closed Timelike Curves: I. Classical Theory," *Physical Review D*, August 15, 1991, pp. 1077–1099. The authors credited the physicist/science fiction writer Robert Forward (who used the same ideas in his 1992 novel *Timemaster*) for motivating their research.

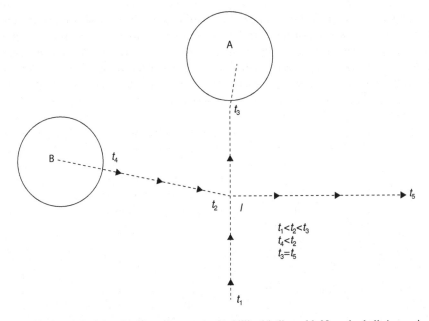

Fig. 6.10 The principle of self-consistency in the billiard ball world. Now the ball, in passing through point *I* on its dead-on center path toward mouth A of the wormhole time machine, is suddenly hit a grazing blow by another ball that has just shot out of mouth B (and into the past) at an angle. The impact knocks the first ball slightly off its original trajectory, and it enters mouth A slightly off-center. Thus, the ball emerges from B into the past slightly off-center and just in time to glace off itself at *I*—which explains *why* it emerged from B slightly off-center!

possibility. The actual results were, however, surprisingly different. It was found, under very general assumptions about the wormhole parameters, that (1) there are no trajectories with zero multiplicity and (2) the multiplicity is *not* one but rather is *always* infinity! Thus, the billiard ball form of the grandfather paradox *was* found to be *not* well-defined, but not for the expected reason that there was no self-consistent solution. Instead, it was because there are *too many* solutions.

This astonishing, completely unexpected result seems to be just what is needed to support the viability of time machines, as it appears to allow a definition of *well-defined* in the Cauchy sense and still permit an answer to the puzzle of free will. The initial conditions of a time traveling ball give rise to an infinity of self-consistent trajectories, each occurring in the same way that a random variable takes on different values with each new performance of the experiment that the random variable is defined on. And yet, there are still unique probability density functions for all sets of measurements that one might make anywhere along these trajectories. Thus, the Cauchy problem is *stochastically* well-defined; at the start of any trajectory, we do not know in detail what will happen except that whatever does happen will be self-consistent. In this probabilistic sense, then, wormhole time travel to the past and the retention of free will both make sense. The Russian physicist Igor Novikov and his colleagues continued the study of time traveling billiard balls,[56] demonstrating that one can *deduce* self-consistency from the long-accepted principle of least action (that is, self-consistency is not an additional assumption to existing physics).[57]

The one-wormhole, two-mouth time machine was actually not the first kind of wormhole time machine described in the physics literature. In their 1988 paper (note 21), Morris and Thorne initially described a time machine constructed from *two* wormholes, but they added a note-in-proof at the end that they had just discovered how to build a time machine using one wormhole (the machine we have been discussing). This reduction in the required number of wormholes was thought to be a technical advance, of course, and so the two-wormhole time machine was put aside.

But not for long. Soon thereafter the concerns about Cauchy horizon stability began to surface, a concern that one-wormhole time machines might destroy themselves just at the instant their mouths were about to be threaded by closed timelike curves. As noted earlier, the negative mass wormhole throat has a defocusing effect on electromagnetic radiation (and so the initial concern, that time traveling photons might be fatal, faded)—but then it was found that vacuum fluctuations of quantum fields are *not* so defocused.[58] That failure to defocus time-traveling vacuum polarizations (as quantum field fluctuations are called) was shown to result in an unphysical divergence of the stress-energy on the Cauchy horizon of a one-wormhole time machine.[59] This sounds bad, but the hope was that the divergence wouldn't actually be fatal: it appeared to be sufficiently sluggish that it was suggested reaching an actual *infinity* of the stress-energy would be precluded by the eventual intercession of quantum gravity. That is, the stress-energy might *try* to become unbounded as spacetime approached the formation of a time machine

---

[56]A. Lossev and I. D. Novikov, "The Jinn of the Time Machine: Nontrivial Self-Consistent Solutions," *Classical and Quantum Gravity*, October 1992, pp. 2309–2321; E. V. Mikheeva and I. D. Novikov, "Inelastic Billiard Ball in Spacetime with a Time Machine," *Physical Review D*, February 15, 1993, pp. 1432–1436; M. B. Mensky and I. D. Novikov, "Three-Dimensional Billiards with a Time Machine," *International Journal of Modern Physics D*, April 1996, pp. 179–192.

[57]A. Carlini, *et al.*, "Time Machines: The Principle of Self-Consistency as a Consequence of the Principle of Minimal Action," October 1995, pp. 557–580, and "Time Machines and the Principle of Self-Consistency as a Consequence of Stationary Action (II): The Cauchy Problem for a Self-Interacting Particle," October 1996, pp. 445–479, both in *International Journal of Modern Physics D*.

[58]V. P. Frolov, "Vacuum Polarization in a Locally Static Multiply Connected Spacetime and a Time-Machine Problem," *Physical Review D*, June 15, 1991, pp. 3878–3894.

[59]S.-W. Kim and K. S. Thorne, "Do Vacuum Fluctuations Prevent the Creation of Closed Timelike Curves?" *Physical Review D*, June 15, 1991, pp. 3929–3947. See also L.-X. Li, "Must Time Machines Be Unstable Against Vacuum Fluctuations?" *Classical and Quantum Gravity*, September 1996, pp. 2563–2568.

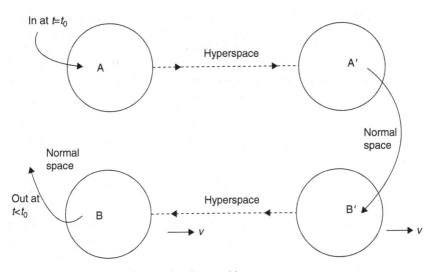

Fig. 6.11 A two-wormhole, Roman ring time machine

but, before it becomes so large as to destroy the time machine, quantum gravity would cut-off the divergence 'in time' (so to speak!) to save the machine.

Hawking disagreed (in his famous Chronology Protection Conjecture—see note 54 in Chap. 1), arguing that Kim and Thorne had made a crucial error in their calculations. According to Hawking, the divergence of the stress-energy may indeed be cut off by quantum gravity, but not before the development of spacetime disturbances representing perhaps a hundred million times the energy levels associated with ordinary chemical binding energies. These would be sufficiently big disturbances to raise serious doubts about the physical survival of a one-wormhole time machine, even in the absence of a true stress-energy infinity. Hence the resurrection of the two-wormhole time machine geometry. Perhaps *it* could avoid the destructive effect of time-traveling vacuum fluctuations.

If a spacetime contains multiple wormholes, then it is called a *Roman spacetime* after the physicist Thomas Roman (at Central Connecticut State University), who was the originator of such spacetimes. Each of these wormholes, individually, is *not* a time machine. Together, however, they form a time machine geometry called a *Roman configuration* (or a *Roman ring*).[60] Here's how.

In Fig. 6.11 two pairs of wormhole mouths are labeled A, A' and B, B'. We imagine that the A, A' wormhole is stationary and that its two mouths are very far apart in normal space—so far apart, in fact, that if a traveler enters A and almost instantly (because the wormhole handle is very short in hyperspace) emerges from A', it will appear to an observer at rest with respect to the wormhole that the traveler has moved faster than light. That is, entering A and exiting A' are events with

---

[60]M. Visser, "Traversable Wormholes: The Roman Ring," *Physical Review D*, April 15, 1997, pp. 5212–5214.

spacelike separation. Now, imagine also that the wormhole with mouths B, B' is moving past the first wormhole at speed v. To an observer in this second, *moving* frame of reference, the spacelike separation of entering A and exiting A' can result in the two events being temporally reversed if v is sufficiently large (but still less than the speed of light). Therefore, upon emerging from A' the traveler crosses normal space to the moving wormhole mouth B', enters the wormhole, and then almost instantly emerges from mouth B, and finally travels again through normal space to mouth A. If the traveler can make the two trips in normal space in less time than the backward time shift achieved by the temporal reversal of entering A and exiting A', then we have a time machine

Two simultaneous analyses of the Roman ring time machine each concluded that, for suitable choices of sizes (the radii of the wormhole mouths, the wormhole lengths in normal space, the lateral offset of the two wormholes, and the relative speed of the wormholes), the stress-energy divergence *can* be limited by quantum gravity to an arbitrarily weak level. That is, the two-wormhole time machine is not necessarily destroyed by an unbounded stress-energy on the Cauchy horizon.[61] But not all was now put right.

Visser, in particular, had some strong reservations about the Roman ring. Although he granted that a quantum gravity cut-off the stress-energy divergence *would* probably occur in the Roman ring, he called the required special sizing conditions "bizarre," and asserted that the resulting time machine would be quite useless for a human traveler in any case. For example, he calculated that only if the mouths of the wormholes are separated in normal space by the radius of the universe (!), and only if the wormhole mouths have radii on the order of that of an atomic nucleus, would the cut-off be sufficient to allow the putative time machine to avoid destruction. When Visser reduced the wormholes from universe size to 'merely' that of the distance between the Sun and the Earth, he concluded that it would require energy at the level of the Superconducting Supercollider accelerator to blast an information-bearing message through the narrow wormholes. And even then the 'short' wormholes would provide a maximum penetration into the past of just eight minutes. As Visser put it, "This does not seem to be a workable recipe for studying tomorrow's *Wall Street Journal*.

Lyutikov, on the other hand, took a far less negative stance. He concluded that although Visser's calculations "make it very inconvenient for time travel [by humans]," nevertheless "the [principal] question of the possibility of transmitting information back in time through traversable wormholes would still remain."

The wormholes we have been discussing so far are *static* in time, but another approach is to allow them to be dynamic structures in spacetime. That is, to allow one or more of their parameters to vary with time (perhaps, for example, the throat diameter could collapse). Then, according to one analysis, it is possible to have a

---

[61]M. Visser, "Van Vleck Determinants: Traversable Wormhole Spacetimes," April 15,1994, pp. 3963–3980, and M. Lyutikov, "Vacuum Polarization at the Chronology Horizon of the Roman Spacetime," April 15, 1994, pp. 4041–4048, both in *Physical Review D*.

traversable wormhole *made of normal matter* and, even though it is collapsing, it would take so long to do so that "a space adventurer will have enough time to pass through the throat of the wormhole from one asymptotically flat region [of spacetime outside the entry mouth of the wormhole] to the other [spacetime region outside the exit mouth of the wormhole] before the radius of the throat shrinks to . . . where the event horizon is developed."[62] Such a dynamic wormhole, it was claimed, satisfies both the weak and the dominant energy[63] conditions, but not the strong energy condition. Thus, gravity would still be repulsive in the throat, but this condition (which would seem to require *exotic* matter) was brushed aside because such a condition is thought to have actually occurred, on a massive scale, during the inflationary stage of the Big Bang[64] (although how that would help in the construction of a wormhole in the future is a bit murky).

Is it reasonable to think 'useable' wormholes, static or otherwise, can be acquired for the purpose of creating a time machine? At one time, Hawking was sure the answer is no, once writing "The philosophy of this paper is . . . to look for vacuum polarization [the divergence of the stress-energy on the Cauchy horizon] to enforce the chronology protection conjecture."[65] It became increasingly apparent, however, that matters would be a great deal more involved and, as Hawking himself came to admit, "the fact that the energy-momentum tensor fails to diverge [in certain special cases of time machine spacetimes] shows that the back reaction does not enforce chronology protection."[66].

I think the best (and most honest) way to respond to the 'reasonable' question that opened the previous paragraph is with words from 20 years ago, by the Russian astrophysicist Serguei Krasnikov, words still valid today: "It may well be that the vacuum fluctuations do make the time machine unstable, but nothing at present

---

[62]A. Wang and P. S. Letelier, "Dynamical Wormholes and Energy Conditions," *Progress of Theoretical Physics*, July 1995, pp. 137–142. See also L. A. Anchordoqui, *et al.*, "Evolving Wormhole Geometries," *Physical Review D*, January 15, 1998, pp. 829–833.

[63]The dominant energy condition is the weak energy condition *plus* the requirement that any observed energy flux is *never* superluminal.

[64]During inflation the universe is thought to have expanded at a rate far beyond human comprehension. It has been estimated that during the first 10^{-35} second of the Big Bang the universe doubled in each spatial dimension by a factor of two each 10^{-37} second; that is, there were about 100 such doublings. Thus, there was an increase by a factor of $2^{100} \approx 10^{30}$ in each linear dimension of the universe, and the volume increased by the cube of that enormous factor. See Alan Guth, *The Inflationary Universe*, Addison-Wesley 1997.

[65]S. Hawking, "Quantum Coherence and Closed Timelike Curves," *Physical Review D*, November 15, 1995, pp. 5681–5686.

[66]M. J. Cassidy and S. W. Hawking, "Models for Chronology Selection," *Physical Review D*, February 15, 1998, pp. 2372–2380. See also L.-X. Li and J. R. Gott, "Self-Consistent Vacuum for Misner Space and the Chronology Protection Conjecture," *Physical Review Letters*, April 6, 1998, pp. 2980–2983.

suggests this. All we have are a few simple examples. In some of them the energy density diverges at the horizon and in some does not. So, the time machine perhaps is stable and perhaps is not."[67]

The daunting level of technology required to *build* a wormhole (with or without exotic matter) doesn't mean we can't search for existing wormholes. Perhaps, for example, vast wormhole networks were formed naturally at Big Bang time, as described in Gregory Benford's 1997 novel *Foundation's Fear*, where wormholes are "leftovers from the Great Emergence [the Big Bang]." Or perhaps "advanced civilizations" long ago constructed a vast, pan-galactic 'subway system' of wormholes like the one described in Carl Sagan's 1985 novel C*ontact* (and dramatically illustrated in the 1997 film).

Of course, any such wormhole, if found (via its double-spike light signature, for example), could be a very long way from Earth. It might even be in another galaxy. So, even if we found a wormhole, what could we *do* with it? Surprisingly, maybe a lot. The Russian physicists Igor Novikov and Andrei Lossev (note 56) suggested that a wormhole might be very useful *even if its location is completely unknown, even if we haven't yet even discovered it!* The only assumption they made was that the wormhole has existed for a "sufficiently long time" (and precisely what that means will be explained in just a bit). With that assumption, they showed how to make an information-creating time loop. Here's how they did that.

They began their analysis by assuming that people have no knowledge of how to build spacecraft that can make the interstellar voyage to the distant wormhole, even if they knew in which direction to go to reach the mouth that leads backward in time (mouth B). Instead, they build an automatic spacecraft construction plant that can follow any detailed sequence of instructions provided to it, and then stockpile it with a supply of raw materials (energy, steel, plastic, computers, and so on). When the spacecraft construction is done (*how* that is done is explained in the next paragraph), the last step before launching the spacecraft toward mouth B *will be* to load the on-board computer with the following three pieces of information:

1. The detailed sequence of instructions to be followed in the construction of the spacecraft;
2. The direction from Earth to mouth B;
3. The direction from mouth A (the wormhole exit mouth in the past) back to Earth.

To summarize, people build the automatic plant, load it up with raw materials, *and then withdraw*. This last step is crucial, because it eliminates human free will from further consideration, that is, it removes any temptation to create a bilking paradox. So, what happens next?

Lossev and Novikov suggest that what happens next is that a *very* old spacecraft suddenly appears in the sky and lands next to the automatic construction plant. In its on-board computer are items a, b, and c. Using item a, the automatic plant makes a

---

[67]S. V. Krasnikov, "Quantum Stability of the Time Machine," *Physical Review D*, December 15, 1996, pp. 7322–7327.

new spacecraft, then loads the new on-board computer with items a, b, and c from the *very* old spacecraft's on-board computer, and then the new spacecraft is launched toward mouth B (using the information of item b). The *very* old spacecraft is given an honored place in a museum.

The new spacecraft arrives at the distant mouth B in the far future, by which time it is, of course, an old spacecraft (but not yet a *very* old spacecraft). It then plunges into mouth B and almost immediately emerges from mouth A, in the past. Indeed, it repeats this process as many times as required until it is in the far distant past, at a time even before it left Earth. (It might seem that to do this, the spacecraft's computer memory needs a fourth piece of information, the direction from mouth A back to mouth B, but in fact items b and c are sufficient for the old spacecraft to find its way from A to B.) It is now clear how long the wormhole must have been in existence. The old spacecraft repeatedly uses the wormhole time machine until it is so far in the past that it can cruise back to Earth at normal speed (it knows the way back because of item c) and arrive as a *very* old spacecraft, just in time to be placed in the museum!

As Lossev and Novikov pointed out, this remarkable, looped sequence of events has increased knowledge from what it was at the time just before the automatic construction plant was built. People now know both how to build an interstellar spacecraft, and the locations of both mouths of the wormhole. They also now possess a *very* old, used spacecraft. It is curious to note that although the information in the *very* old spacecraft's computer memory has traveled on a closed time loop, the *very* old spacecraft itself has not. This is because the spacecraft left Earth when new, but arrived back (before it left) as *very* old, whereupon it promptly entered a museum. There is therefore no question about the origin of the *very* old spacecraft, but where did the information of items a, b, and c come from? Lossev and Novikov say it came from the energy gained by the spacecraft as it interacted (will interact?) with the rest of the universe while on its journey.

Nobody said time travel isn't weird!

6.4 Gott's Cosmic String Time Machine

"It's an amazingly simple solution. It doesn't take much physics to understand it."
—MIT astrophysicist Alan Guth, on Gott's discovery of the cosmic string time machine[68]

"Louise, working out the spacetime geometry of a cosmic string is a hard problem in general relativity. But, given that geometry, all the rest of it is no more than Pythagoras' theorem ..."
—a character in Stephen Baxter's 1994 novel *Ring*, agreeing with Guth.

---

[68]Quoted from J. Travis, "Could a Pair of Cosmic Strings Open a Route Into the Past?" *Science*, April 10, 1992, pp. 179–180.

A new way to gain access to closed timelike curves, without the involvement of the exotic matter needed by negative-mass wormholes, was described in 1991 by the Princeton physicist J. Richard Gott.[69] Gott gave exact solutions to Einstein's gravitational field equations for what are called *cosmic strings*, solutions that (1) unlike wormholes, do not violate any of the energy conditions, (2) unlike black holes have no crushing singularities or event horizons, and (3) are not topologically multiply connected.

Cosmic strings are fantastically thin (10^{-28} cm in radius) filaments of pure energy that are thought to stretch the width of the universe and to have an enormous linear mass-energy density of 10^{28} g/cm. To generate closed timelike paths in spacetime, Gott required that either two fast-moving (which means moving at practically the speed of light) parallel cosmic strings pass each other on a near-collision course, or that there be a closed-loop string that collapses in a slightly non-planar manner so that the opposite, nearly straight sides 'just miss.' The gravitational interaction of the passing strings can 'warp' spacetime enough to produce closed timelike curves.

A hint at the possibility of violating causality with strings had appeared before Gott's work, but those authors didn't take the time travel implications seriously. As they wrote, "We argue ... that any realistic model [for a spinning string with angular momentum[70]] ... will not have closed timelike curves."[71] Gott, however, showed that as two strings pass each other, closed timelike loops *do* encircle the strings.

Gott, who appears to be far less rigid in his view of time travel than are many of his fellow physicists, held out an escape to those who pale at the very thought of time travel to the past. Perhaps, he suggested (following in the footsteps of an analysis by Hawking[72]), as the strings (or string-loop sides) pass, a black hole will form with an event horizon that will seal-off the closed timelike curves from any would-be time traveler. Or perhaps, he further suggested, the more realistic case of non-singular strings (that is, strings with non-zero-filament radii) and possessing

---

[69]J. R. Gott, "Closed Timelike Curves Produced by Pairs of Moving Cosmic Strings: Exact Solutions," *Physical Review Letters*, March 4, 1991, pp. 1126–1129.

[70]The two strings in a Gott-pair are *not* necessarily spinning, and no such assumption was made by Gott. They don't even have to be parallel. If the strings have no spin, then it takes two strings to make a time machine. If spin *is* allowed, however, then just a single string will suffice for time travel: see S. Deser and R. Jackiw, "Time Travel?" *Comments on Nuclear and Particle Physics*, September 1992, pp. 337–354.

[71]D. Harari and A. P. Polychronakas, "Gravitational Time Delay Due to a Spinning String," *Physical Review D*, November 15, 1988, pp. 3320–3322.

[72]S. W. Hawking, "Gravitational Radiation from Collapsing Cosmic String Loops," *Physics Letters B*, August 23, 1990, pp. 36–38.

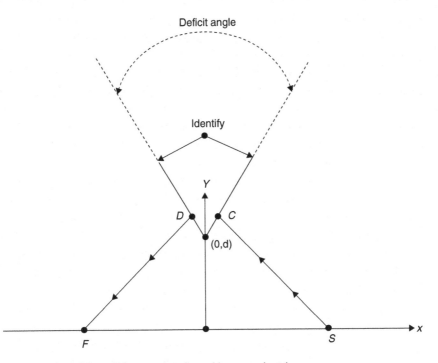

Fig. 6.12   The deficit angle in spacetime formed by a cosmic string

spin would banish the terrifying closed timelike curves. Subsequent analyses along those lines, however, continued to find the time travel implications intact.[73]

Here's how the cosmic string time machine works. In an earlier work,[74] published in 1985, Gott discovered that a cosmic string warps spacetime in a highly characteristic way, as shown in Fig. 6.12. A stationary cosmic string is imagined as perpendicular to the *xy*-plane (the plane of the page) and passing through the page at the point $(0, d)$ on the *y*-axis. The warp produced by the string is as though a wedge of angle 2α (this angle is called the *deficit angle*) were cut out of spacetime and the

[73]See, for example, B. Jensen, "Notes on Spinning Strings," *Classical and Quantum Gravity*, January 1992, pp. L7–L12, H. H. Soleng, "A Spinning String," *General Relativity and Gravitation*, January 1992, pp. 111–117, (the next two are in the *Physical Review D*), B. Jensen and H. H. Soleng, "General-Relativistic Model of a Spinning Cosmic String," May 15, 1992, pp. 3528–3533, and M. Novello and M. C. M. da Silva, "Cosmic Spinning String and Causal Protecting Capsules," January 15, 1994, pp. 825–830.

[74]J. R. Gott, "Gravitational Lensing Effects of Vacuum Strings: Exact Solutions," *The Astrophysical Journal*, January 15, 1985, pp. 422–427. Gott's discovery was independently reported in W. A. Hiscock, "Exact Gravitational Field of a String," *Physical Review D*, June 15, 1985, pp. 3288–3290.

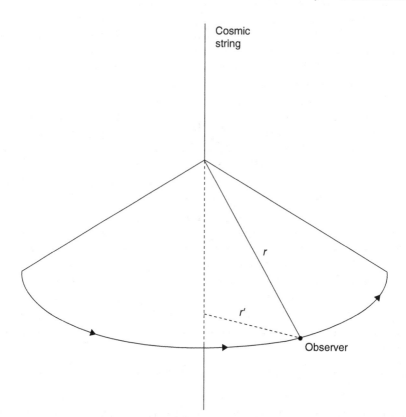

Fig. 6.13 The warped, conical spacetime around a cosmic string, An observer in this spacetime thinks she is distance r from the string, but a 'meta-observer' sees that she is actually distance r' from the string. Thus, if the observer follows a complete circular path around the string, she will travel a distance of $2\pi r' < 2\pi r$. The observer in the spacetime will interpret this result by saying that the angle 2π is really 2π minus 'a deficit'

edges of the cut were then 'glued' together; for example, points C and D are identified as identical. The reason for the term *deficit angle* is that at radius r from the string, a circular path around the string has the reduced length $(2\pi - 2\alpha)r$, and not the usual $2\pi r$ (spacetime around the string, while *locally* flat, is actually 'conical,' as illustrated in Fig. 6.13).

The deficit angle is equal to $8\pi\mu$ radians (in a system of units where G, Newton's gravitational constant, is 1) if the linear mass-energy density μ is expressed in units of Planck masses per Planck length. For example, $\mu = 1$ corresponds to 1.35×10^{28} g/cm (think of something on the order of the mass of the Earth per inch of the string). For 'more typical' values—say a 'mere' $\mu = 10^{22}$ g/cm, $2\alpha = 0.001°$. While Gott's paper had appeared 5 months before Hiscock's (see note 74), it is evident that Hiscock's work was done before he became aware of Gott's. Both papers treat exact derivations of the deficit angle but, in fact, the correct expression had actually been published 4 years earlier (but from a linearized form of the gravitational field equations and so the result was not as 'conclusive' as are Gott's and Hiscock's)[75]

Now, consider the two points S and F on the x-axis at $(x_0, 0)$ and $(-x_0, 0)$ in Fig. 6.12. Suppose we want to send photons from S to F. In normal, 'unwarped' spacetime, the direct path from S to F through the origin has length $2x_0$. There is also another possible path, however, S to C/D to F, that loops out and around the cosmic string. Indeed, this second path is simply the path a gravitationally lensed photon would take (an observer at F would see *two* images of S) and this—*not* time travel—is the issue that originally attracted Gott's attention to cosmic strings.[76] If the deficit angle were zero, then this alternative path would always be longer than $2x_0$, for any value of x_0. For the case of $2\alpha > 0$, however, if x_0 is large enough $(x_0 \gg d)$, then it is possible for 'around the string and over the missing spacetime wedge' path to be shorter than the direct path.

The indirect path provides a way for a *sub*luminal trip (say, by rocket) from S to F to beat a photon traveling on the direct path. That is, the two events of the 'rocket leaving S' and the 'rocket arriving at F' are spacelike separated. Thus, it is possible to find a moving frame of reference in which these two events are *reversed in temporal order*. In that frame of reference, the cosmic string (which is stationary in the reference frame of S and in that of F) will move—at speed v, say—in the $+x$ direction, and in that frame of reference the rocket will arrive at F *before* it leaves S.[77]

Then to complete the construction of a closed timelike path, simply repeat the process as shown in Fig. 6.14. That is, after the rocket arrives at F, have it turn around and fly back to S out-and-around and through the deficit angle spacetime warp due to a *second* cosmic string on the negative y-axis and perpendicular to the xy-plane. This second string is moving at speed $-v$ (that is, opposite to the first string), so the rocket will arrive at S before it leaves F. But that means it arrives at S before it leaves S; that is, the rocket has traveled into the past. In other words, the rocket has traveled into the past. This entire process is precisely the same idea behind the two-wormhole Roman-ring time machine discussion from the previous section.

Now, instead of having two oppositely moving reference frames, one in which the top, stationary string at $(0, d)$ appears to be moving at $+v$ and another frame in

---

[75]A. Vilenkin, "Gravitational Field of Vacuum Domain Walls and Strings," *Physical Review D*, February 15, 1981, pp. 852–857.

[76]Just as discussed earlier in the context of wormholes, gravitational lensing may offer a way to detect cosmic strings. See, for example, the two papers by D. L. Ossipov, "Diffraction of Light by a Cosmic String," November 1995, pp. 765–771, and "Contribution of Strings to the Observed Variability of Extragalactic Sources of Radiation," September 1996, pp. 419–425, both in *JETP Letters*.

[77]For this to happen, however, v must be very close to the speed of light. Gott (see note 69) showed that with $v = \tanh(\theta)$, the condition for the rocket to arrive back at S before it leaves S is $\cosh(\theta) \sin(\alpha) > 1$, where α is one-half the deficit angle. For $\mu = 10^{22}$ g/cm, this gives $v = 0.99999999995$ (times the speed of light).

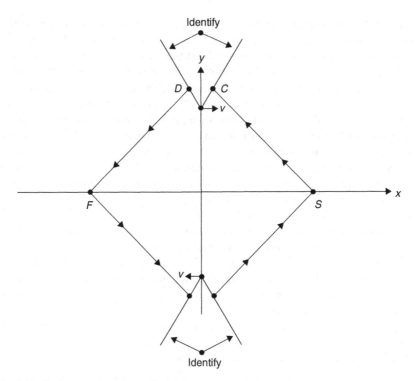

Fig. 6.14 Gott's spacetime, formed by joining two oppositely moving versions of the spacetime in Fig. 6.13

which the bottom, stationary string appears to be moving at $-v$, we can imagine an observer in the *stationary center-of mass* frame watching two strings that are moving at $+v$ and $-v$. This leaves the situation unchanged, so in the center-of-mass frame the rocket does travel into the past, arriving back at S before it leaves S. That is, the rocket has traveled all the way around a closed timelike world line. Note, too, that the geometric condition mentioned earlier of $x_0 \gg d$ immediately implies that for x_0 *not* sufficiently large, there isn't a closed timelike path from S to F and then back to S; that is, there is a region in Gott's spacetime where such time travel journeys *cannot occur*.

Another physicist pursued Gott's analysis in an attempt to see whether these 'time travel paths' are *created* as the strings approach each other or, instead, if the paths exist at other times as well.[78] This important question gets to the idea of whether such time machine paths can be *intentionally created* by humans via a dynamical process (a *strong* time machine), or whether all such paths have existed since the formation of the universe (a *weak* time machine). This issue involves Hawking's

[78]A. Ori, "Rapidly Moving Cosmic Strings and Chronology Protection," *Physical Review D*, October 15, 1991, pp. 2214–2215.

chronology protection conjecture, which you'll recall asserts that the laws of physics will always (somehow) prevent the creation of a time machine. One reason Hawking repeatedly gave for believing the Conjecture is the apparent absence of time travelers from the future among us now (in their past). The only possible exception allowed by the Conjecture is the creation of closed timelike loops at the moment the universe was created (at that moment there *was no past* for time travelers to invade!). Ori proved that the closed timelike loops around Gott's cosmic strings are *always* present: that is, a time machine is *not created* by the near collision of the strings. Thus, Hawking's Conjecture is nor refuted by Gott's spacetime.

One very curious issue is *where* the closed time loops are before the strings pass one another. As mentioned briefly by Ori, and further discussed by others,[79] the time loops are initially at spatial infinity. To this concern, Gott and a colleague made the following very strong reply:

> "[A problem] Deser *et al.* present with respect to the Gott spacetime is that it contains CTC's at spacelike infinity; this is supposed to be an unacceptable boundary condition. We wonder, however, how they know so much about boundary conditions at spacelike infinity. In our own Universe we do not know what spacelike infinity looks like (if it exists) since we have not seen it yet. We certainly have no way of knowing whether or not there are CTC's there. The working physicist is, of course, free to impose simple and convenient boundary conditions (e.g., asymptotic flatness) on a system in order to isolate and understand the processes occurring within it. *But boundary conditions are tools of physicists, and they should not be confused with laws of physics* [my emphasis]. There may be such laws of nature that restrict the possible structure of spacelike infinity, and even prohibit CTC's there, but in the absence of evidence such laws should not be postulated *ad hoc*."[80]

Still, as Ori had observed the year before, having time loops collapsing inward from infinity toward humans who might, fortuitously, wish to use them at *just the instant they so conveniently arrive*, is "a situation which has little to do with the creation of a time machine by a *human being* [my emphasis]."[81]

The most damning objection to Gott's cosmic string time machine came, ironically, from Gott himself. The two-string spacetime of Fig. 6.14 might actually, he and a colleague wrote (see Li and Gott, note 66) be destabilized by the non-zero mass of any would-be time traveler. They suggest that this concern could perhaps be 'solved' by assuming that the time traveler and her spaceship have a spherically symmetric mass distribution surrounded by a negative-mass shell to give zero net mass (and thus a zero net gravitational field that would *not* destroy the closed timelike curves of the strings). But that, they further observed, would negate the crucial advantage—no exotic matter and so no violation of the weak energy condition—that a cosmic string time machine enjoys over a wormhole time machine.

---

[79]See, for example, S. Deser, *et al.*, "Physical Cosmic Strings Do Not Generate Closed Timelike Curves," *Physical Review Letters*, January 20, 1992, pp. 267–269.

[80]M. P. Headrick and J. R. Gott, "(2+1)-Dimensional Spacetimes Containing Closed Timelike Curves," *Physical Review D*, December 15, 1994, pp. 7244–7259.

[81]A. Ori, "Must Time-Machine Construction Violate the Weak Energy Condition?" *Physical Review Letters*, October 18, 1993, pp. 2517–2520.

Does a trip around a pair of cosmic strings present other problems *aside* from the sheer fantastic physics of the strings themselves? Well, I think "turning the rocket around at *F* and flying back to *S*" is a lot easier to write than it would be to actually do! The entire trip has to occur while the strings (moving at essentially light speed) are in a position *to be* flown around. As one character says to another in Stephen Baxter's novel *Ring*, "Louise, the strings are traveling just under the speed of light—within three decimal places of it, actually. [Our ship is] traveling at a little over half-light speed. The turning curves, and the accelerations, are incredible . . ." I think so! And I do wonder who—*or what*(!)—is actually controlling a *maneuvering* rocket traveling faster than $\frac{1}{2}c$?

6.5 Cutting and Warping Spacetime

"The warp drive spacetime of Alcubierre is impossible to set up . . . one needs to transcend the speed of light in order to construct the warp drive in the first place . . . put roughly, you need one to make one!"[82]

In this chapter we've talked about the physics of three specific time machine 'implementations': the rotating cylinder, the wormhole, and the cosmic string. One can also discuss the 'construction' of a time machine in a more geometrical (yet still physical) way by performing what is called *spacetime surgery* to arrive at what is often referred to as a *Deutsch-Politzer spacetime* (after the two physicists who are closely associated with it[83]). With this surgery we arrive at a simple spacetime picture of the grandfather paradox (as you'll soon see).

We start with a flat, two-dimensional Minkowski spacetime, the x,t system in Fig. 6.15, and then imagine that (somehow) two *cuts* in that spacetime come into

Fig. 6.15 Minkowski spacetime transformed into a time machine with two 'cuts'

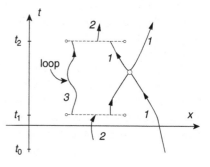

[82]D. H. Coule, "No Warp Drive," *Classical and Quantum Gravity*, August 1998, pp. 2523–2527, offering a pessimistic view of warp drive.

[83]See D. Deutsch, "Quantum Mechanics Near Closed Timelike Lines," November 15, 1991, pp. 3197–3217, and H. D. Politzer, "Simple Quantum Systems with Closed Timelike Curves," November 15, 1992, pp. 4470–4476, both in *Physical Review D*.

existence. These cuts are the two horizontal dashed lines in the figure, one with an arrowhead going into it, and one with an arrowhead coming out of it (each labeled with '2'). We further imagine that each cut has two edges, with the upper edge of the lower cut 'glued' to the lower edge of the upper cut (and the lower edge of the lower cut 'glued' to the upper edge of the upper cut

These 'gluing's' explain why the arrowhead marked 2 into the lower cut at time t_1 emerges from the upper cut at time $t_2 > t_1$, and why the arrowhead marked 1 into the upper cut at time t_2 emerges from the lower cut at time $t_1 < t_2$. The 2-line is the world line of a particle that simply disappears from spacetime during $t_1 < t < t_2$, while the 3-line is the world line of a particle trapped in an endless time loop (remember the 1993 film *Groundhog Day*?). Clearly, the sub-region of spacetime between the two cuts is not 'normal' spacetime. In fact, the 1-line shows that we have encountered a time machine spacetime, as a particle entering it from $t < t_1$ (passing to the right of the lower cut) can enter the lower edge of upper cut at $t = t_2$ and so emerge from the upper edge of the lower cut at the *earlier* time $t = t_1$ and thus interact with itself before it entered the upper edge! And that, of course, sets-up a grandfather paradox situation.

This picture leaves one thing obviously (and glaringly) unexplained—just *how* does one *cut and glue* spacetime? The reason, in spite of that question, that physicists nonetheless study situations depicted in Fig. 6.15, is because it allows them to explore what *could* happen *if* through some (yet unknown) process a time machine spacetime should suddenly appear. Who says physicists aren't optimists?

The idea of modifying spacetime itself to 'make' a time machine (look back in Sect. 3.5, at the discussion there on what it means to solve the gravitational field equations) has also appeared in connection with another of science fiction's favorite ideas, one almost as spectacular as time travel: *the FTL warp drive*. The lure of interstellar FTL travel, for both science fiction enthusiasts and physicists is, of course, simply undeniable. Consider, for example, these words by a Russian physicist:

"Everybody knows that nothing can move faster than light. The regrettable consequences of this fact are also well known. Most of the interesting or promising candidates for colonization are so distant from us that the light barrier seems to make an insurmountable obstacle for any expedition. It is, for example, 200 pc [1 parsec is equal to about 3.2 light-years] from us to the Pole star, 500 pc to Deneb [the brightest star in the constellation Cygnus], and ~10 kpc to the center of the Galaxy, not to mention other galaxies (hundreds of kiloparsecs). It makes no sense to send an expedition if we know that thousands of years will elapse before we receive its report. On the other hand, the prospects of being confined forever to the Solar System without any hope of visiting other civilizations or examining closely black holes, supergiants, and other marvels are so gloomy that it seems necessary to search for some way out."[84]

---

[84]S. V. Krasnikov, "Hyperfast Travel in General Relativity," *Physical Review D*, April 15, 1998, pp. 4760–4766. Possible travel distances have been greatly reduced since Krasnikov wrote. In 2011, for example, astronomers announced the discovery of a red dwarf star with three planets (each of mass comparable to Earth's), all in the star's so-called *habitable zone* (where water can exist on the surface in the liquid state). All three planets are solid (not gaseous as are Jupiter,

In response to that, we might ask if *FTL* trips will someday be made by humans in spaceships? *Maybe*—but only if such journeys can be made in a very *unordinary* spacetime. That is, continuing with Krasnikov's passage:

> "The point ... is that [whereas the light barrier exists in special relativity] in general relativity one can try to change the time necessary for some travel not only by varying one's speed [as in special relativity] but also ... by changing the distance one has to cover."

To understand what Krasnikov was getting at, let's consider the theoretical analysis made 4 years earlier by the Mexican mathematical physicist Miguel Alcubierre on, astonishingly, how to make a *Star Trek* warp drive![85] He did this by demonstrating a spacetime metric that, by literally expanding and contracting the local spacetime of a spaceship and its neighborhood, achieves space travel between any two points, no matter how far apart, in arbitrarily little elapsed time (for both the spaceship, and external non-spaceship observers, there is no time dilation effect[86]).

Alcubierre opened his analysis with words designed to explain how FTL travel is possible, given all that I've told you earlier in this book about how FTL travel is *not* possible (according to *special* relativity). As he wrote,

> "Since our everyday experience is based on a Euclidean space, it is natural to believe that if nothing can locally travel faster than light then given two places that are separated by a proper spatial distance D, it is impossible to make a round trip between them in a time less than $2D/c$ (where c is the speed of light), as measured by an observer that always remains at the place of departure. Of course, from our knowledge of special relativity we know that the time measured by the person making the round trip can be made arbitrarily small if his (or her) speed approaches that of light. However, the fact within the framework of general relativity and without the need to introduce non-trivial topologies (wormholes), one can actually make such a round trip in an arbitrarily short time as measured by an observer that remained at rest will probably come as a surprise to many people."

That last sentence is almost surely a grand understatement, and Alcubierre quickly went on to explain.

> "The basic idea can be more easily understood if we think for a moment of the inflationary phase of the early Universe, and consider the relative speed of separation of two co-moving observers. It is easy to convince oneself that, if we define this relative speed as the rate of change of proper spatial distance over proper time, we will obtain a value that is much larger than the speed of light. This doesn't mean that our observers will be travelling faster

---

Saturn, Neptune and Uranus) and so, as potentially habitable, are candidates for a visit. The star and its planets are 'only' 22 light-years from Earth. In *Star Trek*, FTL speed is described by the *warp factor*, which is the cube-root of the multiple of the speed of light at which the spaceship *Enterprise* travels. So, for example, to make the journey from Earth to the red dwarf in one month of ship time (see ahead also to note 86), the required FTL speed would be 264 times the speed of light, or warp factor 6.4. In science fiction, a *warp drive* is imagined as the means for achieving such speeds.

[85]M. Alcubierre, "The Warp Drive: Hyper-fast Travel Within General Relativity," *Classical and Quantum Gravity*, May 1994, pp. L73–L77.

[86]That is, the passage of time on the spaceship is *identical* with the passage of time on Earth. *With the warp drive, there is no twin paradox.*

than light: *they always move inside their local light-cones* [my emphasis]. The enormous speed of separation comes from the expansion of spacetime itself."

In a similar fashion, a *contraction* of spacetime can result in being able to *approach* an object at FTL speed.

In fact, we've actually already encountered one way to obey special relativity's *local* limit on speeds to that of light, while still achieving superluminal speed on a *global* level. That is, we *could* do that *if* general relativity really does allow wormholes in spacetime (what Alcubierre calls a "non-trivial topology"). That's because we can imagine a wormhole connecting two points in space that are light-years apart in that space, and yet the distance *through* the wormhole itself is quite short. Thus, a spaceship transiting the wormhole could do so at *sub*luminal speed at all times, and yet to an observer in normal space the speed would appear to be far in excess of the speed of light.

Determining just *how* to achieve the spacetime warp, however, is far different from simply demonstrating that such a warp is consistent with the general theory of relativity. The 1996 movie *Star Trek: First Contact*, for example, is about the invention of the warp drive in the twenty-first century. The whole thing fits inside a discarded ICBM which, as you'll soon see, is a *vast* underestimation of the technology required to control the energies associated with a real warp drive. For a spacetime engineer to *build* the warp drive bubble means she has to determine the required mass-energy distribution that results in Alcubierre's assumed spacetime metric. And that brings us to the central problem of the warp drive—the warp drive engine of an FTL starship requires (just like a wormhole) exotic matter (negative energy)—stuff that violates all the usually assumed energy conditions of general relativity.[87]

The weak, strong, and dominant energy conditions are *all* violated because the Alcubierre spacetime warp requires a negative energy density in the 'skin' of the warp bubble. As discussed earlier, in connection with wormholes, negative energy density *can* be achieved on a *micro*scopic scale, but for the Alcubierre warp drive we are talking about a *lot* of exotic matter. In their paper, Pfenning and Ford calculated that, for what they called "a macroscopically useful warp drive" with a radius of 100 m "so that we may fit a ship inside [the warp bubble]," the negative energy required for the warp bubble is on the order of, as they so graphically put it, "roughly ten orders of magnitude greater than the [energy of the] total mass of the entire universe."

Two years later, after making some adjustments to the spacetime metric assumed by Alcubierre (that is, to the distribution of mass-energy to produce the warp bubble), it was shown that the negative energy required by the warp drive could

[87]M. J. Pfenning and L. H. Ford, "The Unphysical Nature of 'Warp Drive'," *Classical and Quantum Gravity*, July 1997, pp. 1743–1751. See also K. D. Olum, "Superluminal Travel Requires Negative Energies," *Physical Review Letters*, October 26, 1998, pp. 3567–3570.

be greatly reduced.[88] The reduction is, in fact, spectacular, but only in a relative sense (you can reduce a mass that is ten orders of magnitude greater than that of the entire universe by a *huge* factor and still be left with a pretty stupendous number). The reduced amount of negative energy required for a warp bubble able to contain a human-sized spaceship is now down to 'only' "of the order of a few solar masses $[-1.4 \times 10^{30}$ kg]."

As mentioned earlier, in connection with the Casimir effect and its theoretical use in a wormhole time machine, although quantum field theory does not preclude negative energy densities, that does not mean it is possible to observe arbitrarily large negative densities for arbitrarily long times. In fact, certain quantum inequalities (QI's), much like Heisenberg's uncertainty principle, have been established that place bounds on the magnitude and duration of observable negative energy density.[89] These QI's have the general form of

$$\widehat{\rho} t_0^4 > -C$$

where C is a positive constant that depends on the nature of the particular quantum field being considered, t_0 is the time duration, and $\widehat{\rho}$ is the integrated energy density along a finite section of a geodesic (free fall) world line. The form of the QI shows that as t_0 increases, $\widehat{\rho}$ must quickly decrease. For example, if t_0 doubles, then $\widehat{\rho}$ must decrease by a factor of *sixteen*, a result that caused Ford and Roman to conclude that it "appears probable that nature will always prevent us from producing gross macroscopic effects with negative energy."

'When it rains it pours,' goes an old saying, and that applies to the warp drive's potential difficulties: in addition to the need for exotic matter, there are two more concerns as well, both *operational* in detail. First, running into any space matter encountered by the leading edges of the warp bubble (where spacetime is shrinking), such as interstellar dust, would certainly generate intense radiation. The ship, then, should carry plenty of shielding which, curiously, would not be a problem because the energy density of the warp, itself, is *independent* of the mass in the bubble's interior. In any case, the warp drive should clearly not be engaged anywhere near any sizeable chunk of matter, like a planet (and, indeed, that constraint was followed in *Star Trek*). Second an even more severe problem was discovered by Krasnikov. In unpublished work he showed that the ship at the center of the bubble is not causally connected to the edges of the bubble. That is, the ship's

---

[88]C. van den Broeck, "A Warp Drive with More Reasonable Total Energy Requirements," *Classical and Quantum Gravity*, December 1999, pp. 3973–3979.

[89]L. H. Ford and T. A. Roman, "Restrictions on Negative Energy Density in Flat Spacetime," *Physical Review D*, February 15, 1997, pp. 2082–2089. For some interesting remarks about the QI's, see J. F. Woodward, "Twists of Fate: Can We Make Traversable Wormholes in Spacetime?" *Foundations of Physics Letters*, April 1997, pp. 153–181.

crew could *not* create a warp bubble on demand and, after it had been created, could not control it on demand.[90]

It is important to both understand what that means, as well as what it does *not* mean. The causality issue does *not* mean that Alcubierre warp bubbles are impossible to create (perhaps they are, but not because of a lack of causality). It only means that whatever action is required to change the spacetime metric to make a warp bubble *has to already have been done before the decision to use the bubble is made*. Thus, a warp bubble wouldn't be of any use for a starship that needs to escape a sudden, unexpected threat. But, as Everett and Roman cautiously observe, the warp bubble might have a more mundane use: "Suppose space has been warped to create a bubble traveling from Earth to some distant star, e.g., Deneb, at superluminal speed. A spaceship, appropriately located with respect to the bubble trajectory, could then choose to enter the bubble, rather like a passenger catching a passing trolley car, and thus make the superluminal journey."

At the end of his paper (note 85) Alcubierre briefly speculated on the possibility of using his superluminal warp drive to build a time machine (showing, again, the intimate connection between the two concepts—look again at note 12 in Chap. 1 for how the connection between FTL and time travel appeared in pulp science fiction), but didn't show *how*. That was done 2 years later by Everett using, not surprisingly, an argument he called "reminiscent of the 'reinterpretation principle' … which played an important role in discussions of the physics of tachyons."[91]

In an attempt to avoid the Alcubierre bubble's causality problem, Krasnikov looked for a different, causal superluminal spacetime metric. This he succeeded in finding[92] but, rather than describing a bubble, Krasnikov's warp is in the shape of a tube. The interior of the tube is flat spacetime, just as in the case of the bubble warp, but unlike the bubble there would be a causal link between the spaceship crew and the tube. Just as the warp bubble requires very thin walls (on the order of a few thousand Planck lengths) of negative energy, so does the Krasnikov tube warp. Unlike the bubble warp, however, the tube warp stretches the entire length of any proposed trip, so the total negative energy in the warp is incredibly huge. For a tube a mere one meter long and one meter wide, for example, the total negative energy is 10^{28} solar masses, and to create a tube from Earth to just the nearest star would require 10^{44} solar masses of negative energy![93]

One curious feature of the Krasnikov warp is that the *outbound* leg of a round trip cannot be made in less time than required by light. *But* on the on the return half

---

[90]A. E. Everett and T. A. Roman, "Superluminal Subway: the Krasnikov tube," *Physical Review D*, August 15, 1997, pp. 2100–2108.

[91]A. E. Everett, "Warp Drive and Causality," *Physical Review D*, June 15, 1996, pp. 7365–7368. Recall the discussion of the RP in Chap. 5.

[92]See notes 84 and 90.

[93]In the same manner as the huge negative energy of the Alcubierre warp drive was later reduced (see note 88), the Krasnikov tube's enormous negative energy requirement was later significantly reduced: see P. Gravel and J. Plante, "Simple and Double Walled Krasnikov Tubes I: tubes with low mass," *Classical and Quantum Gravity*, February 2004, pp. L7–L9.

of the journey, a traveler would find the spacetime metric so altered (because of mass-energy manipulations purposely made on the outbound half) that she would move "backwards in time." The net result is that the round trip could end arbitrarily soon after it started! As Everett and Roman cautiously concluded, the Krasnikov tube is a "very unlikely possibility," but it would make a *wonderful* science fiction gadget, don't you think?

While the Alcubierre warp may seem to be an incredible discovery (it *is*), it was not a unique one. That's because just 8 years later a different warp metric was discovered by the Portuguese mathematician José Natário, in which the expansion/ contraction of spacetime does *not* occur. As Natário wrote, this signature feature of the Alcubierre warp drive "is but a marginal consequence of the choice [for the spacetime metric/mass-energy distribution]."[94] Rather than thinking of the warp bubble as being propelled by the push-pull of spacetime expansion/contraction, Natário wrote that "one could best describe the warp-drive spacetime as 'sliding' the warp-bubble region through space": that is, as analogous to a California surfer riding a wavefront. The surfer is motionless with respect to the water in the immediate vicinity of his board, and yet his speed with respect to the rapidly approaching shore is decidedly non-zero.

The idea of a manipulated or *warped* spacetime allowing time travel was an early arrival in science fiction. Consider, for example, the 1930(!) story in which the narrator (one Thomas Jenkins) walks 18,000 years into the future. In an editorial footnote (a device commonly used in early pulp fiction to inject scientific verisimilitude), we are told that "Jenkins had evidently fallen into a warp in space ... a fault, we might say, borrowing a geologic term, in the curvature of space. Through this warp he had been thrown clear out of our three dimensions into a fourth. There he slid in time over to the other side [of the fault] into the same spot in the three-dimensional world, but into a different era in time."[95]

That was a flawed explanation, with its talk of space rather than of spacetime, but some authors eventually learned to do better. For example, *folded spacetime* as a mechanism for time travel is used in a 1940s cautionary tale on the potential horrors of the atomic bomb. In that story, published 2 years after the atomic bombings in Japan, the world 15 years hence experiences a terrible atomic war. As the time traveler in the tale explains, "During the unprecedented release of atomic energy that arouse during the simultaneous bombings of our cities, something happened to the very continuum in which we exist ... A crook, a twist, a fold—explain it how you will, I accidently stumbled upon an electronic circuit that would create a field that would enable passage from one folded section [of spacetime] to the adjacent section. The fold proved to be about fifteen years in length . . . ."[96]

[94]J. Natário, "Warp Drive with Zero Expansion," *Classical and Quantum Gravity*, March 2002, pp. 1157–1165.

[95]N. Schachner and A. L. Zagat, "In 20,000 A.D.," *Wonder Stories*, September 1930.

[96]R. F. Jones, "Pete Can Fix It," *Astounding Science Fiction*, February 1947.

As Alcubierre and Natário showed, it will take a lot more than a mere electronic circuit to warp spacetime for either FTL or time travel, but at least even the early science fiction pulp writers understood that—*somehow*—a spacetime warping would be required.

The Alcubierre FTL warp drive appeared as the scientific basis for a modern science fiction novel, where at one point we read of a curious optical feature of an FTL spaceship, one I haven't seen mentioned in the physics literature: "This is a ship that traveled faster than light. It's visible as it travels; its warp bubble emits a cascade of exotic radiation . . . but it outruns its own image. So the ship arrives first and the light has to catch up, all the photons it emitted back along its path arriving at mere light speed. The older images arrive last, and you get this effect as if the ship was receding, not arriving."[97]

Another quite interesting feature of Alcubierre's warp drive is that the spaceship crew would experience no acceleration forces, as the ship is always in *free fall*. This may explain why the *Enterprise* crew isn't flattened when Mr. Sulu engages that ship's warp drive. The spaceship is surrounded by a "bubble" of warped spacetime that is swept along by the combined push-pull effect of the expanding spacetime behind the craft and the shrinking spacetime in the front. The ship, itself, resides in the *flat* spacetime interior of the warp bubble. An amusing way to think of this is to imagine a fish (space traveler), inside an aquarium (the warp bubble), which has been tossed into a swiftly flowing river. An observer at the edge of the river sees the aquarium move by her at high speed while, for the fish (swimming in the *still* waters of its aquarium), all is serene because *it is at rest with respect to its local environment*. Thus, the Alcubierre warp drive realizes yet another one of science fiction's wonderful gadgets: the *reactionless* spaceship drive. That is, "the warp bubble moves by interacting with the geometry of spacetime instead of expending reaction mass [as do jet and rocket engines] . . . and the spaceship is simply carried along with it."[98] In picturesque terms, the warp drive starship is like a surfer who makes her own waves.

And so we see, with each passing decade, more and more of science fiction departing from the make-believe to the pages of physics journals.

---

[97]S. Baxter, *Ark*, Gollancz 2009. The spaceship in this work travels at three times the speed of light (warp factor 1.44, as explained in note 84).

[98]F. S. N. Lobo and M. Visser, "Fundamental Limitations on 'Warp Drive' Spacetimes," *Classical and Quantum Gravity*, December 2004, pp. 5871–5892.

6.6   For Further Discussion

The connection between FTL speeds and backward time travel made the jump from theoretical physics to popular culture very quickly. It was in the British humor weekly *Punch*, for example, that the famous (but nearly always misquoted—see note 113 in Chap. 3, which doesn't have it quite right) limerick by A. H. R. Buller (1874–1944) first appeared (December 19, 1923, p. 591):

> *"There was a young lady named Bright Whose speed was far faster than light, She set out one day In a relative way And returned on the previous night."*

Where *Punch* dared to go, Hollywood could not be far behind. Indeed, in this case it was actually there first, with the 1922 one-reel silent comedy film *The Sky Splitter*. This was just a short film (feature pictures generally had at least four reels), so it isn't clear just how widely distributed and viewed it may have been. The story is that of a scientist testing a new spaceship; when it exceeds the speed of light, he begins to relive his life. This all shows that today's fascination, so common in popular culture, of the latest developments in theoretical physics, is nothing new. Why do you think this is so? That is, why (for example) do so many of those who flock to science fiction movies of interstellar invasions (like the 1996 *Independence Day* and its 2016 sequel), nonetheless have no conception of the unlikely possibility of such invasions because of the sheer magnitude of interstellar distances? Distances so immense that, even at the speed of light, it takes 4 *years* to travel to the Sun's nearest stellar neighbor, and *millions* of years to reach the Milky Way's nearest neighboring galaxy? (The vastness of interstellar distances is, as mentioned in the text, the reason for the fascination in warp drives in both science fiction and physics.).

In wormhole and cosmic string time machines, and with warp drives, we encountered the idea of negative mass-energy in the form of 'exotic matter' (see note 39 again). Something like negative mass actually appeared in fiction long ago, in the 1827 novel *A Voyage to the Moon* by "Joseph Atterly," a pseudonym for George Tucker, a professor of moral philosophy at the University of Virginia. (One of Tucker's students was Edgar Allen Poe, who almost surely was influenced by Tucker's book to write his own moon tale, the 1835 "The Unparalleled Adventure of One Hans Pfall.") The trip in Tucker's work was powered by a metal called *lunarium*, which repels Earth.

(continued)

This is *not* the same sort of stuff as Wells' "Cavorite," a metallic alloy that is "transparent" to gravity and that appears in his 1901 *The First Men in the Moon*. Wells' competitor in the 'scientific romance' genre was, of course, Jules Verne. Wells was a visionary who looked far beyond just the next few decades, while Verne was a 'practical engineer' who, for example, got *his* characters to the Moon by the direct method of simply shooting them out of a 900-foot long cannon with 400,000 pounds of guncotton! (Wells' vision *could* sometimes fail him, as it did about the imminent likelihood of airplanes in his 1901 *Anticipations*. He believed they *would* be developed by the year 2000, and maybe even before 1950, but of course just 2 years later …) In a 1903 magazine interview, Verne revealed how he felt about the difference between his and Wells' work: "It occurs to me that his stories do not repose on very scientific bases … He goes to Mars [sic] in an airship, which he constructs of a metal which does away with the law of gravitation. C'*est très joli* [this is all very nice], but show me the metal. Let him produce it." Today the cry from those who dislike wormholes is the Verne-like 'show us the exotic matter!' If Wells and Verne were writing today, how do you think each would respond to that challenge? Would the possible existence (or not) of exotic matter be an issue about which both would *agree*?

Write a time-loop short story based on Lossev and Novikov's idea of a 'very old spacecraft' interacting with a remote wormhole.

Imagine an electronic circuit **A** that has the following behavior: **A**'s input signal is a function of time that has a well-defined maximum value (what electrical engineers call the *peak value*). The circuit's output signal, produced in response to the input, also has a well-defined peak value. Now, imagine further that the output peak occurs *before* the input peak. There is, in fact, nothing paradoxical or impossible about that, and such a circuit can (and has) been constructed, as I'll tell you shortly. Next, suppose that we take **A**'s output signal and use it as the input to another circuit **B** that, when it's input exceeds a certain level, disconnects the input to **A** *before* that input reaches *its* peak value. Circuit **B** can also be constructed in the real world. Indeed, you can read about how to construct **A** and **B** in two papers by M. W. Mitchell and R. Y. Chiao, "Causality and Negative Group Delays in a Simple Bandpass

(continued)

Amplifier," *American Journal of Physics*, January 1998, pp. 14–19, and "Negative Group Delay and 'Fronts' in a Causal System: An Experiment With Very Low Frequency Bandpass Amplifiers," *Physics Letters A*, June 16, 1997, pp. 133–138. What makes all this interesting here is that this "seems to open the way for a variant of the time travel paradox in which the traveler journeys to the past and kills his grandfather before his own father is born," an observation made in Garrison *et al.*, "Superluminal Signals: Causal Loop Paradoxes Revisited," *Physics Letters A*, August 10, 1998, pp. 19–25. This electronic version of the grandfather paradox does indeed follow if one substitutes "input peak" for "grandfather" and "output peak" for "time traveler." But before you think this gadget is a time machine, be assured that its designers also showed that, unlike the causally related grandfather and time traveler, the two peaks are *not* so related. Read these three papers and write a summary report of how circuits **A** and **B** work, and why the two peaks are not so related.

You'll recall that "an advanced civilization" is thought to be required to create a useable wormhole (note 27). The common phrase used by astrophysicists who are interested in the possibility of extraterrestrial life is *arbitrarily advanced civilization*, with a distinction made for at least three progressively higher stages of 'advancement.' Very roughly, Types I, II, and III advanced civilizations are those that, respectively, have the technology to (a) control something like 10^{13} W (ten million megawatts) for massive interstellar radio broadcasts, (b) a technology to control the energy output of the civilization's planet's parent star (10^{27} W), and (c) a technology to control the energy output of the civilization's home galaxy (10^{38} W). We are, today, short of being even a Type I civilization, and it would probably take *at least* a Type III civilization to build a wormhole. Indeed, Stephen Baxter's 1993 novel *Timelike Infinity*, of beings who can manipulate *constellations of galaxies*, seems to assume a Type IV civilization will be required. Since there are typically 10^{11} stars in a galaxy, going from 10^{27} W for a star to 10^{38} watts for a galaxy is consistent. But where do astrophysicists get 10^{27} W for a single star? Here's a calculation for you to perform, to confirm this value for yourself, starting with the experimental fact (not difficult to repeat, as it's at the level of a junior high school science fair project using a solar cell and a few common electrical components) that the solar power level at Earth's equator is 1200 W per m². Then, using the fact that the Earth's orbital radius around the Sun is 93 million miles, compute the total power (energy per second) radiated by the Sun. (You should get a number that is somewhat

(continued)

smaller than 10^{27} W, which is 'explained' by observing that the Sun is really a quite ordinary star, exceeded in size by many other stars in the Milky Way. Next, you'll find, in most books on astrophysics, the statement that the nuclear fusion reactions that power the Sun convert *four million tons* of the Sun's mass to pure energy *every second*. Confirm that your number for the power output of the Sun is consistent with that claim. (Remember Einstein's famous formula $E = mc^2$, that the speed of light is $c = 3 \times 10^8$ m/s, and that 1 kg $\approx$ 2.2 pounds. In the MKS system of units (meters/kilograms/seconds) one watt $=$ one joule (of energy) per second, where to give you some perspective on what a joule is, the chemical energy released by burning a gallon of gasoline is about 100 MJ.

Appendix A
Old Friends Across Time (A Story)[1]

As I sit here in my study, with the photographic evidence spread before me, I can barely comprehend what my eyes tell me must be so. The evidence is incontestable. And yet—I still struggle to believe. Let me try to explain—possibly in the process I will manage to put my tumbling mind to rest.

For as long as I can recall, old photographs have fascinated me. To page slowly through collections of historical pictures, no matter what the theme, was consummate joy. Even when I was quite a small boy I used them as my time machine into the past. They took me up and away from the problems every youngster has while growing up, and let me wonder of people and places long since returned to dust. Matthew Brady's Civil War photos had a particularly strong attraction for me, with the horror (and yes, I will admit it, the *fascination*) of war frozen in the images of young men dead before life had really begun. To look at the fallen youth of more than a century before, and to wonder who they were, and what they had felt and thought—it all sent shivers through my romantic mind.

I suppose I might have become a professional photographer. But somewhere along in the process of looking at pictures, I became aware of the miracle of the *technology* of picture taking. That led me to chemistry and optics, and finally by some wondrous route, I became an electrical engineer. I never lost my love for old pictures, though, but merely turned my interest in them to the photographic history of electrical physics.

To search out and acquire (for by now I had started my own collection) a photograph of Steinmetz, smoldering cigar clamped in his mouth, giving a lecture on AC circuit analysis using the then still mysterious square root of minus one made my heart beat faster. To find a faded picture of Einstein at a long forgotten

---

[1]P. J. Nahin, "Old Friends Across Time," *Analog Science Fiction Magazine*, May 1979. This tale was written with the specific goal of illustrating how a trip into the past *yet to be initiated* could logically influence events in the time traveler's present and future. The story reproduced here is, with only a few very minor alterations, as it originally appeared in *Analog*.

© Springer International Publishing AG 2017
P.J. Nahin, *Time Machine Tales*, Science and Fiction,
DOI 10.1007/978-3-319-48864-6

conference, caught forever in time with his quiet, gentle eyes looking into mine, would send me to the heights of ecstasy.[2]

But it was Maxwell that led me to my incredible discovery. There is no doubt but that James Maxwell was the greatest theoretical physicist of the nineteenth century. Together with Einstein, he was the best of *any* century. Could it possibly be more than mere chance that the same year saw the death of one and the birth of the other? It was Maxwell who gathered together all the then known, but fragmented, experimental bits and pieces of knowledge about electricity and magnetism, and stirred in his own contribution of the displacement current. There was no physical evidence then to justify that last step, but the genius of Maxwell knew it *had* to be. And then, from his soaring mathematical insight and physical intuition, he took it all and wrote down the four magnificent equations for the electromagnetic field![3]

No one who has seen and understood those beautiful equations can come away without a quickening of the pulse and a flush of the blood. They're not long—you can write all four vector differential equations on the back of a postcard, but oh, what they tell us! With them, Maxwell, showed light was electrical in nature, predicted radio waves *two decades* before Hertz discovered them in the lab, explained energy propagation in space, and radiation pressure, and laid the scientific basis for today's television, radar, lasers, giant electric motors, generators, transmission lines and—well, why go on? The equations are the work of a level of genius we may not see again for a millennium. We have hardly begun to discover the marvels wrapped inside the electromagnetic field equations. With their aid, and that of quantum mechanics, the very secret of life, itself, may someday be unraveled.

And so I searched for old photographs of Maxwell. He died at his family's Scottish home in 1879, before the art of picture taking was barely 40 years old. But I knew in my heart that somewhere there *must* be photographs, yet undiscovered, of such a great man. Anyone who has seen the best examples of prints from wet glass collodion negatives knows they are, in the faithfulness of their rendition of detail, better than what we commonly expect today. Working against me was the fact that the process was slow, laborious, and unforgiving of mistakes. The taking of a picture was not a minor decision in Maxwell's time. But still I searched.

I searched for one photo, in particular. When Einstein died, a famous picture was taken of his office, just the way he left it for the last time. On the blackboard behind his desk are the last thoughts he had in his long quest for a Unified Field Theory,

---

[2]The first reference is to Charles Steinmetz (1865–1923), the German-born American electrical engineer and mathematician who became the wunderkind of General Electric. Einstein, of course, needs no introduction!

[3]I wrote this for story effect, but it's not really *quite* true. When Maxwell wrote his theory in mathematical form, he did so using *twenty* (!) equations in as many variables. The equations, as physicists and electrical engineers use them today, were first written in 1885 by the English self-taught eccentric Oliver Heaviside, who considered Maxwell to be his hero (see note 6 in Chap. 6). Modern electrical engineers and physicists write the Maxwell equations as *four* partial differential *vector* equations.

a 'theory of everything.' The writing on the papers covering the desk is clearly legible, and with modern blowup methods, easily readable.

At the time of his death, Maxwell was the Einstein of his times. Surely, I reasoned, a similar photograph of Maxwell's study must have been taken. Even though none has come down through the decades to us, it *must* exist! Gathering dust in an old trunk, or buried in a long forgotten album, it had to be somewhere. I vowed to find it.

I began by writing to all of Maxwell's living descendants, asking that they search through family holdings for any pictures concerning Maxwell that they might possess. For the most part all were cooperative, even though more than just a few thought I was somewhat deranged. Still, it was in vain. I did receive a few old pictures never before seen by other than the family, including a poignant one taken in 1901, showing Maxwell's grave in Parton Churchyard at Glenlair, Scotland. A forlorn, wintry scene, with only what seemed to be three men in the far distance, it brought tears to my eyes. Alas, there were no photos of Maxwell's study.

But then late last year, while on a business trip to London, I stopped off for a few hours at the historical archives maintained by the British Institute of Electrical Engineers. On a chance, I looked through their massive files on Maxwell and was rewarded within the hour! What I found will haunt me throughout the remainder of my life.

There it was, stuck through its border with a rusty pin, between two pieces of yellowed paper covered with what appeared to be some simple, rough lecture notes. An ordinary looking photo of a study. Obviously overlooked through the years, or at best unappreciated for what it was, it was the almost illegible, penciled notation on the back that convinced me of my find—just a date: November 9, 1879. Exactly 4 days after Maxwell's death, precisely when some unknown, yet inspired person (a family member, a neighbor, a local scientist?) would take such a picture!

I am ashamed to admit it, but there was no hope the Institute would let me have the picture. And there was no time to copy it, for I was to return home to America that very night. No, that's not true. The *real* reason for what I did was simply that I *had* to have that original, *old* photo. I took it! It was my undoing, for that dishonorable act destroyed the picture's tie to verified, legitimate historical records. But *I* know what I found is true.

I could barely control my wild emotions on the flight home. Several times I removed the picture from my briefcase, and looked with fascination at the papers lying on Maxwell's desk, and at the tightly written lines of mathematics on the blackboard in the background. My hands trembled with what can only be called lust—once home, reunited with my well-equipped photo lab, I would learn every secret hidden in that picture.

There are no words I know that can convey the thrill I felt as I began the processing of that priceless photo. Alone in my lab, with all the modern equipment a well-off amateur can buy (a Caesar Saltzman 8×10 enlarger with mercury vapor point light source and a $10\times$ Plan Achromat Nikon enlarging lens), I carefully cropped and blew up selected views of the blackboard and desk. Printing the enlargements on ultra-fine grain AGFA Brovira paper, I could scarcely restrain myself from peering at them with a magnifying glass while I waited for them to dry.

Then, at last, I had them spread out across my study desk. I tried to force myself to examine each slowly, carefully, in turn, and not to skip from one to another like a child let loose in a candy store with a dollar. The first three were of the desk papers, including what seemed to be a diary. It must have been lost after Maxwell's death since no trace of it exists in the historical records. I experienced a stunning thrill as I gazed upon the scrawled words, but as they were not easily read at once, I moved on. It was the sixth enlargement, of the upper right corner of the blackboard that sent me reeling back to my chair. An equation that shouldn't, no, *couldn't*, be there. But it was.

To understand my reaction, there is one astounding thing you must realize about Maxwell's field equations. When Einstein turned the world of physics on its head in 1905 with his famous paper, "On the Electrodynamics of Moving Bodies," all the old ideas about absolute motion and simultaneity of events went out the window. Even Newton's laws of mechanics had to be modified. But *not* Maxwell's! His equations, just the way he published them in 1873, are the same ones studied today[4]—they need *no* relativistic corrections.

How can that be, you wonder, as they predate Einstein's by 32 years? The mystery of this has bedeviled the experts down through the years. Oh, they have an explanation, alright. They say that all of electromagnetics is actually relativistic phenomena to begin with, and the laboratory work of Faraday, Ampere, Henry, and the other great experimentalists were studies of relativistic electron interactions in matter (although they, of course, didn't know that). Thus, it is only 'natural' that Maxwell's equations need no correction. So goes the 'expert' explanation, but it isn't right![5] I know Maxwell knew about relativity, and understood it perfectly. He knew all about time paradoxes and the equivalence of mass and energy.

Because how else can you explain the equation visible in my enlargement: $E = mc^2$!

Why, you must wonder (just as I did), didn't Maxwell publish this remarkable result? At first, I believed it was because of a lack of faith in his results. Who would have believed any of it in those Victorian times, so sure of its absolute view of nature? I thought of how Newton, 200 years before Maxwell, had suffered from a similar hesitancy when he wrote the *Principia*. There, when explaining his theory of gravitation, Newton did *not* employ his new invention of the calculus (which he *had used* to make his discoveries), but instead fell back on laborious arguments based on the accepted mathematics of algebra and geometry. Who would have believed him, otherwise?

But then I realized that couldn't be right. Maxwell was a strong man intellectually and he wouldn't have held back for fear of disbelief. No, it had to be that he

---

[4]Don't forget note 3.

[5]Alas, I think it *is* right. Don't forget, this is science *fiction*. When there is a conflict between the needs of a story, and a rigid adherence to physics, the 'needs' wins!

discovered relativity and the mass-energy law just before his death, with no time to make his work known. I was still wrong.

It was later, when I returned to the enlargement of Maxwell's lost diary and read those painfully cramped notes, that I learned the truth. What I saw there showed me Maxwell had thought long and hard about his final discoveries and had purposely withheld them. For clearly visible, after I had slowly deciphered the writing, were the following words:

I have seen monstrous events. My blood has run cold at the sight of two great cities leveled to the ground, their inhabitants cruelly put to death instantly, or left to die slowly from a strange, lingering disease. Other trips, further on, have shown me the root of all these evils is the mass-energy equation, a result I at first believed to be my crowning glory. It will be my crown of thorns unless I ban it from my very being. Another will discover it for himself, but my *soul shall be free! I have dismantled my machine, and shall never look upon or think of those horrible scenes again.*

This passage was dated just 1 month before Maxwell died a savage death from cancer. The reference to 'two cities' can only be that of Hiroshima and Nagasaki. His own death was surely caused by lingering too long among their atomic ruins.

Think of what this *means*. Quite simply, Maxwell knew the secret of time travel! But even more incredible is that it must be *easy*, if one only knows how, to build a time machine! Think about it—Maxwell had no gigawatt power stations at his disposal, no high technology machine shops, or nanosecond computers. He was not a gifted experimentalist, and once he had predicted radio waves, for example, it took others 20 years to finally generate them. And yet, *he* built a time machine. Somehow, with just the puny power sources available to him, and a limited mechanical capability, he wrested free the *simple* implementation of a time machine from his dynamical field equations.

Yes, yes, I know what you must be thinking. How can I really conclude such an incredible thing from a single equation on a blackboard, and a few words written by a man dying a painful death? A man, clearly suffering dearly, and possibly not in complete possession of his once marvelous mind.

This very evening the last bolt of evidence slid into place. Attempting to escape from the emotional maelstrom into which I had fallen, I turned to my old love of picture gazing. I took down from my library shelf a tattered yet cherished volume of the Meserve Collection of Lincoln pictures. My slow paging through the images stopped when I came to the famous photograph by Alexander Gardner of Lincoln's second inauguration. This incredible picture shows John Wilkes Booth looking down on Lincoln from behind a buttress high on the steps of the Capitol, while below in the crowd are the five men who, 41 days later, conspired with him in the assassination.

The following page demonstrated the extraordinary quality of Gardner's work, as it showed an enlargement of Booth's face in which the circular line between the pupil and the white of each eye is sharp and crisp! This impressive picture fascinated me, and I wondered if I could create a similar enlargement. It was then I remembered the old picture of Maxwell's grave, sent to me from Scotland, and the three distant figures in the background. They would present my photo-lab skills with a challenge, and the effort would distract my mind.

I finished the enlargement just 20 min ago. Those faces! Two of them I can now finally accept as being there—it must have been a pilgrimage for one, and for the other, it couldn't have been anything but a mocking, ironic gesture. But I wonder if the youngest one really knew who his two companions were? I don't know the answer to that—yet. But there they are, two men with faces my years of study have made as familiar to me as my own. One is a youthful Albert Einstein. The other, with the signs of death clearly written across his features, is James Clerk Maxwell. The face of the third man is familiar, too, for the third man is me!

Oh, I'm a bit older in the photo than I am now. But it's me, alright. A distinctive, jagged scar across the left cheek, a mark from a childhood accident, is sharply visible, and I can run a finger over my face and match it perfectly with the image in the enlargement. I'd say I'm about 45 or so in the image, no more than 10 years older than I am now. That doesn't leave me much time to keep my appointment, does it? I don't know, right now, how I'm going to do it, but I've got to rediscover Maxwell's secret of time travel. I'm sure I'll succeed—after all, there I am in the picture. Somehow, I'll be going back to pick James and Albert up so we can have our picture taken. Ten years—not much time.

I'm really looking forward to meeting my two new friends from across time.

For Further Discussion

When "Old Friends Across Time" originally appeared in *Analog*, it opened with a quotation from Richard Feynman's famous 1961–1963 Caltech undergraduate course (published in 1964 as *The Feynman Lectures on Physics*): "Ten thousand years from now, there can be little doubt that the most significant event of the nineteenth century will be judged as Maxwell's discovery of the laws of electrodynamics. The American Civil War will pale into provincial insignificance in comparison." This is almost certainly true, but could Maxwell *really* have built a time machine from just a knowledge of electromagnetic theory and special relativity (which is all that is needed to derive $E = mc^2$), if he didn't also have a deep understanding of general relativity (and probably of quantum mechanics, too)? How likely do you think *that* is?

The narrator in "Old Friends Across Time" knows he is going to live long enough to eventually build a time machine; discuss the implications of this knowledge. For example, is he at least *temporarily* invulnerable to committing suicide (or, for that matter, to any other variation of dying?) That is, do we have a 'future' version of the grandfather paradox? This issue has never (to my knowledge) been considered by physicists, and not by philosophers either until recently. See, for example, S. Keller and M. Nelson, "Presentists Should Believe in Time-Travel," *Australasian Journal of Philosophy*, September 2001, pp. 333–345, and M. H. Slater, "The Necessity of Time Travel (On Pain of Indeterminacy)," *The Monist*, July 2005, pp. 362–369. More generally, if we assume that the past is unchangeable then the scenario in "Old Friends Across Time" seems to force at least some level of inevitability on the future as well. Or does it? In the 2007 story by Ted Chiang, "The Merchant and the Alchemist's Gate," that you were asked to read in a *For Further Discussion* at the end of Chap. 4, there is the following exchange between the narrator and the inventor of "the Gate" (a wormhole): "So if you learn that you are dead 20 years from now, there is nothing you can do to avoid your death?" He nodded. This seemed to me very disheartening, but then I wondered if it could not also provide a guarantee. I said, "Suppose you learn that you are alive 20 years from now. Then nothing could kill you in the next 20 years. You could then fight in battles without a care, because your survival is assured." "That is possible," he said. "It is also possible that a man who would make use of such a guarantee would not find his older self alive when he first used the Gate." "Ah," I said, "Is it then the case that only the prudent meet their older selves?" Comment on this issue, with particular attention to free-will.

Appendix B
Newton's Gift (A Story)[6]

Wallace John Steinhope was a sensitive human being, a person deeply concerned about the welfare of his fellow creatures. Any act of injustice, however slight, made his breast pound with righteous indignation. He was a champion of fair play, and his motto in life was taken from the ancient English rule of law—'Let right be done!'

Even while still a lonely, reclusive child, Wallace's heart ached mightly when he read of the laborious, boring, mind-dulling calculations endured by the great mathematicians of old. Just knowing, *thinking*, of Gauss's marvelous mind wasting literally months of its precious existence grinding out tedious mathematics that even a present-day dullard could do in a minute, on a home computer, was sheer agony for Wallace. Contemplation of the God-like Newton suffering endless delays in his gravity research, all because of a simple miscalculation of the length of a degree of longitude, was almost unbearable.

Indeed, Newton played a special role in Wallace's life (and he in Newton's, as we shall soon see). While the other great mathematical physicists had merely been hindered in their work by the lack of modern computational aids, Newton had squandered so much valuable time in other, nonscientific pursuits! His quasireligious writings alone, over half a million words, exceeded his scientific writings. What a waste! Wallace wondered endlessly over the reason for this strange misdirection of talent and bored his friends to the edge of endurance with his constant brooding on the mystery. Still, they all liked and admired Wallace enormously and so put up with it. But more than one of them had sworn to throw up the next time Wallace mentioned Newton during a wedding (but that's another story).

So deep was Wallace's anguish for his predecessors that even as he grew older and his own tremendous talents as a mathematical physicist (the result of a lucky

---

[6]P. J. Nahin, "Newton's Gift," *Omni*, January 1979. This tale was written with the specific goal of illustrating casual loop time paradoxes. The story reproduced here is, with only a few very minor alterations, as it originally appeared in *Omni*.

© Springer International Publishing AG 2017
P.J. Nahin, *Time Machine Tales*, Science and Fiction,
DOI 10.1007/978-3-319-48864-6

genetic mutation induced in a male ancestor some centuries earlier) gained him an international reputation, thoughts of the unmeasurable misery of his scientific ancestors were never far from his mind. It was most appropriate, then, that his greatest discovery gave him an opportunity to *do* something! And Wallace John Steinhope vowed to *help*. He became convinced that it was his purpose on earth— he could not, he *would* not hesitate. As he strapped the knapsack-size time machine to his chest, his excitement was, therefore, easy to understand.

"It is done! And I am ready. I will travel back and bestow this gift of appreciation, this key to mental relief, on the great Newton himself!" Wallace cradled a small, yet powerful hand-calculator in his palm. It was a marvel of modern electronics. Incorporating large-scale integrated circuitry and a Z-8000 microprocessor solid-state chip, the calculator required only a small, self-contained nuclear battery for its power. It could add, subtract, multiply, divide, do square and cubic roots, trig and hyperbolic functions, take powers, find logarithms, all in mere microseconds. It was programmable, too, able to store up to 500 instructions in its micro-memory. The answers it displayed on its red, light-emitting diode readouts would liberate young Isaac from the chains of his impoverished heritage of mathematical calculation. No more Napier's bones for Newton!

But Wallace John Steinhope was no fool. He understood, indeed feared, time paradoxes. He knew Newton could be trusted with the secret, but it wouldn't do for the calculator to survive Newton's time. So Wallace had incorporated a small, self-destructing heat mechanism into it. After 5 years of use, it would automatically melt itself into an unrecognizable, charred slag mass. But that would be enough time for its task to be completed. The emancipation of Newton's mighty brain from tedium! Pleased enormously at the thought of the great good he was about to confer, Wallace set the time and space coordinates for merry old England, flipped the power switch on, and vanished.

Materializing in the Lincolnshire countryside in the spring of 1666, he began his rendezvous with destiny. It was the second and final year of the great bubonic plague, and Newton, seeking refuge from the agony and death plundering London and threatening his college of Trinity at Cambridge, had returned home to work in seclusion. The years of the Black Death were Newton's golden years, when the essentials of calculus would be worked out, when the colored spectrum of white light would be explained, and when the principle of the law of gravitation would be grasped. But how much easier it would be if Newton were released from the binding chains of dreary calculation. Wallace's gift would slip the lock on those chains! Accelerate genius!

It was early evening when, guided by a map of the area prepared by a friend who was both a cartographer and amateur historian, Wallace reached the quiet little town of Woolsthrope-by-Colsterworth. It was here, in a small farmhouse, that Wallace would meet his hero of the ages. A cold, gentle rain was falling as he approached the door. The soft, hazy light of an oil lamp glowed inside, revealing through the translucent glass the form of a man bent over a table. The fragrant smoke of well-dried wood curled from the chimney, announcing a warm fire within.

With his heart about to burst from excitement, Wallace rapped upon the door. After a pause, the shadow rose and moved away from the window. The door opened, and there stood Isaac Newton, a young man of 23 with an intellect that Hume and Voltaire considered "the greatest and rarest genius that ever rose for adornment and instruction of the species." But for the importance of his self-appointed mission, Wallace would have fainted dead away from the thrill of it all. "Is this the home of Isaac Newton?" he asked in a voice quavering with the trembling tones normally used by lovers about to reveal their deepest feelings.

The young man, of medium height and thick hair already showing signs of gray, swung open the door and replied, "My home it is, indeed, stranger. Come into the parlor, please, before the wetness takes you ill."

Isaac followed Wallace into the room and stood quietly watching as his visitor removed his soaked coat and hat. The portable time machine was gently placed on the floor next to a wall. The calculator was snug and safe in its plastic case in Wallace's shirt pocket. "Thank you, Master Newton. May we sit while we talk? I am afraid you may wish to take some time to consider my words." Motioning to a chair near the table, Isaac pulled a second chair from a darkened corner and joined Wallace. "You have a strange sound to your speech, stranger. Are you from hereabouts, or have you traveled far? Please commence slowly your tale."

Wallace laughed aloud at this question, a response prompted by his nervous excitement, and it quite surprised him. "Please forgive me. It is just that I *have* traveled so very, *very* far to see you. You see, I am from the future." Wallace was not one to play his cards close to his chest. Now it was Isaac's turn to laugh. "Oh, this is most ridiculous. Are you a friend of Barrow's at Trinity? It would be so like him to play such a trick.[7] From the future, indeed!"

Wallace's eyes ached at the sight of the papers on the table where Isaac had been working. What wonders must be there about to be born! In any other situation, Wallace would have asked their contents, but the die had been cast. He had to convince Isaac of the truth of his tale. But he had to walk a tight line, too. It just wouldn't do to misdirect Isaac's interest away from the calculator and toward the time machine itself! He must do something dramatic, something that would rivet his idol's attention and hold it.

"Yes, yes, I understand your reluctance to believe me. But, look here. This will convince you of the honesty of my words." Wallace pulled the shiny black plastic-cased calculator from his shirt pocket and flipped the power switch on. The array of LEDs glowed bright in the gloomy room as they flashed on in a random, sparkling red burst. Isaac's eyes widened, and he pushed his chair back. Was he frightened?

"As the Lord is my Savior, is it a creation of Lucifer? The eyes of it shine with the color of his domain. Are you one of his earthly agents?"

---

[7]The reference is to Isaac Barrow (1630–1677), who was the first Lucasian Professor of Mathematics at Cambridge. Barrow resigned that position to allow it to pass to Newton. Centuries later, Hawking became the 17th Lucasian Professor.

"Oh my, no! Look here, Master Newton, let me show you that there is no black magic or chicanery involved. It is all perfectly understandable in terms of the laws of Nature. What I have here is an automatic calculator, a device to perform all of your laborious mathematical labors."

So saying, Wallace squeezed the sides of the calculator case together, releasing pressure snap-fittings, and flipped the case open on a hinge at the top. Revealed to Isaac were the innards of the electronic marvel—a tightly packed interior of printed circuit boards, a mass of integrated circuitry, the small LED display, and the sealed nuclear battery. Isaac stared intently at the sight, and Wallace could see the natural curiosity of Newton's great mind begin to drive away the initial apprehension.

"But where are the gears, levers, springs, and ratchets to carry out the calculations? All I see is a black box with lights that glow red—how is *that* done; where is the lamp or candle to provide the light!—and many little isolated fragments of strange shapes. There is clearly nothing in your box that moves!"

"Oh, it is all done with electronics, Master Newton! The central processing unit has access to a solid-state memory that contains the decoding logic necessary to implement the appropriate algorithmic processes to provide the answers to the specific requests entered through these buttons. The actual performance of the box is achieved by the controlled motion of electrons and holes in suitably doped semiconductor material under the influence of electric fields induced—" Wallace, still overcome by his excitement, had rambled on wildly without thought of the essentially infinite technological gap that separated himself from Newton.

"Stop, stop," cried Isaac. "I understand only a few of the words you use and nothing at all of their meaning! But it is obvious that for calculations to be performed, mechanical work must be done, and that implies motion. Pascal's adding machine has shown the veracity of that. I say again, nothing moves in the box. How *can* it work?"

Wallace was embarrassed. The mistake of overlooking the hundreds of years of progress after Newton's time was one a child might make. "I am sorry, Master Newton. I'm going too fast for you." Isaac looked at Wallace with a frown, but Wallace failed to see the pricked vanity of the proud Newton. Going too fast, indeed!

Wallace prepared to lay a firmer technological foundation for Newton, but then he froze. It couldn't be done! Newton was a genius, certainly, but the task was still impossible. Wallace would have to tell him all about Maxwell's equations, Boolean algebra and computer structure, electronics, and solid-state device fabrication technology. It was just too much, and besides, there was the danger! The potential time paradoxes of all that knowledge out of its proper time sequence! What if Newton, in innocence, revealed some critical bit of knowledge out of its natural place in history? So, Wallace hesitated, but seeing the suspicion grow again in Isaac's eyes, he realized he had to do something, *anything*, immediately.

"You cannot deny your own eyes," answered Wallace. "Let me *show* you how it works. I'll divide two numbers for you with just the punch of a few buttons. Watch this." And, at random, he entered 81,918 divided by 123. Poor Wallace, of all the numbers to use, they were the worst. Within milliseconds the answer glowed brightly in fiery red characters. Wallace looked with pride at the result and then, already enjoying in his mind what he knew would be Isaac's amazement, he turned his eyes to the great man. What he saw made his spine tingle, and the gooseflesh stand high on his neck. Newton had fallen to his knees, with eyes bulging and hands raised as if in prayer.

"The mark of the Beast, it is the mark of the Beast! It is so written in the Book of Revelations—Here is wisdom. Let him that hath understanding count the number of the beast; for it is the number of man; and his number is six hundred three-score and six!" Rising to his feet, Newton fell back into his chair. "Your cursed box bears the brand of its master. There can be no doubt now, it *is* the creation of the fallen archangel!" Wallace was aghast at Isaac's violent reaction. The seventeenth century genius had now stumbled backward from his chair and had grasped a poker from the hot coals of the fireplace.

"Wait, please wait! Watch this; I'll multiply two other numbers together for you, watch!" Wallace quickly punched in the data, and then the answer gleamed steadily in burning red characters on the LEDs. Isaac's eyes first went wide with fear as he again saw the wizard electronics do their marvelous assignment, and then he shut them tight. Wallace was becoming desperate—this wasn't the way it was supposed to be! "Don't you see—imagine the tedious work, the mind-deadening labor this machine will save you from. And it is yours."

"Yes? But only for the exchange of my soul! That is always the Devil's price for his seductive gifts from Hell!" As Isaac shrieked these last words at Wallace, he raised the poker over his head. "Begone, you emissary of the Dark World! I know now you must be in the employ of the Father of the Antichrist, but the Lord God Almighty will protect me if I do not waver in my resolve. Begone, or I'll strike your brains out on the floor where you stand!"

Isaac's eyes were wide with fear, nearly rolling back to show all white spittle sprayed from his mouth as he yelled at Wallace, who stared in shock at the wild man who threatened him with death. "Please, please, *listen* to me, please! I beg you to understand—I am a scientist, just like you. The concept of the devil, and all it stands for, is contrary to everything I believe. How *could* I be in the devil's employ, when I don't even accept his existence? You *must* believe me!"

"Blasphemy!" screamed Isaac. "Your own words condemn you. To deny the reality of Satan in a sinful world is to deny that of God, too. Now leave my home, you dark beast from hell, or by the heavens above, *I shall destroy you!*" As he shrilled these words, Isaac brought the poker down in a wild swing that barely missed Wallace's head.

Struck dumb with confusion at the uncontrolled outburst, Wallace stuffed the calculator into his shirt, grabbed his hat, coat, and time machine and rushed from the house. As he hurried into the cold, wet night, he turned back, just once, to see Isaac Newton framed in the light of the open door. "Go, go, you foul messenger from the Lord of Evil! Back to your stinking pit of burning hell-fire! This is a house that honors the Divine Trinity and is no haven for the likes of you!"

Wallace rushed away into the blackness, the time machine bouncing unheeded upon his chest.

He ran, for how long he couldn't recall, until he fell exhausted next to a stream running heavy with the rain. Tears of rage, frustration, and shock streamed from his eyes. Rejected by the great Newton! Well, damn him! Wallace flung the calculator into the stream in his terrible anger and activated the return coordinates. He faded from Newton's world as quickly and quietly as he had come.

As for Isaac Newton, after having chased the Devil's messenger from his house, he returned on shaking legs to his desk. Pushing aside his rough calculations on the orbit of the moon around the earth, he swore to redeem himself in the eyes of the Savior. Somehow, he had been found lacking and had been tested. And the test was surely not over! He began to reapply his marvelous mind to determine the origin of his failure before the Lord God Jehovah. Taking quill in hand, he wrote the first of the many hundreds of thousands of words that his numerous religious tracts would devour from his allotted time.

Five years later, long after Newton had returned to Cambridge, a group of picnicking children were frightened when a nearby stream suddenly erupted into a geyser of steam. Moments later, the bravest (or most foolhardy) of the boys— who, by an astonishing coincidence that befits any good time travel paradox, would be Wallace's great-grandfather nine times removed—cautiously examined the streambed. All he found were some twisted, hot pieces of what he thought was a hard, black rock, and he tossed them back. They were all that was left of the calculator's nuclear battery. He did receive a tiny radiation dose from them, which caused a recessive genetic mutation that centuries later would suddenly appear as the cause of Wallace's genius, but otherwise the lad was unaffected. The incident was soon forgotten.

Well over 300 years later, Wallace John Steinhope reappeared in his own time. He was essentially the same man as before he left—kind, generous, and sensitive, and ready to come to the aid of any man or beast that might need help. As far as his friends were concerned, in fact, he was even improved (naturally, they didn't know what had brought about the welcome change but, if they had, they would have applauded it).

Wallace John Steinhope, you see, never again had another kind word for Newton, or for that matter, any words for him at all.

For Further Discussion

In his book *Travels in Four Dimensions: the enigmas of space and time* (Oxford 2003), the philosopher Robin Le Poidevin writes (p. 176) "But, as everybody knows, when a time machine leaves for another time it *disappears*." This is, indeed, how the time machine in "Newton's Gift" works; however, after reading *Time Machine Tales* do you think such behavior is in agreement or in conflict with general relativity? Defend your position.

"Newton's Gift" contains causal loops. Identify two of them, and discuss their role in the story (that is, are they central to the story or merely incidental?).

The idea of a time traveler visiting famous people in the past occurs fairly frequently in science fiction. In Ian Watson's "Ghost Lecturer" (*Isaac Asimov's Science Fiction Magazine*, March 1964), for example, the inventor of the "Roseberry Field" uses it to yank geniuses out of time to supposedly honor them, to let them know their lives had been worthwhile in the eyes of the future. But then he goes on to tell them—oh so kindly—where they had gone wrong or had fallen short of the mark, and of how much more we know nowadays. "You almost got it right, boy! You were on the right track, and no mistake. Bravo! *But* . . ." Watson makes the interesting observation that one can easily imagine playing this pathetic game of 'second-guessing' history with scientists, but what could even the most talented modern do to upstage a Mozart or a Shakespeare? Most similar to "Newton's Gift," however, are (for example) Gregory Benford's "In the Dark Backward" (*Science Fiction Age*, June 1994) where Shakespeare and Hemingway are visited, and Jack McDevitt's "The Fort Moxie Branch" (*Full Spectrum*, October 1988) where Hemingway and Thomas Wolfe appear. Read these stories, and then compare/contrast their descriptions of how story characters react to the appearances of time travelers, to Newton's behavior in "Newton's Gift."

Appendix C
Computer Simulation of the Entropic Gas Clock

%**gasclock.m**/created by PJNahin for *TIME MACHINE TALES*(6/27/2015)
%This MATLAB m-file simulates the diffusion of gas molecules in a sealed
%container by using the Ehrenfest ball exchange rules. The simulation
%starts with n molecules (i.e., balls) of one type (i.e., black) on
%one side of the container, and n more molecules of another type (i.e.,
%white balls) on the other side. The two urns play the roles of the
%two sides of the container. To simulate the ball (molecule)
%movements, the program selects two random numbers from 0 to 1, which
%are then compared to the current probabilities of selecting a black
%ball from urn I and a white ball from urn II. If BOTH random numbers
%are greater than these two probabilities then a white ball has been
%selected from urn I and a black ball has been selected from urn II,
%and so the number black balls in urn I is increased by one while the
%number of white balls in urn II is increased by one. If BOTH random
%numbers are less than or equal to these two probabilities then a
%black ball has been selected from urn I and a white ball has been
%selected from urn II and so the number of black balls in urn I is

© Springer International Publishing AG 2017
P.J. Nahin, *Time Machine Tales*, Science and Fiction,
DOI 10.1007/978-3-319-48864-6

```
%decreased by one while the number of white balls in urn II is
decreased
%by one. If one of the random numbers is greater than its
corresponding
%probability while the other random number is less than its
%corresponding probability, then no action is taken because then a
%white (black) ball moves from urn I to urn II at the same time a
white
%(black) ball moves in the opposite direction. That is, there is
no
%net change. Then, the ball selection probabilities are recalcu-
lated and
%another ball exchange is simulated.
rand('state',100*sum(clock))        %new seed for the random number
generator;
n=100;                              %number of balls in each urn;
nb1=n;                              %number of black balls INITIALLY
in urn I;
nw2=n;                              %number of white balls INITIALLY
in urn II;
pb1=nb1/n;                          %probability of selecting a black
ball from urn I;
pw2=nw2/n;                          %probability of selecting a white
ball from urn II;
for trials =1:1000;
   system(trials)=pb1;
   ball1=rand;
   ball2=rand;
   if(ball1>pb1&ball2>pw2)          %white ball selected from urn I
      nb1=nb1+1;                    %and black ball selected from
      nw2=nw2+1;                    %urn II;
   elseif(ball1<=pb1&ball2<=pw2)    %black ball selected from urn I
      nb1=nb1-1;                    %and white ball selected from
      nw2=nw2-1;                    %urn II;
   end
   pb1=nb1/n;
   pw2=nw2/n;
end
plot(system)
axis([1 trials 0 1])
grid
xlabel('time, in arbitrary units')
ylabel('fraction of balls in urn I that are black')
figure(1)
```

Epilogue

[Science fiction] cannot be good without respect for good science . . . This does not include time machines, space warps and the fifth dimension; they will continue to exist in the hazy borderland between [science fiction] and fantasy.[8]

In many science-fiction stories, the trip into the past is by way of some futuristic machine that can take you through time at will . . . That, however, is totally impossible on theoretical grounds. It can't and won't be done.[9]

The opening quotations, particularly the second one from Asimov who was one of the great modern writers of science fiction, is a gloomy one indeed for fans of time travel, but it is not difficult to find inconsistency in Asimov's own tales dealing with the concept. Asimov is famous, in particular, for his stories of robots, and the very last such tale that he wrote combines robotics with time travel, with a robot sent two centuries into the future.[10] At the start of the story, the narrator tells us that time travel to the past is impossible because the past is unchangeable and (of course) a time traveler would necessarily disturb history. (That is (of course) simply a failure to distinguish between the difference of *changing* the past and *affecting* the past, as well as a failure to see how the principle of self-consistency negates the issue of paradoxes.) Then, when the robot returns from the future (and so backward time travel is *not* impossible!), he reports that his arrival had been expected, that history had recorded that he would appear. At the end of the story we learn how the future knew this—it had read "Robot Visions"! So now Asimov *uses*

---

[8]Harry Harrison, in his essay "With a Piece of Twisted Wire . . .," *SF Horizons* (no. 2), 1965. Harrison (1925–2012) was a well-known (if little appreciated outside the SF community) writer, whose 1966 novel *Make Room! Make Room!* was the inspiration for the excellent (if somewhat depressing) 1973 film *Soylent Green* (a movie about future over-population of the Earth that will make you think twice about ever eating a cookie again).

[9]From an essay Isaac Asimov wrote on the time travel movie *Peggy Sue Got Married* for the *New York Times*, October 5, 1986.

[10]I. Asimov, "Robot Visions," *Isaac Asimov's Science Fiction Magazine*, April 1991.

© Springer International Publishing AG 2017
P.J. Nahin, *Time Machine Tales*, Science and Fiction,
DOI 10.1007/978-3-319-48864-6

the principle of self-consistency, with the narrator realizing that he *must* preserve his story so the future can read it.

Not a very consistent story! Asimov, was, of course, writing a story for entertainment's sake, so perhaps it's unreasonable to hold him *scientifically* accountable (although logic wouldn't seem to be too much to ask for).

In any case, was Asimov right? Lots of his fellow science fiction writers certainly thought so. One, for example, bluntly asserted that

Time travel is inconceivable.[11]

Other critics agreed:

In science fiction we find the lunatic fringe more often than not trying to perfect time-travel mechanisms.[12]

and

Scientifically, time travel can't stand inspection.[13]

and

Time travel is . . . scientific nonsense.[14]

and

It would be untrue . . . to present the idea of a time machine as anything but what it is, an intriguing literary device, part of the bag of tricks of the science fiction writer . . . There is no such thing as a 'science' of time travel.[15]

You'll notice that these pronouncements are from decades ago: Conklin (1904–1968), Gold (1914–1996), and Oliver all wrote just 3 years after Gödel, and so perhaps it was simply too soon for his work to be widely known outside of the physics community. But physicists have learned a lot since 1952! Have they learned enough to make Asimov and his fellow SF skeptics (if they were still alive) change their minds, or at least reconsider? I suspect not.

I say that because, even 25 years after Conklin, Gold, and Oliver wrote, while we do find an awareness of Gödel starting to appear in the science fiction world, a feeling of skepticism was still in the air. In a fascinating analysis[16] of the first half-century of the science fiction magazines, Paul Carter admitted that there *is* a rationality to time travel because of Gödel but, nonetheless, the conventional view remained that backward time travel is simply impossible. Then, citing the work of Tipler, Carter wrote "Only as recently as 1974 (see note 130 in Chap. 1), in

---

[11]Kingsley Amis, *New Maps of Hell*, Harcourt 1960.

[12]Groff Conklin, *Science Fiction Adventures in Dimension*, Vanguard 1953.

[13]H. L. Gold, *The Galaxy Reader of Science Fiction*, Crown 1952.

[14]Alexei Panshin, *The Mirror of Infinity*, Canfield 1970.

[15]Chad Oliver, "The Science of Man," a non-fiction essay included in Oliver's 1952 time machine novel *Mists of Time*. Chad Oliver (1928–1993) was a scientist by profession (anthropology), and his opinion carried weight among SF writers *and* (non-physicist) scientists.

[16]P. A. Carter, *The Creation of Tomorrow*, Columbia University Press 1977.

the sober pages of the *Physical Review*, has a physicist been more bold ... For 70 years in the meantime, however, without waiting for Professor Tipler to solve his equations ... writers had happily helped themselves to Mr. Wells' invention and sent their characters through time in every direction, forward, backward, and sideways."[17]

In the 1980s writers were apparently just as unaware of Gödel's time travel analyses (and of the much later ones of Tipler) as had been the 1950s commentators. In his marvelous 1985 book *The Past Is a Foreign Country*, for example, David Lowenthal repeatedly refers to time travel as "fantasy" and to science fiction stories about time travel as "unbridled by common sense." And for another example from the start of the 1980s, consider the case of James Gunn (born 1923), professor of English at the University of Kansas, past president of both the Science Fiction Writers of America and the Science Fiction Research Association, author of *The Immortals* (inspiration for the 1970–1971 TV series of the same name), and eminent scholar (see his 1975 book *Alternate Worlds*). His literary credentials are impeccable and his critical influence profound. And yet, 30 years after Gödel and 5 years after Tipler, Professor Gunn wrote in *The Road to Science Fiction*, "Time travel has been an anomaly in science fiction. Clearly fantastic—there is no evidence that anyone has ever traveled in time and *no theoretical basis for believing that anyone ever will* [my emphasis]." If you've read this book carefully, however, of the analyses by Gott, Krasnikov, Thorne, Alcubierre, Novikov, Natário, and others, you know that what Gunn claims in those last words is actually not necessarily so.

The British-born American theoretical physicist Freeman Dyson of the Institute for Advanced Study has commented[18] on that sort of narrow mindset, with words quoted from the 1979 physics Nobel prize winner Steven Weinberg, words reminding us that rigidity concerning time travel is not limited to science fiction writers: "This is often the way it is in physics—our mistake is not that we take our theories too seriously, but that we do not take them seriously enough. It is always hard to realize that these numbers and equations we play with at our desks have something to do with the real world. Even worse, there often seems to be a general agreement that certain phenomena are just not fit subjects for respectable theoretical and experimental effort." The words *time travel* and *time machine* are never mentioned, but could they have been far from either Weinberg's or Dyson's thoughts?

All through this book we have seen how people have argued against time travel to the past (Tipler's cylinder is unphysically long, Gödel's universe requires an unphysical rotation, wormholes and warps require unphysical energy conditions, what about all those paradoxes ... and on and on). These arguments remind me of

---

[17]Given that *The Time Machine* was published in 1895, it is not clear how Carter arrived at the value of 70 until Tipler's work (he should have written 79), and of course it was only 54 years between Wells' time travel fiction and Gödel's time travel mathematical physics.

[18]F. J. Dyson, "Time Without End: Physics and Biology in an Open Universe," *Reviews of Modern Physics*, July 1979, pp. 447–460.

the debate in the 1930s between the illustrious British astrophysicist Sir Arthur Eddington and the young Indian astrophysicist Subrahmanyar Chandrasekhar (1910–1995), winner of the 1983 Nobel prize in physics. In his analyses of the life history of stars, Chandrasekhar had arrived at an astonishing conclusion, one that Eddington simply could not accept. As Eddington sarcastically explained in an address at Harvard University in the summer of 1936, "Above a certain critical mass (two or three times that of the sun), the star could never cool down, but must go on radiating and contracting until heaven knows what becomes of it. That did not worry Chandrasekhar, he seemed to like the stars to behave that way, and believes that is what really happens."[19] Eddington then went on to declare such 'unbelievable' behavior to be nothing less than "stellar buffoonery."

As far as Eddington was concerned, Chandrasekhar had simply made an error in combining relativity theory with non-relativistic quantum theory. Indeed, so appalled was Eddington at the thought of a star contracting "until heaven knows what becomes of it" (that is, until it gravitationally collapses into a black hole) that he had earlier, in 1935, stated "There should be a law of nature to prevent a star from behaving in this absurd way!" Today, of course, no astrophysicist feels the need for a 'star protection conjecture'—which perhaps reminds you of another, more recent 'protection conjecture.'

What can one conclude from all the similar controversy concerning time travel, time machines, and spacetime warps? Not much, I think, except that these are open issues and will remain the subjects of on-going study for a long time yet to come. The one thing I am fairly certain of is that if time travel is ever achieved, it will be by means that we cannot today even begin to guess. It will almost certainly require *at least* a mutant child genius with an IQ of 270 to fix the slightly broken time machine found abandoned in a cellar![20] But that view isn't uniformly shared across all of science fiction. I very much doubt, for example, that things will be quite so elementary as depicted in the story[21] where the time machine was so simple that "If it were taken apart or put together before you, your wife, or the man across the street, you would wonder why you didn't think of it yourselves." Not only that, but its power source was just two dry cell batteries!

The time machine in an earlier story is almost as simple, requiring (besides a piece of strange crystal) only a "little stack of dry cells, a Ford [automotive ignition] coil, a small brass switch, a radio 'B' battery, an electron tube, and a rheostat."[22] Even Wells' *Time Machine* couldn't resist making it all look easy: as one critic put it, "The time machine, like all products of supreme inventive genius, was a

---

[19]See S. Chandrasekhar, *Eddington: The Most Distinguished Astrophysicist of His Time*, Cambridge University Press 1983, p. 48.

[20]F. B. Long, "A Guest in the House," *Astounding Science Fiction*, March 1946.

[21]R. Abernathy, "Heritage," *Astounding Science Fiction*, June 1947.

[22]J. Williamson, "In the Scarlet Star," *Amazing Stories*, March 1933.

remarkably simple affair. A few rods, wires, some odd glass knobs—nothing more!"[23] That sort of simplistic fictional description of a time machine reminds me of the reaction of the great Polish science fiction writer Stanislaw Lem to the general treatment of time travel in the genre: "There have been mountains of nonsense written about traveling in time, just as previously there were about astronautics—you know, how some scientist, with the backing of a wealthy businessman, goes off in a corner and slaps together a rocket, which the two of them—and in the company of their lady friends, yet—then take to the far end of the Galaxy. Chronomotion, no less than Astronautics, is a colossal enterprise, requiring tremendous investments, expenditures, planning ..."[24]

An example of what Lem was talking about is the 1956 novella *Arcturus Landing* by Gordon R. Dickson (1923–2001). There we read of aliens who have confined humans to the solar system—until (if) Earth scientists discover the secret of FTL travel. So, a genius physicist does just that (with no mention of spacetime engineering, but rather we encounter a lot of mumbo-jumbo gibberish as the 'explanation'), and uses it to instantly transport himself and some friends to a planet orbiting Arcturus.[25] And when they get there the friendly aliens speak perfect English.

Lem would have snorted in derision, too, at this statement made to a prospective graduate student by the head of a college physics department, that the college "has been awarded a million dollars to build [a time machine]. It means ... a raise for me and maybe a doctorate for you, so we'll build one and have some fun doing it."[26] Is it any wonder that Lem so readily dismissed stories that reduce space (and time) travel to weekend adventures in a home laboratory? As Lem wrote in another essay, time travel and its close relation, FTL space travel, have reduced much of science fiction to "a bastard of myths gone to the dogs."[27] Because of precisely that, Harry Harrison wrote (note 1) of the early science fiction magazines that published so much nonsense, "I used to moan over the fact that pulp magazines were printed on pulp paper and steadily decompose back towards the primordial from which they sprang. I am beginning to feel that this is a bit of a good thing."

I don't know whether time travel to the past can actually be accomplished, but I do know that speculations once thought to be as outlandish as finding the Philosopher's Stone for turning base elements into gold, *have* eventually been realized (and, come to think of it, with modern nuclear physics we *have* learned how to turn lead into gold, if only a few atoms at a time). Television, nuclear power, home computers that run at multi-gigahertz clock rates in the bedrooms of high school

---

[23]W. B. Pitkin, "Time and Pure Activity," *Journal of Philosophy, Psychology and Scientific Methods*, August 27, 1914, pp. 521–526.

[24]S. Lem, "The Twentieth Voyage of Ijon Tichy," in *The Star Diaries*, Seabury Press 1976.

[25]A journey incorrectly given in the story as 120 light years, when in fact it is less than 40 light years.

[26]W. West, *River of Time*, Avalon Books 1963.

[27]S. Lem, "Cosmology and Science Fiction," *Science-Fiction Studies*, July 1977, pp. 107–110.

students, even faster computers that animate our movies and simulate the formation of black holes and galaxies, voyages to the Moon and back—all these amazing developments would be pure magic to nineteenth century science. The ghosts of not just a few Victorian scientists who had poo-pooed the possibility of such things, have watched their reputations eat a lot of posthumous crow during the last 150 years.

My personal position on the question of time travel leans towards the rejoinder made to the skeptic in one science fiction story who, even after having done some time traveling, *still* argues against it by invoking paradoxes. He is sharply rebuked with "Oh, for heaven's sake, shut up, will you? You remind me of the mathematician who proved that airplanes couldn't fly."[28] I subscribe to the optimistic philosophy of the British writer Eden Phillpotts (1862–1960), who wrote in his 1934 novel *A Shadow Passes* "The Universe is full of magical things, patiently waiting for our wits to grow sharper." Perhaps he had a famous saying by the British-born Indian scientist J. B. S. Haldane (1892–1964) in mind, words from his 1928 *Possible Worlds*: "Now my suspicion is that the universe is not only queerer than we suppose, but queerer than we can suppose."

Still, even if time travel is possible, the engineering phase will surely be tough going. I am certain that before we see a working time machine, there will be many, *many* episodes like the one described in a very funny, novel-length spoof of academic research.[29] All physicists and engineers who have tried to get some stubborn piece of apparatus to work, apparatus that *should* work and simply won't, will appreciate Professor Demetrious Demopoulos' frustration and will, I am sure, forgive him his intemperate language:

> ... the distinguished physicist took a step back and, arms akimbo, surveyed the complex and sophisticated machine that was the culmination of years of dedicated scientific research and pains-taking technological development.
>
> "What a pile of ****," he said.
> "Oh, no, Dr. Demopoulos, don't say that!"
> "Well, it is." A sneer formed on the professor's thin lips. "Time machine, my ****. This thing couldn't give you the time much less travel in it."
> "But we haven't incorporated all our latest test data yet," the pretty research assistant reminded him. "These last few adjustments might do it, Professor."
> "Hell, we've been tinkering with it for 2 years," Demopoulos complained.
>
> "We've tried everything and it's all come to dog ****."

[28]R. Heinlein, "By His Bootstraps," *Astounding Science Fiction*, October 1941. As discussed at the end of Chap. 1 (in "For Further Discussion") the mathematician was the American astronomer Simon Newcomb.

[29]J. DeChancie and D. Bischoff, *Dr. Dimension*, ROC 1993.

That scene probably won't actually happen for a long time to come but, even before the practical nuts-and-bolts bugs in the Professor's time machine are worked out, I think some adjustments are called for in our thinking about time travel. I believe that present-day philosophers and science fiction writers are going to have to become knowledgeable about the work by physicists on time travel. It simply won't do any longer for Philosophy Professor X to invoke the grandfather paradox during a discussion of causality and free will and then airily declare them to be 'obviously' incompatible with time travel to the past. And it simply won't do any longer for Famous SF Writer Y to send his hero into the past to kill Hitler as a baby and thereby change recorded history. One might as well keep watching a video recording of the 9/11 destruction of the World Trade Center, in the vain hope that maybe, on the next viewing, the planes will miss.

The principle of self-consistency around closed timelike curves is going to have to become as much a part of the science fiction writer's craft (or else she will be a writer of fantasy) as it will have to become part of the fundamental philosophical axioms.[30] The 'time police,' like the "operatives of the Bureau of Time Exploration and Manipulation" that appeared in the science fiction of Andre Norton (1912–2005), will have to be put out to pasture with the unicorns and telepathic dragons of fantasy fiction. Just as the recent physics literature on time machines has displayed a growing awareness of what science fiction writers and philosophers have had to say on the subject of time travel, so too are writers and philosophers going to have to learn some more physics. Most people can enjoy a good fantasy tale now and then, but the use of 'magic mirrors' to see through time is *not* physics. Such devices were popular and acceptable in medieval times—see "The Squire's Tale" in Chaucer's *The Canterbury Tales*, and later (see Act IV of *Macbeth*)—but good science fiction needs much more than that today.

Time travel to the past is a beautiful, romantic idea, and some words written by two physicists in a technical paper—words embedded in the midst of swirls of tensor equations—show that even hard-nosed physicists can share this dream: "In truth, it is difficult to resist the appealing idea of traveling into one's own past . . ."[31] The appeal of that dream is explained in Ray Bradbury's Foreword to a beautiful little 1989 book by Charles Champlin (*Back There Where the Past Was*). In it Bradbury clearly illuminated *why* we want to go back into the past. It is for the same reason that we go, time and again, to see *Hamlet*, *Othello*, and *Richard III*: "We don't give a hoot in hell who poisoned the King of Denmark's semicircular canal.

---

[30] Bud Foote (1930–2005), late professor of English at Georgia Tech, wrote (in his book *The Connecticut Yankee in the Twentieth Century: Travel to the Past in Science Fiction*, Greenwood Press 1991) that consistency is simply a well-used plot device: "The attempt of the time traveler to prevent something or take advantage of it [and so causing] the event in question, is so popular and so ubiquitous that it seems to be about worn out." Worn out or not, I believe that plot device to be correct science.

[31] A. J. Accioly and G. E. A. Matsas, "Are There Causal Vacuum Solutions with the Symmetries of the Gödel Universe in Higher-Derivative Gravity?" *Physical Review D*, August 15, 1988, pp. 1083–1086.

We already know where Désdemona lies smothered in bedclothes and that Richard goes headless at his finale. We attend them to toss pebbles in ponds, not to see the stones strike, but the ripples spread."

That's why a visit to the past is so mysteriously and marvelously fascinating. It would let us watch ripples spread through time. Our own visit to the past, in fact, might even be the pebble in the pond of history that starts an interesting ripple or two that will one day sweep over—us! (Take a look at Appendices A and B) Who would want to miss that? Indeed, if modern philosophers are right, if the analyses discussed earlier in this book are correct, you *can't* (didn't/won't) miss it.

I think time travel appeals, irresistibly, to the romantic in the soul of anyone who is human.[32] A time traveler does not exist either *here* or *then*, but rather *everywhen*. For a time traveler passing back and forth through the ages, history would be the ultimate puzzle, a chronicle described in one novel as beginning "not in one place, but everywhere at once ... It might be begun at any point along the infinite, infinitely broken coastline of time."[33] Romanticism doesn't preclude there also being a dark side to visiting the past, of course, as one time traveler from 1989 learns when he takes up residence in 1962. Falling asleep on a hot summer night in that long-ago year, he thinks "JFK slept. Oswald slept. Martin Luther King slept. [I sleep and dream] of Chernobyl ... *I am a cold wind from the land of your children*."[34]

But, I must admit, I personally am more attracted by happier descriptions of time travel. In his marvelous 1996 book *1939: The Lost World of the Fair*—which is proof that there are not enough Pulitzers to go to all the books that deserve one—Yale professor David Gelernter caught just the right spirit in his Prologue: "The best of all reasons to return to the fair is that travel is broadening, and time travel most of all ... The 1939 New York World's Fair is one amazing show. It still stands, undisturbed on Flushing Meadow, just over the edge of time; it would be an unforgivable shame to miss it." Trust me—if you read Gelernter's book, you'll come as close as you can in today's world to taking a ride in a 'time machine'!

The eminent philosopher Sir Karl Popper opens his biography with a wonderful story about his apprenticeship as a young man in 1920s Vienna to a master cabinetmaker.[35] After winning the old man's confidence, the student learned his mentor's great, secret desire: For years the master had been looking for the solution to perpetual motion. He knew what physicists thought of such machines, but nonetheless he had never given up his dream: "They say you can't make it; but

---

[32] How else to explain the pleasure, for modern children *and* adults too, in watching rebroadcasts of the 1960s animated TV cartoon program 'starring' Mr. Peabody, a nice but slightly stuffy, professorial white beagle. (Don't all dogs wear glasses and a bow tie?) Mr. Peabody, with his brainy adopted son Sherman, routinely travels into the past in the "Way-Bac" machine to see what *really* happened in history.

[33] John Crowley, *Great Work of Time*, Bantam 1991.

[34] R. C. Wilson, *A Bridge of Years*, Doubleday 1991.

[35] See volume 1 of *The Philosophy of Karl Popper* (P. A. Schilpp, editor), The Library of Living Philosophers, Open Court 1974, p. 3.

once it's been made they'll talk differently." Popper's master sounds just a bit like the American writer Gertrude Stein (1874–1946) in her 1938 essay "Picasso," where she writes "It is strange about everything, it is strange about pictures, a picture may seem extraordinarily strange to you and after some time not only it does not seem strange but it is impossible to find what there was in it that was strange." Might we one day say the same thing about time travel?

An alternative point of view can be found in a discussion of time travel via cosmic strings that makes this assessment: "While there is still hope that one day a sufficiently clever design may make building a time machine possible, it is beginning to seem more and more improbable. Like the perpetual motion machines of the nineteenth century, the designs have an elegant simplicity (as well as enormous commercial potential), but it seems that Nature also may abhor them just as much."[36] Of course, at one time it was thought that Nature abhorred a vacuum, but then we learned that she must actually love a vacuum because else why did she make so much of it?!

The theoretical basis for time travel is very different from that of perpetual motion (there is *more* reason to accept time travel as a plausible possibility). And so maybe one day, *just maybe*, the first time traveler will receive a toast such as the one in a story telling us about the arrival of the inventor of the first time machine and his no longer skeptical friend in the Civil War past:

"To you, Mac," I said.
McHugh loosened his tie. "To the Creator," he said, "who has given us a Universe with such marvelous possibilities."[37]

---

[36]B. Allen and J. Simon, "Time Travel on a String," *Nature*, May 7, 1992, pp. 19–21.

[37]J. McDevitt, "Time's Arrow," *Isaac Asimov's Science Fiction Magazine*, November 1991.

Glossary[38]

Action the integral over a **world line** of a quantity called the *Lagrangian*. When a massive particle is moving at non-relativistic speed through a gravitational field, for example, the instantaneous value of the Lagrangian is the difference between the kinetic and potential energies of the particle. For other types of fields (such as the electromagnetic) and/or relativistic motion in any type of field, the Lagrangian is different. In any case, however, the actual world line of the particle is the one for which the integrated Lagrangian, that is, the action, is minimized. See **least action**.

Action at a distance the direct interaction of two separated objects, without concern for the details of what (if anything) occurs in the region between the objects (see also **field**). Newton's theory of gravity is action at a distance, whereas Einstein's theory of gravity is a *field* theory.

Advanced solution the prediction, by Maxwell's electromagnetic field equations, of radio waves that travel into the past (see also **Dirac radio**).

Anti-matter quantum mechanical prediction (experimentally verified) that all fundamental particles of matter come in two forms (the 'normal' version and the 'anti-matter' version). The positron, for example, is the anti-matter version of the electron, differing only in the sign of its electric charge. The photon, on the other hand, is its own anti-particle. A **subluminal** anti-particle traveling forward in time can be thought of as its 'normal' version traveling backward in time.

Arbitrarily advanced civilization for time travel discussions, a civilization with a technology sophisticated enough to construct a traversable **wormhole** in spacetime. More generally, Types I, II, and III of such civilizations are, respectively, those that can control 10^{13} W, 10^{27} W (the total power output of their home star), and 10^{38} W (the total power output of their home galaxy).

---

[38] "I hate definitions." (Usually attributed to writer and British Prime Minister Benjamin Disraeli (1804–1881) but, more precisely, they are the words of one of the characters in his 1826 novel *Vivian Grey*.)

—but they *can* be useful

© Springer International Publishing AG 2017
P.J. Nahin, *Time Machine Tales*, Science and Fiction,
DOI 10.1007/978-3-319-48864-6

Arrow of time the statement the time appears to have a direction, that there is a difference between the past and the future. There are several different arrows: the psychological (we remember the past, we anticipate the future), the thermodynamic (organized systems evolve toward disorganization, that is, **entropy** increases as time increases), the electromagnetic (radio waves propagate *away* from their generators), and the cosmological (the expansion of the universe is directed toward the future).

Asymptotically flat if the geometry of a curved spacetime is such that, as one moves ever further away from all matter and energy, the spacetime **metric** becomes that of flat **Minkowski spacetime**, then the curved spacetime is said to be *asymptotically flat*. As a counter-example, the spacetime of a **Tipler cylinder** time machine is *not asymptotically flat*.

Autoinfanticide paradox see **grandfather paradox**.

Averaged null energy condition the claim that the *averaged* value of the observed mass-energy density along the entirety of any **null geodesic** is non-negative.

Averaged weak energy condition the claim that the *averaged* value of the observed mass-energy density along the entirety of any **timelike world line** is non-negative.

Back reaction the tendency of spacetime to resist the formation of **closed timelike lines** (see also **stress-energy divergence**).

Bell's theorem an inequality that either holds or does not hold, depending on whether quantum mechanics is non-local or local, respectively.

Big Bang the singular beginning of spacetime.

Big Crunch the singular end of spacetime.

Bilking paradox what would happen if a **causal loop** were disrupted. For example, suppose a time traveler builds a time machine using plans he received years earlier from a mysterious stranger. He now realizes that the stranger was himself, using the time machine to travel back into the past to give his younger self the plans. A bilking paradox would be created if the time traveler builds the time machine, verifies that it works, and then decides *not* to visit his younger self to hand over the plans. See also **bootstrap paradox**.

Black hole a region of spacetime where gravity is so strong that nothing can escape, including light. Black holes are thought to be created when sufficiently massive stars burn out (see **white dwarf** and **neutron star**) and undergo *gravitational collapse*. A black hole of ten solar masses would have a radius of about twenty miles. Black holes might have been created at the Big Bang singularity and, if so, could theoretically come in any mass and size (a black hole with the mass of the Earth would have a diameter of less than half an inch).

Block universe a spacetime in which all world lines are completely determined from beginning to end (a fatalistic universe). There is no free will in such a spacetime.

Boost matrix matrix formulation of the **Lorentz transformation**.

Bootstrap paradox the puzzle of the origin of *information* on a closed loop in time. The classic example is that of a time traveler from the future giving his younger-self the plans for the time machine the time traveler has just used to visit the past so that he can then build the time machine to visit the past. The time machine plans appear not to have been *created* by anyone! The plans just *are*. See also **bilking paradox**.

Cauchy horizon a **spacelike** hypersurface in spacetime that intersects, exactly once, every **timelike** world line that has no end point. Knowledge of the conditions on such a surface uniquely determines the spacetime at all other points.

Causal loop a time loop containing an event caused by a *later* event that, itself, is caused by the earlier event (see the example in **bilking paradox**).

Causality the metaphysical claim that every event is caused by a prior event. Time travel to the past inherently violates causality.

Chronal regions those parts of spacetime that have no closed timelike curves.

Chronology horizon a (hyper)surface in spacetime that separates **chronal** and non-chronal regions. It is a special case of a **Cauchy horizon**.

Chronology protection the claim, as yet unproved, that time machines and time travel to the past are impossible because of the **back reaction** of spacetime will lead to **stress-energy divergence**. Popularized among physicists as the *Hawking chronology protection conjecture* (1992), Hawking has since admitted that stress-energy divergence is *not* sufficient to enforce his conjecture.

Chronon science fiction name for **Planck time**.

Closed timelike line (or curve) a **timelike** world line of finite length that has no ends, i.e., that forms a *closed loop* in spacetime. A region of spacetime containing closed timelike lines is said to be a **time machine**.

Conservation law physical quantities in interacting systems that remain unchanged are said to be conserved. Total energy, total momentum (linear and angular), and electric charge are conserved quantities.

Cosmic string hypothetical, threadlike spacetime structures with enormous mass-energy and density that may have formed during the **Big Bang**. Cosmic strings may have been initially formed either as infinitely long, or as closed loops, and it is the former that are thought to be physically meaningful in the present-day universe. Cosmic strings do not violate the **weak energy condition** (as do **wormholes**), and they can theoretically create **closed timelike lines**.

Cosmological constant an extra term specifically added by Einstein to the general theory of relativity to keep that theory from predicting the expansion of the universe (which was later observationally found to actually be the case). Einstein subsequently said that his failure to believe the general theory's original prediction of the expansion of the universe was the greatest mistake of his life. The constant (which today is believed to be almost zero, if not exactly zero) appears in Gödel's rotating time travel spacetime as a determining factor in the minimum radius of a **closed timelike line**.

Determinism the metaphysical belief that effects are uniquely determined by causes (this is *not* **fatalism**).

Dirac radio science fiction gadget for sending information at infinite speed, which thus travels backward in time (see also **ultraluminal**).

Dominant energy condition the **weak energy condition** *plus* the claim that the observed energy flux is never **superluminal**.

Electron fundamental particle of mass that possesses one quantum of negative electric charge. *Bound* electrons orbit the nuclei of atoms and plays a central role in determining the chemical properties of the elements and of their compounds. *Free* electrons carry electric current, either in conductors (wires) or through space.

Elsewhen the collection of spactime events that cannot be reached from the **here-now** with a **timelike world line**.

Entropy a measure of the randomness of a system that plays a central role in the thermodynamic **arrow of time**.

Ether a substance once thought to fill all space to allow radiation 'something to propagate through' (as opposed to simply a vacuum). The special theory of relativity showed that the ether is an unnecessary concept because it has no observable effects (physicists argue that if something is impossible to detect, then it is meaningless to talk about it being part of *science*).

Event a point in spacetime.

Event horizon the spacetime surface of a black hole or of a non-traversable wormhole, at which light can *just* escape to the outside universe. It is called a *horizon* because, by definition, an external observer can't see beyond it and into the interior of the hole. To see the inside of a hole you must enter the hole by crossing the horizon (but then you can't get out).

Exotic matter matter that violates one or both of the **weak/strong energy conditions**. Exotic matter appears in the theories of wormholes and warp drives.

Fatalism the metaphysical belief that all events have been *pre*determined from the beginning of time.

Field the concept that if a physical law is local, then it is describable by differential equations that relate what is 'happening' at every point in spacetime to what is 'happening' at its closely located neighboring points. Electromagnetism and general relativity are field theories, for example, described by sets of partial differential equations called *Maxwell's equations* and *Einstein's gravitational field equations*, respectively.

Fourth dimension either time or a fourth spatial dimension.

Frame of reference a spacetime coordinate system.

Free will the condition that prevails when we can *choose* to do what we do. There is no free will in a **block universe**.

Future the collection of spacetime events that can be reached from the **here-now** via a **timelike world line** directed toward a later time (for each individual, the future is what hasn't yet been experienced).

Gamma ray very high-energy, very high-frequency electromagnetic radiation. Gamma rays have frequencies on the order of *ten trillion* (10^{13}) times greater than those of AM radio broadcast radio waves.

General theory of relativity Einstein's theory of curved spacetime, which explains gravity in terms of nothing but geometry. Its fundamental premise is that *all* the laws of physics should appear the same to all observers in *any* **frame of reference**. It is believed the theory will fail when the local mass-energy density reaches a level of about 10^{94} g/cm^3, a density so enormous (the density of water is just 1 g/cm^3) that there is no known mechanism for achieving it anywhere in the universe except in another **Big Bang**. See also **Planck density**.

Geodesic the shortest path connecting two points in space (if the space is spacetime, the world line of a particle in free-fall).

Global in the large.

Gödel universe a spacetime that, unlike the one we live in, is rotating so fast that it automatically generates *closed timelike lines* and thus constitutes a weak **time machine**. In such a universe, time travel to the past would be a natural phenomenon.

Grandfather paradox *the* classic time travel paradox, of a time-traveler killing, while in the past and *before* the time traveler has been conceived, an ancestor directly linked to the future birth of the time traveler. A more direct form of this sort of paradox is simply the time traveler killing his own younger self (called the **autoinfanticide paradox**).

Gravitational field equations a set of coupled, partial differential, non-linear tensor equations, considered to be the most complicated equations in all of mathematical physics. They show how the local curvature of spacetime depends on the local mass-energy of spacetime. The equations are independent of the **topology** of spacetime.

Gravitational lensing the ability of gravitational fields to bend and focus light.

Graviton the quantum particle of gravity.

Hawking radiation the emission of particles (energy) by a **black hole** into the region *outside* its **event horizon**, which results in the eventual evaporation of the hole. This is a quantum mechanical effect.

Here-now the point or **event** (for each observer) in spacetime that separates the **past**, the **future**, and **elsewhen**.

Hyperspace any space of four or more dimensions (for example, four-dimensional spacetime is a hyperspace).

Inertial frame any frame of reference in which Newton's laws of mechanics are true (there are no *acceleration forces* in inertial frames, and so *rotating* or 'merry-go-round' frames are not inertial).

Invariance a quantity that remains the same in any frame of reference is an invariant. Two examples are the distance between any two points on a piece of paper (because it is independent of any particular coordinate system), and the speed of light.

Kerr-Newman black hole a *rotating* black hole, which may (or may not) be electrically charged.

Krasnikov tube a particular spacetime **metric** (or *warp*) allowing **superluminal** travel, with the great difficulty of requiring *enormous* negative energy. Two Krasnikov tubes can be made into a time machine. Named after its Russian inventor.

Least action general principle in physics that asserts the world line of a particle is the one that *minimizes* the **action**.

Light cone the **lightlike** surface in spacetime that, at each point in spacetime, separates the **past** from the **future** from **else-when** from the **here-now**.

Lightlike the world line of a photon (or of any other form of mass-energy traveling *at* the speed of light).

Li mirror a perfectly reflecting, spherical surface that can be used to stabilize a wormhole against energy loops circulating through a **wormhole** time machine (thus creating unbounded energy levels that destroy the time machine). Named after its Chinese inventor.

Local in the small.

Lorentz factor the ubiquitous square-root expression that appears in so many relativistic calculations, such as time dilation, length contraction, and the variation of mass with speed. For example, the mass m of a moving body is not independent of it speed v but rather varies as $m = \dfrac{m_0}{\sqrt{1-\left(\frac{v}{c}\right)^2}}$, where m_0 is the rest mass (that is, the mass when $v = 0$) and c denotes the speed of light (186,210 miles per second). The denominator is the Lorentz factor.

Lorentz-FitzGerald contraction the conclusion from special relativity that the appearance (to a stationary observer) of a moving object will be shortened in length along the direction of motion. Many years after Einstein's work, it was shown that the object will also appear to be *rotated*.

Lorentz transformation equations from the special theory of relativity that describe how the space and time measurements of two relatively moving observers are related.

Many-worlds interpretation quantum mechanical view of splitting universes.

Mass-energy the famous $E = mc^2$, the equation behind atomic fission and nuclear fusion weapons.

Metric the measure of the separation between any two events in a spacetime.

Minkowski spacetime the flat spacetime of the special theory of relativity. In this spacetime there is no gravity, no spacetime curvature (hence it is *flat*) and no backward time travel.

Neutron star the end state of a star with one to three solar masses that has collapsed to a density of up to 10^{17} g/cm^2.

Non-Euclidean geometry the geometry of spacetime, whether curved or flat. Spacetime is non-intuitive precisely because it is always hard to resist thinking in terms of high school Euclidean geometry, which is simply the *wrong* geometry.

Null geodesic the world line of a photon in spacetime.

Observer physicist's term for 'somebody' equipped with recording instruments (such as a clock, a pencil and notepad, and the like).

Parallel transport a procedure for moving a vector around any closed curve in a space to determine whether that space is flat or curved.

Parallel worlds simultaneous existence of multiple (perhaps) infinite versions of reality.

Past the collection of spacetime events that can reach the **here-now** via timelike world lines directed from an earlier time (for each individual, the past is what has already been experienced).

Photon the quantum particle of electromagnetism. A photon of frequency f has energy hf, where h is **Planck's constant**.

Planck density the density of mass-energy that distinguishes classical from quantum spacetimes; about 10^{94} g/cm^3, equal to the **Planck mass** divided by the cube of the **Planck length**.

Planck length the non-zero length in quantum theory (about 1.6×10^{-33} cm) below which quantum gravity effects will become important.

Planck mass the fundamental mass in quantum theory (about 22×10^{-6} g), but *not* the smallest non-zero mass in quantum theory.

Planck's constant fundamental constant in quantum theory, h, associated with the discrete nature of quantum effects. (If h had the value of zero, rather than its actual value of about 6.6×10^{-34} joule-seconds, then the microworld would appear to be continuous.)

Planck time the time interval in quantum theory (about 5.3×10^{-44} s) below which quantum gravity effects become important. The time required to travel the **Planck length** at the speed of light.

Positron the electron's anti-particle (see **anti-matter**).

Proper time the timekeeping of an observer's clock.

Pulps the old science fiction magazines, through the 1940s and into the early 1950s or so, published on inexpensive, wood-pulp paper.

Quantum foam see **topology**.

Quantum gravity the yet-to-be-discovered theory that unifies quantum field theory with the curved spacetime of general relativity.

Quantum mechanics the exact physics of the very small (atoms and things smaller).

Quantum theory any theory in which physical quantities are not continuous but rather assume their values in discrete jumps (the size of the jump is the *quantum*).

Recurrence paradox the claim that if you wait long enough, then every system will return to every previous state infinitely often.

Red dwarf small (less than about half a solar mass) star with a very long life (hundreds of times that of the Sun). They are 'cool' stars, with a surface temperature less than 4000 °C, and are thought to be the most common type of star in the universe.

Red shift the *down* shift in frequency of light received from all distant stars due to the Doppler effect induced by the expansion of the universe. The opposite effect is called a *blue shift*.

Reinterpretation principle asserts that negative mass-energy traveling forward in time is positive mass-energy traveling backward in time, and vice-versa.

Reissner-Nordström black hole a spherically symmetric, non-rotating electrically charged black hole.

Reversibility paradox based on the fact that the equations of physics contain no **arrow of time**; that is, they work equally well with time running forward or backward.

Roman ring a time machine made of two or more traversable wormholes connected in a closed sequence.

Schwarzschild black hole a spherically symmetric, non-rotating, uncharged black hole.

Self-consistency the assertion that the events on a closed **timelike** line must never be in contradiction; generally attributed to the Russian physicist Igor Novikov, who with his colleagues showed that it is not an independent assumption but rather an implication of the principle of **least action**.

Sexual paradox a special type of causal loop, where the connected events on a time loop are 'coupled' (pun intended!) through reproductive sex. An example is a time traveler to the past who becomes her own ancestor.

Singularity either a region in spacetime where the curvature becomes infinite and the laws of physics fail, or a point in spacetime beyond which world lines cannot be extended. Singularities of the first kind are called *curvature* or *crushing* singularities, and those of the second kind are called incomplete singularities. The **Big Bang** was a curvature singularity, as is the center of a black hole. In a **Schwarzchild black hole** the curvature singularity is a point, whereas in a **Kerr-Newman black hole** it is an extended region in the form of a ring.

Spacelike a world line on which propagating mass-energy would exceed the speed of light.

Spacetime the 'stuff' out of which reality is built. Everything there is—the universe—is the total collection of events in spacetime. A *flat* spacetime has no gravity, whereas a curved spacetime is the *origin* of gravity.

Special theory of relativity Einstein's theory of *flat* spacetime, which assumes that gravity is absent (gravity is the result of the geometry of *curved* spacetime). Its fundamental premise is that the laws of physics should appear the same to observers in different **inertial frames**.

Splitting universes the idea that every decision causes reality to split into separate copies, identical in every respect except for each of the different possible results of the decision.

Stargate science fiction name for the mouth of a traversable **wormhole**.

Stress-energy divergence the unbounded growth of the **general theory of relativity**'s measure of the density of mass-energy in spacetime.

Strong energy condition the claim that gravity is always (that is, locally) attractive. A traversable **wormhole** violates this condition.

Subluminal slower than light.

Superluminal faster than light.

Tachyon a particle (hypothetical, so far) that always travels faster than light, so its **world line** is always spacelike.

Temporally orientable spacetime any spacetime in which the direction of time at every point agrees with the direction of time at its local neighboring points.

Tensor mathematical generalization of the scalar and vector concepts. Einstein's **gravitational field equations** are tensor-differential equations (for example, the metric tensor contains information about the curvature of spacetime), whereas Newton's and Maxwell's equations are vastly less complex vector-differential equations.

Tidal force force experienced by a non-point mass (one with spatial extension) in a non-uniform gravitational field. Such forces tend simultaneously to compress and stretch spatially extended masses. Black holes and wormhole mouths can generate enormous tidal forces on extended masses as small as a human body. Interestingly, the *more* massive a black hole, the *less* severe its tidal forces are at distances outside the **event horizon**. However, no matter what the black hole mass is, the tidal forces are infinite at the central curvature **singularity**.

Time dilation the altering of the rate of timekeeping by a clock, either by motion or by gravity.

Time machine (in the weak sense) a machine able to traverse **closed timelike world lines** inherent in a spacetime (e.g., a rocket in Gödel spacetime) but unable to *create* such world lines; (in the strong sense) a machine able to manipulate mass-energy in a finite or compact region of spacetime in such a way as to *create* closed timelike world lines.

Time police story characters in science fiction charged with the (unnecessary!) job of preventing time travelers from changing the past.

Time warp science fiction name for a **time machine**.

Tipler cylinder an infinitely long cylinder, made of super-dense matter, rotating so fast around its long axis that it warps spacetime enough to create closed timelike lines that encircle the cylinder. It can be used as a strong sense time machine to travel both into the future and into the past (but *not* to a time before the creation of the cylinder).

Topology the structure of a *space* (including *spacetime*) without regard to a metric. That is, topology is concerned only with how a space is connected together and not with how far apart points in the space are. Topologists consider stretching or compressing a space to be irrelevant, just as long as one doesn't *tear* it and so put holes in the space. The simplest topology is that of a *simply connected* space, in which if you construct any closed surface that lies totally in the space around any point in the space, then every other point inside the surface is also in the space. A space with a hole in it fails this test, and so is said to be *multiply connected*. A *quantum foam* spacetime has a multiply connected topology. The classical spacetime of general relativity is simply connected *until* the appearance of wormholes.

Twin paradox the conclusion from special relativity that a clock's rate of time keeping slows with motion.

Ultraluminal motion sufficiently **superluminal** that mass-energy appears to travel backward in time (see also **Dirac radio**).

Uncertainty principle the statement in quantum mechanics that says certain pairs of quantities cannot simultaneously be measured with arbitrarily small error. The position and momentum of a particle are one such pair, and energy and time are another.

Vacuum fluctuation the particle/anti-particle creation and annihilation processes allowed, even empty space, by the **uncertainty principle** of quantum mechanics.

Warp drive science fiction name for the propulsion mechanism of a faster-than-light spaceship, now commonly used by physicists, too.

Weak energy condition the claim that the observed mass-energy density is always (locally) non-negative. Quantum mechanics predicts (and it has been experimentally confirmed) that there are exceptions.

White dwarf a burnt-out star with a mass less than 1.4 solar masses, of planetary size with a density up to 10^7 g/cm^3. The ultimate fate of our Sun.

World line the trajectory of mass-energy in spacetime.

Wormhole a spacetime structure (violating the **weak and strong energy conditions**, if traversable) connecting two points of the same spacetime (or even two *different* spacetimes) with a timelike path that requires less time to travel along than does a photon traveling *outside* the wormhole between the two points. A wormhole is *traversable* if it has no **event horizons**, and such wormholes can apparently be made into a **time machine** (sometimes called a *time tunnel*) using a time shift (see **time dilation**) between the two mouths of the wormhole *unless* quantum effects forbid time machines (still an open question).

Index

© Springer International Publishing AG 2017
P.J. Nahin, *Time Machine Tales*, Science and Fiction,
DOI 10.1007/978-3-319-48864-6

CPSIA information can be obtained
at www.ICGtesting.com
Printed in the USA
LVOW02s0902270317
528553LV00019B/2/P